高等学校教材

化妆品专业英语

徐洪伍 主编

COSMETIC ENGLISH

化学工业出版社

·北京·

内容简介

《化妆品专业英语》共八个单元。第一单元包括两章，主要介绍化妆品的定义、历史及中国化妆品市场情况；第二单元包括四章，主要介绍无机化合物的命名方法、有机化合物的命名方法、高分子化合物的命名方法及INCI命名；第三单元包括四章，主要介绍装饰性化妆品、护肤化妆品、发用化妆品及香水；第四单元包括九章，主要介绍油性原料、表面活性剂、增稠剂、保湿剂、着色剂、防晒原料、美白剂、抗衰老剂、防腐剂和抗氧化剂；第五单元包括三章，主要介绍中国化妆品监督管理条例、欧盟化妆品安全的法律法规及美国化妆品安全的相关法律法规；第六单元包括两章，主要介绍化妆品的配方、技术、分析和包装；第七单元包括三章，主要介绍化妆品的安全性测试、功效评价和感官评价；第八单元主要介绍生物技术在化妆品中的应用研究进展。

本书可作为本科化妆品技术与工程、化妆品科学与技术等专业的教材，也可作为高职化妆品相关专业的教材，还可供化妆品生产、管理、经营、销售及美容等从业人员阅读。

图书在版编目（CIP）数据

化妆品专业英语/徐洪伍主编. —北京：化学工业出版社，2022.10

ISBN 978-7-122-42630-7

Ⅰ.①化… Ⅱ.①徐… Ⅲ.①化妆品-英语-教材 Ⅳ.①TQ658

中国版本图书馆CIP数据核字（2022）第229046号

责任编辑：提 岩 张双进　　文字编辑：曹 敏
责任校对：李雨函　　装帧设计：李子姮

出版发行：化学工业出版社（北京市东城区青年湖南街13号 邮政编码100011）
印　　装：三河市延风印装有限公司
787mm×1092mm 1/16 印张16 字数370千字
2023年8月北京第1版第1次印刷

购书咨询：010-64518888　　售后服务：010-64518899
网　　址：http://www.cip.com.cn
凡购买本书，如有缺损质量问题，本社销售中心负责调换。

定　价：49.80元

前言

随着我国经济的持续发展，居民生活水平不断提高，化妆品已由奢侈品变成了不可或缺的日用品，我国已经是全球第二大化妆品消费市场。近年来电商行业的飞速崛起进一步带动了化妆品行业的发展，我国化妆品行业规模迅速扩大。资料显示，2021年我国化妆品类商品零售额达4026亿元，同比增长18.4%。化妆品行业的发展需要大量化妆品专业人才的支撑。

近年来，全国许多高校相继开设了化妆品相关本科专业。随着化妆品行业的国际化趋势日趋加强，化妆品专业英语成为化妆品相关本科专业的一门重要课程。但是，由于中国高校前期没有开设过化妆品相关本科专业，所以化妆品专业英语方面的本科教材相对缺乏。为此，肇庆学院化妆品技术与工程专业组织从事化妆品专业一线教学的教师编写了这本《化妆品专业英语》。

化妆品专业英语课程的目的是培养学生阅读英文专业文献的能力，使其在今后的工作中能够熟练运用。考虑到大多数本科同学在大学英语的学习过程中已经熟练掌握了英语的各种基础知识，本书直接选取原汁原味的英文素材给同学们阅读，教师在授课过程中可对一些重点或特殊的语法和词法进行介绍。本书收录了最新的化妆品行业信息、最常用的化妆品专业词汇和用语，适合化妆品类本科专业学生作为教材使用，也适合化妆品企业技术人员作为参考书或培训教材。

本书由徐洪伍担任主编，吴利欢、高爱环、张博担任副主编。具体编写分工为：第一单元、第五单元～第八单元由徐洪伍编写；第二单元由吴利欢、李倩编写；第三单元由张博、王亮编写；第四单元由高爱环、刘莉莉编写。全书由徐洪伍统稿，潘勇主审。

本书的出版得到了肇庆学院精品教材建设项目的资助，同时也融入了化学工业出版社各位编辑的心血，在此深表感谢！

由于作者水平所限，书中不足之处在所难免，欢迎各位读者提出宝贵意见和建议，以便不断完善！

编 者

2022年8月

目录

Unit 6 Formulations, Technologies, Analyses and Packaging of Cosmetics / 180

Unit 7 Safety Testing, Efficacy and Sensory Evaluation of Cosmetic Products / 199

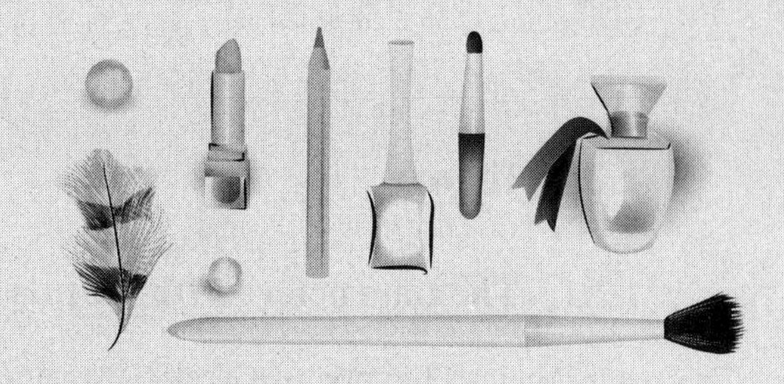

Unit 1
Definition, History and China's Cosmetics Market

In this unit, you will see:
① Definitions for the cosmetics in different countries;
② A brief history about cosmetics;
③ China's cosmetics market.

Chapter 1 Definition and History

1.1 Definition

Different countries have different legal definitions for cosmetics.

In China, cosmetics refer to daily chemical industrial products that are applied to any part of the human body surface (skin, hair, nails, lips, *etc.*) by brushing, spraying or other similar methods to achieve the purpose of cleaning, eliminating bad smell, skincare, beauty and modification.

In the United States, the Food and Drug Administration (FDA), which regulates cosmetics, defines cosmetics as products "intended to be applied to the human body for cleansing, beautifying, promoting attractiveness, or altering the appearance without affecting the body's structure or functions". This broad definition includes any material intended for use as an ingredient of a cosmetic product, with the FDA specifically excluding pure soap from this category.

In European Union, cosmetic product means any substance or mixture intended to be placed in contact with the external parts of the human body (epidermis, hair system, nails, lips and external genital organs) or with the teeth and the mucous membranes of the oral cavity with a view exclusively or mainly to cleaning them, perfuming them, changing their appearance, protecting them, keeping them in good condition or correcting body odours.

1.2 History

1.2.1 The time before 19th century

Ancient Egyptians, Greeks, Romans, Incas and Chinese frequently used personal care products. Cleopatra's baths in asses' milk and the conspicuous makeup and malachite eye shadow of Nefertiti and of the pharaohs still appeal to our imagination. The embalmment of the deceased pharaohs and the makeup articles found in their tombs are evidence of their interest in and knowledge of cosmetics. They already tried to counteract the formation of wrinkles by means of an ointment based on the fenugreek plant. The Greeks and the Romans also applied massage oils on a large scale. The Greek Theophrastus (372 B.C.) and the Roman Galen (130-200 A.D.) wrote the first systematic books about the preparation of cosmetics through the introduction of distillation and extraction processes. Via the crusaders and the Portuguese and Spanish explorers, cosmetics gradually became known in Europe. During the Renaissance, cosmetics and perfumes were still mainly used at the royal courts. Not until 1900 did they come within reach of ordinary citizens.

1.2.2 The 19th century

During the late 1800s, the Western cosmetics industry began to grow due to a rise in "visual self-awareness," a shift in the perception of color cosmetics, and improvements in the safety of products. Prior to the 19th century, limitations in lighting technology and access to reflective devices stifled people's ability to regularly perceive their appearance. This, in turn, limited the need for a cosmetic market and resulted in individuals creating and applying their own products at home. Several technological advancements in the latter half of the century, including the innovation of mirrors, commercial photography, marketing and electricity in the home and in public, increased consciousness of one's appearance and created a demand for cosmetic products that improved one's image.

Face powders, rouges, lipstick and similar products made in home were found to have toxic ingredients, which deterred customers from their use. Discoveries of non-toxic cosmetic ingredients and the distribution of cosmetic products by established companies such as Rimmel, Guerlain, and Hudnut helped popularize cosmetics to the broader public. Skincare, along with "face painting" products like powders, also became in-demand products of the cosmetics industry. The massive advertisements of cold cream brands such as Pond's through billboards, magazines, and newspapers created a high demand for the product. These advertisement and cosmetic marketing styles were soon replicated in European countries, which further increased the popularity of the advertised products in Europe.

1.2.3 The 20th century

During the early 1900s, makeup was not excessively popular. In fact, women hardly wore makeup at all. Makeup at this time was still mostly the territory of prostitutes, those in cabarets and on the black & white screen.

Around 1910, makeup became fashionable in the United States of America and Europe owing to the influence of ballet and theatre stars such as Mathilde Kschessinska and Sarah Bernhardt. Colored makeup was introduced in Paris upon the arrival of the Russian Ballet in 1910, where ochers and crimsons were the most typical shades. The Daily Mirror beauty book showed that cosmetics were now acceptable for the literate classes to wear. With that said, men often saw rouge as a mark of sex and sin, and rouging was considered an admission of ugliness.

In the 1920s, the film industry in Hollywood had the most influential impact on cosmetics. Stars such as Theda Bara had a substantial effect on the makeup industry. Helena Rubinstein was Bara's makeup artist; she created mascara for the actress, relying on her experiments with kohl. Others who saw the opportunity for the mass-market of cosmetics during this time were Max Factor Sr and Elizabeth Arden. Many of the present-day makeup manufacturers were established during the 1920s and 1930s. Lipsticks were one of the most popular cosmetics of this time, more so than rouge and powder, because they were colorful and cheap. In 1915, Maurice Levy invented the metal container for lipstick, which gave license to its mass production. The Flapper style also influenced the cosmetics of the 1920s, which embraced dark eyes, red lipstick, red nail polish, and the suntan, invented as a fashion statement by Coco Chanel. The eyebrow pencil became vastly popular in the 1920s, in part because it was technologically superior to what it had been, due to a new ingredient: hydrogenated cottonseed oil (also the key constituent of another wonder product of that era Crisco Oil). The early commercial mascaras, like Maybelline, were simply pressed cakes containing soap and pigments. A woman would dip a tiny brush into hot water, rub the bristles on the cake, remove the excess by rolling the brush onto some blotting paper or a sponge, and then apply the mascara as if her eyelashes were a watercolor canvas. Eugène Schueller, founder of L'Oréal, invented modern synthetic hair dye in 1907 and he also invented sunscreen in 1936. The first patent for a nail polish was granted in 1919. Its color was a very faint pink. Previously, only agricultural workers had sported suntans, while fashionable women kept their skins as pale as possible. In the wake of Chanel's adoption of the suntan, dozens of new fake tan products were produced to help both men and women achieve the "sun-kissed" look. In Asia, skin whitening continued to represent the ideal of beauty, as it does to this day.

In the time period after the First World War, there was a boom in cosmetic surgery. During the 1920s and 1930s, facial configuration and social identity dominated a plastic surgeon's world. Face-lifts were performed as early as 1920, but it wasn't until the 1960s when cosmetic surgery was used to reduce the signs of aging. During the twentieth century, cosmetic surgery mainly revolved around women. Men only participated in the practice if they had been disfigured by the war. Silicone implants were introduced in 1962. In the 1980s, the American Society of Plastic Surgeons made efforts to increase public awareness about plastic surgery. As a result, in 1982, the United States Supreme Court granted physicians the legal right to advertise their procedures. The optimistic and simplified nature of narrative advertisements often make the surgeries seem hazard-free, even though they were anything but. The American Society for

Aesthetic Plastic Surgery reported that more than two million Americans elected to undergo cosmetic procedures, both surgical and non-surgical, in 1998, liposuction being the most popular. Breast augmentations ranked second, while numbers three, four, and five went to eye surgery, face-lifts, and chemical peels.

During the 1920s, numerous African Americans participated in skin bleaching in an attempt to lighten their complexion as well as hair straightening to appear whiter. Skin bleaches and hair straighteners created fortunes worth millions and accounted for a massive thirty to fifty percent of all advertisements in the black press of the decade. Oftentimes, these bleaches and straighteners were created and marketed by African American women themselves. Skin bleaches contained caustic chemicals such as hydroquinone, which suppressed the production of melanin in the skin. These bleaches could cause severe dermatitis and even death in high dosages. Many times, these regimens were used daily, increasing an individual's risk. In the 1970s, at least 5 companies started producing makeup for African American women. Before the 1970s, makeup shades for black women were limited. Face makeup and lipstick did not work for dark skin types because they were created for pale skin tones. These cosmetics that were created for pale skin tones only make dark skin appear grey. Eventually, makeup companies created makeup that worked for richer skin tones, such as foundations and powders that provided a natural match. Popular companies like Astarté, Afram, Libra, Flori Roberts and Fashion Fair priced the cosmetics reasonably due to the fact that they wanted to reach out to the masses.

From 1939 to 1945, during the Second World War, cosmetics were in short supply. Petroleum and alcohol, basic ingredients of many cosmetics, were diverted into war supply. Ironically, at this time when they were restricted, lipstick, powder, and face cream were most desirable and most experimentation was carried out for the post war period. Cosmetic developers realized that the war would result in a phenomenal boom afterwards, so they began preparing. Yardley, Elizabeth Arden, Helena Rubinstein, and the French manufacturing company became associated with "quality" after the war because they were the oldest established. Pond's had this same appeal in the lower price range. Gala cosmetics were one of the first to give its products fantasy names, such as the lipsticks in "lantern red" and "sea coral".

During the 1960s and 1970s, many women in the western world influenced by feminism decided to go without any cosmetics. In 1968 at the feminist Miss America protest, protestors symbolically threw a number of feminine products into a "Freedom Trash Can". This included cosmetics, which were among items the protestors called "instruments of female torture" and accouterments of what they perceived to be enforced femininity.

Cosmetics in the 1970s were divided into a "natural look" for day and a more sexualized image for evening. Non-allergic makeup appeared when the bare face was in fashion as women became more interested in the chemical value of their makeup. Modern developments in technology, such as the high-shear mixer facilitated the production of cosmetics which were more natural looking and had greater staying power in wear than their predecessors. Though the prime cosmetic of the time was eye shadow, women also were interested in new lipstick colors such as lilac, green, and silver. These lipsticks were often mixed with pale pinks and whites, so women could create their own individual shades. "Blush-on" came into the market in this decade, with Revlon giving them wide publicity. This product was applied to the forehead, lower cheeks, and chin. Contouring and highlighting the face with white eye shadow cream also became popular. Avon introduced the lady saleswoman. In fact, the whole cosmetic industry in general opened opportunities for women in business as entrepreneurs, inventors, manufacturers, distributors, and promoters.

1.2.4 The 21st century

Beauty products are now widely available from dedicated internet-only retailers, who have more recently been joined online by established outlets, including the major department stores and traditional bricks and mortar beauty retailers.

Although modern makeup has been used mainly by women traditionally, gradually an increasing number of males are using cosmetics usually associated with women to enhance their own facial features. Concealer is commonly used by cosmetic-conscious men. Cosmetics brands are releasing cosmetic products especially tailored for men, and men are using such products more commonly. There is some controversy over this, however, as many feel that men who wear makeup are neglecting traditional gender, and do not view men wearing cosmetics in a positive light. Others, however, view this as a sign of ongoing gender equality and feel that men also have rights to enhance their facial features with cosmetics if women could.

Today the market of cosmetics has a different dynamic compared to the 20th century. For example, China's cosmetics market became the second largest one in the world.

Reference

Wikipedia org. History of cosmetics[EB/OL]. 2014[2023-05-17]. https://encyclopedia.thefreedictionary.com/History+of+cosmetics.

Key Words & Phrases

cosmetic *n.* 化妆品；美容品
skincare *n.* 护肤品
epidermis *n.* 表皮
mucous *adj.* 黏液的；黏的；分泌黏液的
personal care 个人护理；个人护理用具；个人护理用品
makeup *n.* 化妆品；天性；性格；组成；构成
malachite *n.* 孔雀石
eye shadow 眼影膏
embalmment *n.* 尸体防腐法；（尸体的）防腐处理；薰香
ointment *n.* 药膏；软膏；油膏
fenugreek *n.* 葫芦巴（种子用于南亚食物调味）
perfume *n.* 香水；芳香；香味；馨香
color cosmetic 彩妆
face powder 扑面粉；敷面粉
rouge *n.* 胭脂
lipstick *n.* 口红；唇膏
cold cream 洁面乳；润肤膏
ocher *n.* 赭石
crimson *n.* 深红色
kohl *n.* 黑色眼影粉（尤指东方人用的）
flapper *n.* 新潮女郎
eyebrow pencil 眉笔
mascara *n.* 睫毛膏；染睫毛油
hair dye 染发剂
sunscreen *n.* 防晒霜；防晒油
nail polish 指甲油
cosmetic surgery 整容手术

face-lift 美化；翻新；面部皱纹切除术；面部拉皮术
implant *n.*（植入人体中的）移植物；植入物
plastic surgery 整形手术；整形外科
liposuction *n.* 吸脂术；脂肪抽吸（术）
breast augmentation 隆胸术；丰胸；隆胸；隆乳；隆乳手术
chemical peel 化学换肤；化学脱皮术；果酸换肤；化学剥脱
skin bleaching 漂白皮肤
hair straightening 拉直头发
melanin *n.* 黑色素
dermatitis *n.* 皮炎
regimen *n.* 养生；养生之道；生活规则
face cream 面霜；雪花膏
concealer *n.* 遮瑕膏

Chapter 2 China's Cosmetics Market

2.1 Market Overview

China's cosmetics sector has been growing rapidly in recent years, in line with the growth of the economy. According to Euromonitor, retail sales of skincare products in China reached RMB260.4 billion in 2020, while sales of makeup products totalled RMB52.5 billion according to Shenzhen-based Forward Industry Research Institute. Though the trade through physical stores was affected by the COVID-19 pandemic, online sales of skincare products and makeup conducted via the Alibaba group of e-commerce platforms grew respectively 31.8% and 32.5% year-on-year. The table below shows the retail sales of cosmetics by wholesale and retail enterprises above a designated size in recent years.

Year	Retail sales/billion RMB
2016	222.2
2017	251.4
2018	261.9
2019	299.2
2020	340.0

Source: National Bureau of Statistics of China.

The current structure of China's consumer market of cosmetic products is as follows:

- Skincare products: the largest and fastest-growing segment in the cosmetics market.
- Shampoos and haircare products: a market niche becoming saturated, with slowing growth.
- Makeup products: this market is far from saturated, particularly for enhancement items such as colour correcting (CC) and blemish balm (BB) cream. Sales of eye makeup products have recorded significant growth in recent years.
- Products for children: sales of products designed for use by children continue to soar.
- Sunscreen products: seasonal demand ensures sales do not slow down during traditionally quiet periods.
- Anti-aging products: cosmetic products that help consumers stay looking youthful and fight aging are increasingly popular.
- Sports cosmetics: many consumers who love sports and fitness pursuits are keen to look good as well. They use sports cosmetics that help prevent moisture loss and are anti-odour, anti-sweat, and anti-bacterial, packaged in compact, portable sizes.

Cosmeceuticals: consumers are increasingly aware of products that combine cosmetic and pharmaceutical features, such as spot lightening cream, acne treatment lotion and acne ointment. Cosmeceuticals can be roughly classified into three types: pharmaceutical cosmetics that are sideline cosmetic products from pharmaceutical companies, medical skincare products that provide supplementary therapeutic functions and functional cosmetics that deal with specific skin problems.

Green/natural cosmetics: these contain natural or nutritional ingredients, such as aloe and vitamins.

China's demand for makeup products is maturing and diversifying. Previously, mainland consumers mostly bought basic products, but now demand has expanded to include products such as liquid foundation, eye makeup and lipstick. Eyeshadow was previously considered a western style makeup and was not sought by Chinese consumers. Now, thanks to online video tutorials about makeup, more consumers have come to appreciate eyeshadow and consumption has shotted up.

As pointed out in a mainland research report on rational skincare in the post-pandemic period, people have been wearing facemasks constantly since the beginning of the pandemic. As a result, allergies and acnes have become the main skin concerns of consumers. The demand for cosmetics has fallen significantly, particularly for makeup removal and sunscreen products. As consumers stay home more often, they spend more time on skincare. This trend is particularly evident among the post-90s generation. In the post-pandemic period, the ingredients, efficacy and safety of makeup and skincare products are the key reasons for choosing a product.

Natural ingredients and safety matter more in cosmetics for children than for adults. To ensure consumer safety, the former State Food and Drug Administration issued its guidance on application and review of children's cosmetics in 2012. According to Euromonitor, the size of the Chinese market for children's cosmetics reached RMB 26.8 billion in 2020. Statistics from kaola.com show that sales of children's makeup products shot up 300% year-on-year in 2020. There is a good chance that the market could reach RMB 45.7 billion by 2025.

The Chinese market is moving towards more high-end products, especially foreign brands. Chinabaogao.com figures show that China's market for high-end skincare products was RMB 91.7 billion in 2020, up 9.4% year-on-year. Nevertheless, China's domestic cosmetics brands are performing well. According to the analysis on new domestic beauty brands released by mainland research firm CBNData, these domestic brands are promoting their products both online and offline, using a host of celebrities and KOLs as spokespersons and investing heavily in marketing to raise brand recognition. They have also been maintaining high value-for-money ratios to maximise the chance of consumers trying out their products.

Male consumers are increasingly receptive to skincare and makeup products designed especially for men. When it comes to skincare, men are mainly concerned with cleansing and managing oily skin. While facial cleansers make up the lion's share of the male skincare market, demand for specialty products such as masks, sun-blocks and whitening and moisturising agents is also on the rise. This demonstrates that male consumers are beginning to pay more attention to skin conditions such as ageing and coarseness. Data from AskCI Research reveal that 59.5% of the men buying cosmetics are aged under 25.

Cosmeceuticals, especially traditional Chinese herbal medicine cosmetics, are relatively new in the market. According to iResearch, sales of cosmeceuticals currently comprise only some 24% of China's cosmetics market, so there is plenty of potential for growth. Following the introduction of new regulations related to cosmetics in 2021, consumer confidence in cosmeceuticals has increased markedly. According to the Regulation on the Supervision and Administration of Cosmetics which came into force on 1 January 2021, no medical jargon or claims of medical efficacy should be used in these items' packaging or instructions to avoid misleading consumers.

China's imports of major cosmetic products in 2020 are summarized below:

China's imports of major cosmetic products in 2020

HS Code	Description	2020/ million USD	Y-o-Y change/%
33030000	Perfumes and toilet waters	745	33.6
33041000	Lip makeup preparations	827	–11.3
33042000	Eye makeup preparations	288	3.2
33049100	Powders, whether or not compressed	324	7.5
33049900	Other (including preparations for the care of the skin, suntan preparations, *etc.*)	15899	35.7
33051000	Shampoos	394	17.0
33052000	Preparations for permanent waving	4.6	–30.3
33053000	Hair lacquers	18.9	10.4
33072000	Personal deodorants and antiperspirants	23.8	15.0

Source: Global Trade Atlas.

2.2 Market Competition

According to statistics from the National Medical Products Administration (NMPA), which is now under the State Administration for Market Regulation (SAMR), there were 5447 enterprises qualified to produce cosmetics in China in 2020. Most of these producers are concentrated in Guangdong and East China: in 2020, there were 2967 or 54.5% of all qualified cosmetics producers in Guangdong, followed by 1434 or 26.3% of qualified producers in East China. Domestic brands are mostly concentrated in the mid- to low-end market segments, while joint ventures and enterprises with foreign investment dominate the high-end segment. In 2020, there were 2.77 million registered cosmetics-related enterprises, up 12% from the 2.47 million enterprises registered in 2019. Among these, 74% were small enterprises (less than RMB 1 million in capital) and only 1% were large enterprises (more than RMB 30 million in capital). Evidently, competition among small cosmetics enterprises is fierce.

In recent years slowing growth has characterised the pharmaceutical industry. Consequently, pharmaceutical enterprises, which are under NMPA control, have started to expand into the cosmeceuticals market where they see bright prospects. Biotech and pharmaceutical companies are accustomed to meeting rigorous quality standards with advanced manufacturing technology. Their raw materials also tend to meet higher standards than those used by a typical cosmetics company, and so they have a competitive edge in the cosmetics industry. In April 2021, for example, NanHua Bio-Medicine Co Ltd, which aims to become a world-leading biotech company, was granted a patent to apply its technology in cosmetics production by the China National Intellectual Property Administration. Some domestic brands, including Tongrentang and Herborist, have also ventured into the cosmeceuticals market and are achieving growing recognition from consumers. In the report on brand competition and investment opportunities in China's cosmeceutical industry released by Forward Industry Research Institute, it is projected that, as more overseas firms enter the market and domestic brands become more prominent, the size of China's cosmeceutical market will reach RMB125 billion by 2025.

The high-end cosmetic market has traditionally been the exclusive territory of international cosmetic brands, but with world-renowned fashion brands such as Hermes and LV now entering the market, competition is growing increasingly intense. With the Chinese people enjoying ever higher quality of life and income, the prospects for the high-end cosmetic market look promising.

With traditional style and culture now in vogue, domestic companies have successfully targeted the cosmetics market with such items as Chinese-style vanity gift boxes and limited festival editions. Domestic brands are actively applying traditional Chinese medicine concepts and natural extraction methods in their development of skincare products, such as the Tai Ji and Yu Wu Xing series from Herborist.

2.3 Sales Channels

The main sales channels for cosmetics on the mainland include integrated e-commerce platforms, wholesale markets, supermarkets and department stores, dedicated counters, speciality chain stores, drugstores, beauty parlours and direct selling. Integrated e-commerce platforms, department stores and speciality chain stores are now the top three sales channels.

National Bureau of Statistics figures reveal that sales of cosmetics in November 2020 and March 2021 respectively hit peaks which were 32.3% and 42.5% greater than that in the previous year, thanks to promotional activities such as the "Double 11" and "Goddess Day" (on 8 March) shopping festivals. According to a mainland research report on market demand forecast and strategic investment planning in China's cosmetic sector, online sales channels accounted for 38% of total cosmetic sales, a sharp increase from the 5.3% share 10 years ago. On the other hand, because of the closing of stores during the pandemic, there was a 0.9% drop in physical store sales year on year, to RMB 281.3 billion in 2020.

E-commerce on social media platforms such as Xiaohongshu, TikTok and Kuaishou has further stimulated spending on cosmetics. At first these platforms shared graphics and video clips, but they have evolved into livestream marketing. To whet the appetite of consumers, livestream anchors interact with them and offer discounts. The makeup livestream report 2020 released by TikTok shows that the number of people watching makeup livestreaming in May 2020 soared 122% from January 2020, and the average time they spent watching increased by 257%. Clearly consumers are receiving product information from livestream platforms, and the large increase in shopping cart click rates during livestreaming is testimony to its effectiveness.

The 'dedicated counter' is a major traditional sales channel for cosmetics, adopted by most international cosmetic brands. Dedicated counters play a huge role in brand image building. According to iiMedia Research, brand and word-of-mouth recommendation are the two factors that mainland consumers care about most when buying cosmetics. Top global brands such as Lancôme, Estée Lauder, Chanel and Dior dominate the sales of cosmetics through dedicated counters on the mainland. Only a few domestic brands, such as Herborist, are able to compete with these giants.

Some brands seek to expand their business by opening specialty stores, either directly operated or as franchises. Many multinational cosmetics giants prefer directly operating specialty stores, where they can control the brand image, maintain quality of service and ensure unified, stable pricing. Franchise chain stores, however, are generally regarded to be the most effective format, requiring minimum input and achieving the highest success rates.

Cosmetic products are also distributed through pampering and therapeutic beauty parlours, large and medium-sized high-end beauty spas, franchise chain stores, and grooming and hairdressing parlours.

Fairs held in China provide an ideal channel for industry players to gather the latest information and to meet dealers.

Reference

Hong Kong Trade Development Council. China's Cosmetics Market[EB/OL].(2022-08-26)[2023-05-17]. https://research.hktdc.com/en/article/MzA4Nzg0MTgw.

Key Words & Phrases

shampoo *n.* 洗发水；香波；洗发剂
haircare *n.* 头发护理
blemish balm 修护霜；伤痕保养霜；遮瑕霜
anti-aging product 抗衰老产品
sports cosmetic 运动化妆品
cosmeceutical *n.* 药妆品；药妆；药用化妆品
spot lightening cream 祛斑霜
acne treatment lotion 痤疮治疗乳液
acne ointment 痤疮膏
aloe *n.* 芦荟
liquid foundation 粉底液；粉底；液体粉底；粉底霜；湿粉
ingredient *n.* 成分；原料；因素；要素
efficacy *n.* 功效；（尤指药物或治疗方法的）效力
high-end product 高端产品
facial cleanser 洗面奶；洁面乳
mask *n.* 面具；面罩；假面具；护肤膜；面膜
sun-block 防晒霜

whitening agent 美白成分；增白剂
moisturising agent 保湿剂
jargon *n.* 行话
toilet water 花露水
permanent waving 烫发
hair lacquer 毛发定型剂
personal deodorant 个人除臭剂
antiperspirant *n.* 止汗剂
mid-end market 中端市场
low-end market 低端市场

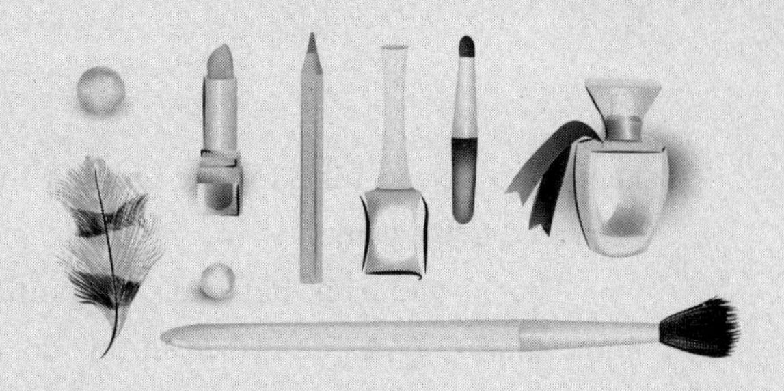

Unit 2

Nomenclatures of Cosmetic Ingredients

Generally, cosmetics contain a variety of ingredients, which may be chemically synthesized or natural. As cosmetics related practitioners, you must know the names of them. Therefore, in this unit, you will see:

① Nomenclature of inorganic chemical compounds;

② Nomenclature of organic compounds;

③ Nomenclature of polymers;

④ International nomenclature cosmetic ingredients (INCI for abbreviation).

Chapter 3 Brief Guide to the Nomenclature of Inorganic Chemistry

The universal adoption of an agreed chemical nomenclature is a key tool for communication in the chemical sciences, for computer-based searching in databases, and for regulatory purposes, such as those associated with health and safety or commercial activity. The International Union of Pure and Applied Chemistry (IUPAC) provides recommendations on the nature and use of chemical nomenclature. The basics of this nomenclature are shown here, and in companion documents on the nomenclature systems for organic chemistry and polymers, with hyperlinks to the original documents. An overall summary of chemical nomenclature can be found in *Principles of Chemical Nomenclature*. Greater detail can be found in the *Nomenclature of Inorganic Chemistry*, colloquially known as the Red Book, and in the related publications for organic compounds (the Blue Book) and polymers (the Purple Book). It should be noted that many compounds may have non-systematic or semi-systematic names (some of which are not accepted by IUPAC for several reasons, for example because they are ambiguous) and IUPAC rules allow for more than one systematic name in many cases. IUPAC is working towards identification of single names which are to be preferred for regulatory purposes (Preferred IUPAC Names, or PINs). Note: In this document, the symbol '=' is used to split names that happen to be too long for the column format, unless there is a convenient hyphen already presented in the name.

The boundaries between 'organic' and 'inorganic' compounds are blurred. The nomenclature types described in this document are applicable to compounds, molecules and ions that do not contain carbon, but also to many structures that do contain carbon (Section 3.2), notably those containing elements of Groups 1-12. Most boron-containing compounds are treated using a special nomenclature.

3.1 Stoichiometric or Compositional Names

A stoichiometric or compositional name provides information only on the composition of an ion, molecule, or compound, and may be related to either the empirical or molecular formula for that entity. It does not provide any structural information. For homoatomic entities, where only one element is present, the name is formed (Table 3.1) by combining the element name with the appropriate multiplicative prefix (Table 3.2). Ions are named by adding charge numbers in parentheses, *e.g.* (1+), (3+), (2–), and for (most) homoatomic anion names 'ide' is added in place of the 'en' 'ese' 'ic' 'ine' 'ium' 'ogen' 'on' 'orus' 'um' 'ur' 'y' or 'ygen' endings of element names[9]. Exceptions include Zn and Group 18 elements ending in 'on', where the 'ide' ending is added to the element names. For some elements (*e.g.*, Fe, Ag, Au) a Latin stem is used before the 'ide' ending (*cf.* Section 3.2.3). Certain ions may have acceptable traditional names (used without charge numbers). Binary compounds (those containing atoms of two elements) are named stoichiometrically by combining the element names and treating, by convention, the element reached first when following the arrow in the element sequence (Figure 3.1) as if it were an anion. Thus, the name of this formally 'electronegative' element is given an 'ide' ending and is placed after the name of the formally 'electropositive' element followed by a space (Table 3.3).

Table 3.1 Examples of homoatomic entities

Formula	Name	Formula	Name
O_2	dioxygen	Cl^-	chloride(1–) or chloride
S_8	octasulfur	I_3^-	triiodide(1–)
Na^+	sodium(1+)	O_2^{2-}	dioxide(2–) or peroxide
Fe^{3+}	iron(3+)	N_3^-	trinitride(1–) or azide

Table 3.2 Multiplicative prefixes for simple and complicated entities

No.	Simple	Complicated	No.	Simple	Complicated
2	di	bis	8	octa	octakis
3	tri	tris	9	nona	nonakis
4	tetra	tetrakis	10	deca	decakis
5	penta	pentakis	11	undeca	undecakis
6	hexa	hexakis	12	dodeca	dodecakis
7	hepta	heptakis	20	icosa	icosakis

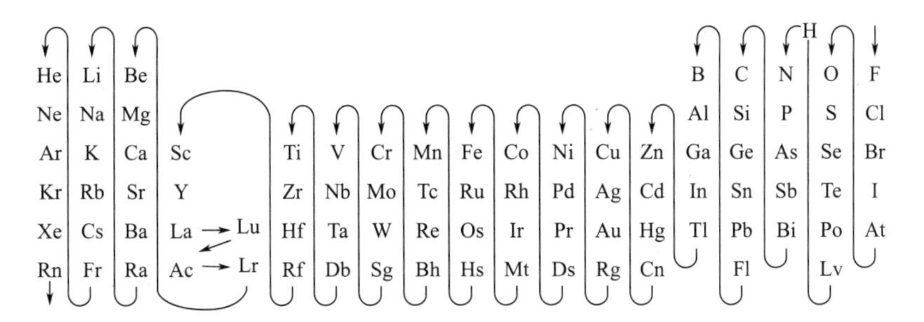

Figure 3.1 Element sequence

Table 3.3 Examples of binary compounds

Formula	Name	Formula	Name
GaAs	gallium arsenide	$FeCl_2$	iron dichloride or iron (II) chloride
CO_2	carbon dioxide	$FeCl_3$	iron trichloride or iron (III) chloride
CaF_2	calcium difluoride or calcium fluoride	H_2O_2	dihydrogen dioxide or hydrogen peroxide

Again, multiplicative prefixes (Table 3.2) are applied as needed, and certain acceptable alternative names may be used. Stoichiometry may be implied in some cases by the use of oxidation numbers, but is often omitted for common cases, such as in calcium fluoride. Heteropolyatomic entities in general can be named similarly using compositional nomenclature, but often either substitutive or additive nomenclature (Section 3.2) is used. In the latter case information is also provided about the way atoms are connected. For example, $POCl_3$ (or PCl_3O, compositional name phosphorus trichloride oxide) is given an additive name in Table 3.10.

Certain ions have traditional short names, which are commonly used and are still acceptable (*e.g.*, ammonium, NH_4^+; hydroxide, OH^-; nitrite, NO_2^-; phosphate, PO_4^{3-}; diphosphate, $P_2O_7^{4-}$). Inorganic compounds in general can be combinations of cations, anions and neutral entities. By convention, the name of a compound is made up of the names of its component entities: cations before anions and neutral components last (see examples in Table 3.4). The number of each entity present has to be specified in order to reflect the composition of the compound. For this purpose, multiplicative prefixes (Table 3.2) are added to the name of each entity. The prefixes are 'di' 'tri' 'tetra', *etc.*, for use with names for simple entities, or 'bis( )' 'tris( )' 'tetrakis( )', *etc.*, for names for most entities which themselves contain multiplicative prefixes or locants. Care must also be taken in situations when use of a simple multiplicative prefix may be misinterpreted, *e.g.*, tris(iodide) must be used for $3I^-$ rather than triiodide (which is used for I_3^-), and bis(phosphate) rather than diphosphate (which is used for $P_2O_7^{4-}$). Examples are shown in Table 3.4. There is no elision of vowels (*e.g.*, tetra-aqua, pentaoxide), except in the special case of monoxide.

Table 3.4 Use of multiplicative prefixes in compositional names

Formula	Name
$Ca_3(PO_4)_2$	tricalcium bis(phosphate)
$Ca_2P_2O_7$	dicalcium diphosphate

(Continued)

Formula	Name
BaO_2	barium(2+) dioxide(2−) or barium peroxide
$MgSO_4 \cdot 7H_2O$	magnesium sulfate heptahydrate
$CdSO_4 \cdot 6NH_3$	cadmium sulfate—ammonia (1/6)
$AlK(SO_4)_2 \cdot 12H_2O$	aluminium potassium bis(sulfate)−water (1/12) or aluminium potassium bis(sulfate) dodecahydrate
$Al_2(SO_4)_3 \cdot K_2SO_4 \cdot 24H_2O$	dialuminium tris(sulfate)−dipotassium sulfate−water (1/1/24)

Names of neutral components are separated by 'em' dashes without spaces. Inorganic compounds may themselves be components in (formal) addition compounds (last four examples in Table 3.4). The ratios of component compounds can be indicated, in general, using a stoichiometric descriptor in parentheses after the name (see the last three examples in Table 3.4). In the special case of hydrates, multiplicative prefixes can be used with the term 'hydrate'.

3.2 Complexes and Additive Nomenclature

3.2.1 Overall approach

Additive nomenclature was developed in order to describe the structures of coordination entities, or complexes, but this method is readily extended to other molecular entities as well. Mononuclear complexes are considered to consist of a central atom, often a metal ion, which is bonded to surrounding small molecules or ions, which are referred to as ligands. The names of complexes are constructed (Table 3.5) by adding the names of the ligands *before* those of the central atoms, using appropriate multiplicative prefixes. Formulae are constructed by adding the symbols or abbreviations of the ligands *after* the symbols of the central atoms (Section 3.2.7).

Table 3.5 Producing names for complexes: simple ligands

Structure to be named	$[NH_3, H_3N, NH_3, Co, H_3N, OH_2, NH_3]^{3+}$ $3Cl^-$	$2Cs^+$ $[Cl, Cl, Cl, Cl, Re{\equiv}Re, Cl, Cl, Cl, Cl]^{2-}$
Central atom(s)	cobalt (Ⅲ)	2×rhenium
Identify and name ligands	ammonia → ammine water → aqua	chloride → chlorido
Assemble name	pentaammineaqua=cobalt (Ⅲ) chloride	cesium bis(tetrachlorido=rhenate)(Re—Re)(2−)

3.2.2 Central atom(s) and ligands

The first step is to identify the central atom(s) and thereby also the ligands. By convention, the electrons involved in bonding between the central atom and a ligand are usually treated as belonging to the ligand (and this will determine how it is named).

Each ligand is named as a separate entity, using appropriate nomenclature–usually substitutive nomenclature for organic ligands and additive nomenclature for inorganic ligands. A small number of common molecules and ions are given special names when presented in complexes. For example, a water ligand is represented in the full name by the term 'aqua'. An ammonia ligand is represented by 'ammine', while carbon monoxide bound to the central atom through the carbon atom is represented by the term 'carbonyl' and nitrogen monoxide bound through nitrogen is represented by 'nitrosyl'. Names of anionic ligands that end in 'ide' 'ate' or 'ite' are modified within the full additive name for the complex to end in 'ido' 'ato' or 'ito', respectively. Note that the 'ido' ending is now used for halide and oxide ligands as well. By convention, a single coordinated hydrogen atom is always considered anionic and it is represented in the name by the term 'hydrido', whereas coordinated dihydrogen is usually treated as a neutral two-electron donor entity.

3.2.3 Assembling additive names

Once the ligands have been named, the name can be assembled. This is done by listing the ligand names in alphabetical order before the name of the central atom(s), *without* regard to ligand charge. If there is more than one ligand of a particular kind bound to a central atom in the same way, the number of such identical ligands is indicated using the appropriate multiplicative prefix for simple or complicated ligands (Table 3.2), not changing the already established alphabetical order of ligands. The nesting order of enclosing marks, for use in names where more than one set of enclosing marks is required, is: ( ), [( )], {[( )]}, ({[( )]}), *etc*. Any metal-metal bonds are indicated by placing the central atom symbols in parentheses, in italics and connected by an 'em' dash, after the name of the complex (without spaces). The charge number of the complex or the oxidation number of the central atom is appended to the name of the complex. For anions that are named additively, the name of the central atom is given the 'ate' ending in a similar way to the 'ide' endings of homoatomic anions (Section 3.1). In some cases, by tradition, the Latin stem is used for the 'ate' names, such as in ferrate (for iron), cuprate (for copper), argentate (for silver), stannate (for tin), aurate (for gold), and plumbate (for lead). Finally, the rules of compositional nomenclature (Section 3.1) are used to combine the additive names of ionic or neutral coordination entities with the names of any other entities that are part of the compound.

3.2.4 Specifying connectivity

Some ligands can bind to a central atom through different atoms under different circumstances. Specifying just which ligating (coordinating) atoms are bound in any given complex can be achieved by adding κ-terms to the name of the ligand. The κ-term comprises the Greek letter κ followed by the italicised element symbol of the ligating atom. For more complicated ligands the κ-term is often placed within the ligand name following the group to which the κ-term refers. Multiple identical links to a central atom can be indicated by addition of the appropriate numeral as a superscript between the κ and element symbols (see Table 3.6). These possibilities are discussed in more detail in the Red Book. If the ligating atoms of a ligand are contiguous (*i.e.*, directly bonded to one another), then an η-term is used instead, for example, for many organometallic compounds (Section 3.2.6) and the peroxido complex in Table 3.6.

Table 3.6 Producing names for complexes: complicated ligands

Structure to be named	Ba^{2+} 2 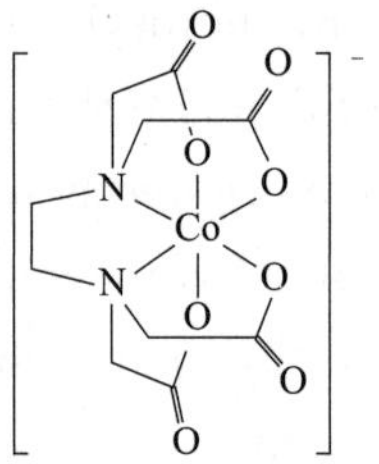	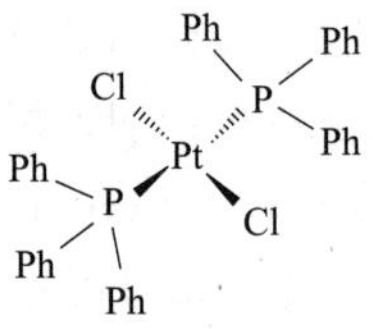
Central atom(s)	cobalt (Ⅲ) → cobaltate (Ⅲ)	platinum (Ⅱ)
Identify and name ligands	2,2′,2″,2‴-(ethane-1,2-diyl=dinitrilo)tetraacetate → 2,2′,2″,2‴-(ethane-1,2-diyl=dinitrilo)tetraacetato	chloride → chlorido triphenylphosphane
Specify ligating atoms	2,2′,2″,2‴-(ethane-1,2-diyl=dinitrilo-κ^2N)tetraacetato-κ^4O	*not required for chloride* triphenylphosphane-κ*P*
Assemble name	barium [2,2′,2″,2‴-(ethane-1,2-diyldinitrilo-κ^2N)tetra=acetato-κ^4O]cobaltate (Ⅲ)	dichloridobis(triphenyl=phosphane-κ*P*)platinum (Ⅱ)
Structure to be named	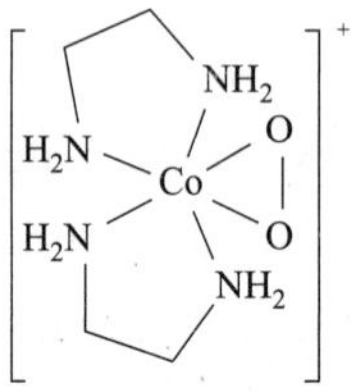	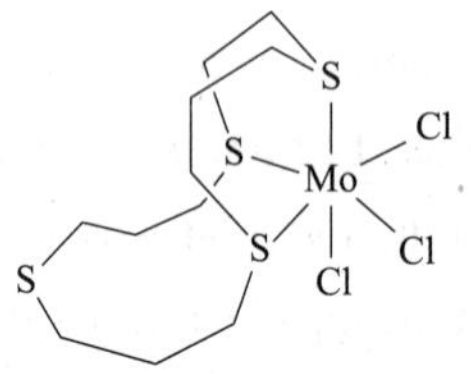
Central atom(s)	cobalt (Ⅲ)	molybdenum (Ⅲ)
Identify and name ligands	ethane-1,2-diamine peroxide → peroxido	chloride → chlorido 1,4,8,12-tetrathiacyclopentadecane
Specify ligating atoms	ethane-1,2-diamine-κ^2N η^2-peroxido	*not required for chloride* 1,4,8,12-tetrathiacyclo=pentadecane-κ^3S^1,S^4,S^8
Assemble name	bis(ethane-1,2-diamine-κ^2N)=(η^2-peroxido)cobalt (Ⅲ)	trichlorido(1,4,8,12-tetrathiacyclopentadecane-κ^3S^1,S^4,S^8)molybdenum (Ⅲ)

A κ-term is required for ligands where more than one coordination mode is possible. Typical cases are thiocyanate, which can be bound through either the sulfur atom (thiocyanato-κ*S*) or the nitrogen atom (thiocyanato-κ*N*), and nitrite, which can be bound through either the nitrogen atom (M-NO_2, nitrito-κ*N*), or an oxygen atom (M-ONO, nitrito-κ*O*). The names pentaammine(nitrito-κ*N*)cobalt(2+) and pentaammine(nitrito-κ*O*)cobalt(2+) are used for each of the isomeric nitrito complex cations. More examples of constructing names using κ-terms to specify the connectivity of ligands are shown in Table 3.6. A κ-term may also be used to indicate to which central atom a ligand is bound if there is more than one central atom (Section 3.2.5).

3.2.5 Bridging ligands

Bridging ligands are those bound to more than one central atom. They are differentiated in names by the addition of the prefix 'μ' (Greek mu), with the prefix and the name of the bridging ligand being separated from each other, and from the rest of the name, by hyphens. This is sufficient if the ligand is monoatomic, but if the ligand is more complicated it may be necessary to specify which ligating atom of the ligand is attached to which central atom. This is certainly the case if the ligating atoms are of different kinds, and κ-terms can be used for this purpose.

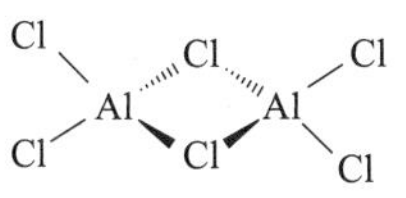

di-μ-chlorido-bis[di=chloridoaluminium(III)]
$[Cl_2Al(\mu\text{-}Cl)_2AlCl_2]$

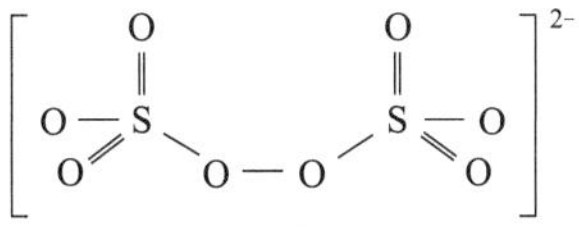

μ-peroxido-1κO^1, 2κO^2-bis(tri=oxidosulfate)(2−)
$[O_3S(\mu\text{-}O_2)SO_3]^{2-}$

3.2.6 Organometallic compounds

Organometallic compounds contain at least one bond between a metal atom and a carbon atom. They are named as coordination compounds, using the additive nomenclature system (see above). The name for an organic ligand binding through one carbon atom may be derived either by treating the ligand as an anion or as a neutral substituent group. The compound $[Ti(CH_2CH_2CH_3)Cl_3]$ is thus named as trichlorido(propan-1-ido)titanium or as trichlorido(propyl)titanium. Similarly, 'methanido' or 'methyl' may be used for the ligand—CH_3.

When an organic ligand forms two or three metal-carbon single bonds (to one or more metal centres), the ligand may be treated as a di- or tri-anion, with the endings 'diido' or 'triido' being used, with no removal of the terminal 'e' of the name of the parent hydrocarbon. Again, names derived by regarding such ligands as substituent groups and using the suffixes 'diyl' and 'triyl' are still commonly encountered. Thus, the bidentate ligand $-CH_2CH_2CH_2-$ would be named propane-1,3-diido (or propane-1,3-diyl) when chelating a metal centre, and μ-propane-1,3-diido(or μ-propane-1,3-diyl) when bridging two metal atoms.

Organometallic compounds containing a metal-carbon multiple bond are given substituent prefix names derived from the parent hydrides which end with the suffix 'ylidene' for a metal-carbon double bond and with 'ylidyne' for a triple bond. These suffixes either replace the ending 'ane' of the parent hydride, or, more generally, are added to the name of the parent hydride with insertion of a locant and elision of the terminal 'e', if present. Thus, the entity CH_3CH_2CH=as a ligand is named propylidene and $(CH_3)_2C$=is called propan-2-ylidene. The 'diido'/'triido' approach, outlined above, can also be used in this situation. The terms 'carbene' and 'carbyne' are not used in systematic nomenclature.

$P(C_6H_{11})_3$
Cl
Cl
Ru
Ph
$P(C_6H_{11})_3$

dichlorido(phenylmethylidene)bis(tricyclohexylphosphane-κ*P*) ruthenium,
dichlorido(phenylmethanediido)bis(tricyclohexylphosphane-κ*P*) ruthenium,
or (benzylidene)dichloridobis(tricyclohexylphosphane-κ*P*)ruthenium

The special nature of the bonding to metals of unsaturated hydrocarbons in a 'side-on' fashion *via* their π-electrons requires the eta (η) convention. In this 'hapto' nomenclature, the number of contiguous atoms in the ligand coordinated to the metal (the hapticity of the ligand) is indicated by a right superscript on the eta symbol, *e.g.*, η^3 ('eta three' or 'trihapto'). The η-term is added as a prefix to the ligand name, or to that portion of the ligand name most appropriate to indicate the connectivity, with locants if necessary. A list of many π-bonding unsaturated ligands, neutral and anionic, can be found in the Red Book.

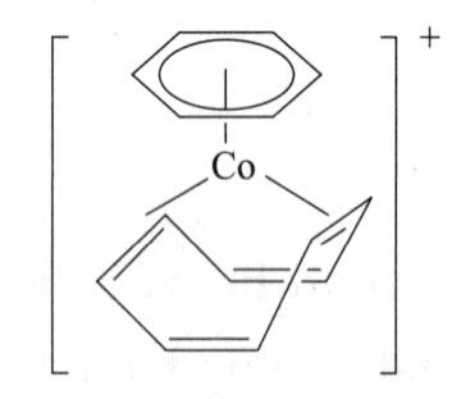

(η^6-benzene)[(1, 2, 5, 6-η)-cycloocta-1, 3, 5, 7-tetraene]cobalt(1+)

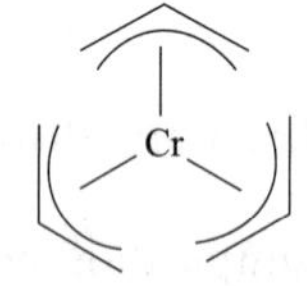

tris(η^3-prop-2-en-1-ido)chromium,
tris(η^3-prop-2-en-1-yl)chromium,
or tris(η^3-allyl)chromium

Note that the ubiquitous ligand η^5-C_5H_5, strictly η^5-cyclopenta-2,4-dien-1-ido, is also acceptably named η^5-cyclopentadienido or η^5-cyclopentadienyl. When cyclopenta-2,4-dien-1-ido coordinates through one carbon atom *via* a σ bond, a κ-term is added for explicit indication

H
Fe
OC
OC

dicarbonyl(η^5-cyclopentadienido)(cyclopenta-2,4-dien-1-ido-κC^1)iron
or dicarbonyl(η^5-cyclopentadienyl)(cyclopenta-2,4-dien-1-yl-κC^1)iron

of that bonding. The symbol η^1 should not be used, as the eta convention applies only to the bonding of contiguous atoms in a ligand.

Discrete molecules containing two *parallel* η^5-cyclopentadienido ligands in a 'sandwich' structure around a transition metal, as in bis(η^5-cyclopentadienido)iron, $[Fe(\eta^5\text{-}C_5H_5)_2]$, are generically called metallocenes and may be given 'ocene' names, in this case ferrocene. These 'ocene' names may be used in the same way as parent hydride names are used in substitutive nomenclature, with substituent group names taking the forms 'ocenyl' 'ocenediyl' 'ocenetriyl' (with insertion of appropriate locants).

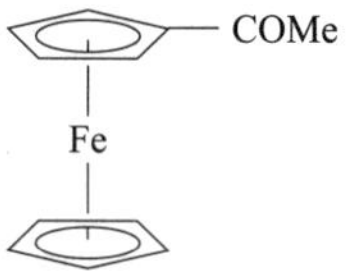

COMe
Fe

1-ferrocenylethan-1-one

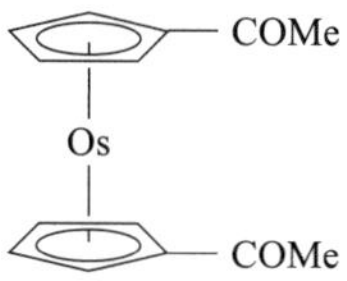

COMe
Os
COMe

1,1'-(osmocene-1,1'-diyl)di=(ethan-1-one)

By convention, 'organoelement' compounds of the main group elements are named by substitutive nomenclature if derived from the elements of Groups 13-16, but by additive nomenclature if derived from the elements of Groups 1 and 2. In some cases compositional nomenclature is used if less structural information is to be conveyed. More detail is provided in the Red Book.

3.2.7 Formulae of coordination compounds

Line formulae for coordination entities are constructed within square brackets to specify the composition of the entity. The overall process is shown in Table 3.7. The symbol for the central atom is placed first and is then followed by the symbols or abbreviations for the ligands (in alphabetical order according to the way they are presented in the formula). Where possible the coordinating (ligating) atom should be placed nearer the central atom in order to provide more information about the structure of the complex. If possible, bridging ligands should be placed between central atom symbols for this same reason (see examples in Section 3.2.5).

Table 3.7 Producing line formulae for complexes

Structure	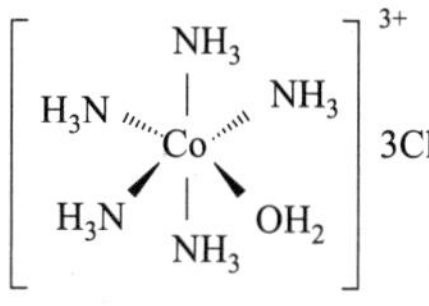 $[Co(NH_3)_4(H_3N)(OH_2)]^{3+}$ $3Cl^-$ (NH₃, H₃N, NH₃, H₃N, NH₃, OH₂ around Co)	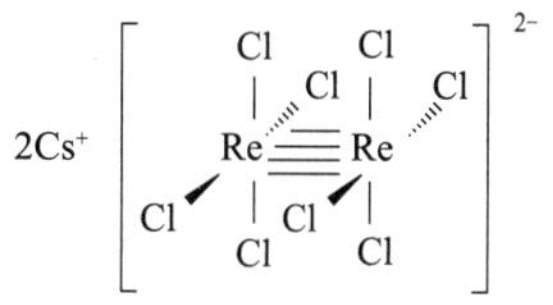 $2Cs^+$ $[Cl_4Re\equiv ReCl_4]^{2-}$
Central atom(s)	Co	2 × Re
Ligands	NH_3, OH_2	Cl
Assemble formula	$[Co(NH_3)_5(OH_2)]Cl_3$	$Cs_2[Cl_4ReReCl_4]$

(Continued)

Structure	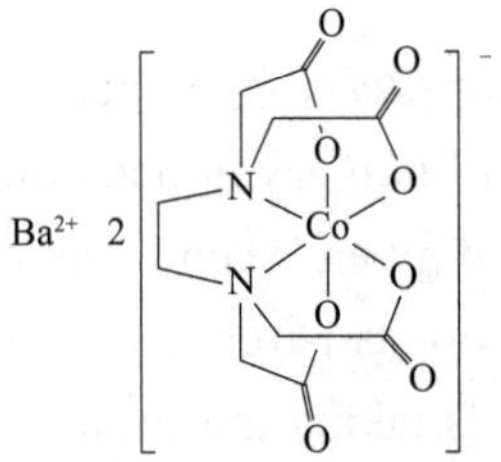 Ba^{2+} 2	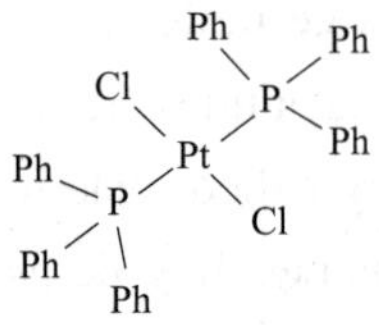
Central atom(s)	Co	Pt
Ligands	2,2′,2″,2‴-(ethane-1,2-diyldinitrilo) tetraacetate → edta	Cl triphenylphosphane → PPh_3
Assemble formula	$Ba[Co(edta)]_2$	$[PtCl_2(PPh_3)_2]$

Generally, ligand formulae and abbreviations are placed within enclosing marks (unless the ligand contains only one atom), remembering that square brackets are reserved to define the coordination sphere. Multiple ligands are indicated by a right subscript following the enclosing marks or ligand symbols.

3.2.8 Inorganic oxoacids and related compounds

Inorganic oxoacids, and the anions formed by removing the acidic hydrons (H^+) from them, have traditional names, many of which are well-known and can be found in many textbooks: sulfuric acid, sulfate; nitric acid, nitrate; nitrous acid, nitrite; phosphoric acid, phosphate; arsenic acid, arsenate; arsenous acid, arsenite; silicic acid, silicate; *etc.* These names are retained in IUPAC nomenclature, firstly because they almost invariably are the names used in practice, and secondly because they play a special role in organic nomenclature when names are needed for organic derivatives. However, all the oxoacids themselves and their derivatives may be viewed as coordination entities and named systematically using additive nomenclature (Table 3.8).

Table 3.8 Examples of inorganic oxoacids and derivatives

Formula	Traditional or organic name	Additive name
H_2SO_4 or $[S(O)_2(OH)_2]$	sulfuric acid	dihydroxido dioxido sulfur
$(CH_3)_2SO_4$ or $[S(O)_2(OMe)_2]$	dimethyl sulfate	dimethoxido dioxido sulfur or dimethanolato dioxido sulfur
H_2PHO_3 or $[P(H)(O)(OH)_2]$	phosphonic acid ①	hydrido dihydroxido oxido=phosphorus
$PhP(O)(OH)_2$	phenylphosphonic acid	dihydroxidooxido(phenyl)=phosphorus

① The term ‘phosphorous acid’ has been used in the literature for both the species named phosphonic acid in Table 3.8 and that with the formula $P(OH)_3$, trihydroxidophosphorus. It is used in organic nomenclature in the latter sense.

The traditional oxoacid names may be modified according to established rules for naming derivatives formed by functional replacement: thus ‘thio’ denotes replacement of =O by=S;

prefixes 'fluoro' 'chloro', *etc.*, and infixes 'fluorid' 'chlorid', *etc.*, denote replacement of OH by F, Cl, *etc.*; 'peroxy'/ 'peroxo' denote replacement of O by OO; and so forth (Table 3.9).

Table 3.9 Examples of derivatives of inorganic oxoacids and anions formed by functional replacement

Formula	Name indicating functional replacement	Additive name
H_3PS_4 or $[P(S)(SH)_3]$	tetrathiophosphoric acid or phosphorotetrathioic acid	tris(sulfanido)sulfido=phosphorus
H_2PFO_3 or $[PF(O)(OH)_2]$	fluorophosphoric acid or phosphorofluoridic acid	fluoridodihydroxido=oxidophosphorus
$S_2O_3^{2-}$ or $[S(O)_3(S)]^{2-}$	thiosulfate or sulfurothioate	trioxidosulfido=sulfate(2–)
$[O_3S(\mu\text{-}O_2)SO_3]^{2-}$	peroxydisulfate	see Section 3.2.5

If all hydroxy groups in an oxoacid are replaced, the compound is no longer an acid and is not named as such, but will have a traditional functional class name as, *e.g.*, an acid halide or amide. Such compounds may again be systematically named using additive nomenclature (Table 3.10).

Table 3.10 Examples of functional class names and corresponding additive names

Formula	Functional class name	Additive name
PCl_3O	phosphoryl trichloride	trichloridooxido=phosphorus
SCl_2O_2	sulfuryl dichloride	dichloridodioxidosulfur
$S(NH_2)_2O_2$	sulfuric diamide	diamidodioxidosulfur

A special construction is used in hydrogen name, which allows the indication of hydrons bound to an anion without specifying exactly where. In such names, the word 'hydrogen' is placed at the front of the name with a multiplicative prefix (if applicable) and with no space between it and the rest of the name, which is placed in parentheses. For example, dihydrogen(diphosphate)(2–) denotes $H_2P_2O_7^{2-}$, a diphosphate ion to which two hydrons have been added, with the positions not known or at least not being specified. One may view the common names for partially dehydronated oxoacids, such as hydrogenphosphate, HPO_4^{2-}, and dihydrogenphosphate, $H_2PO_4^-$, as special cases of such hydrogen names. In these simplified names, the charge number and the parentheses around the main part of the name are left out. Again, these particular anions may be named systematically by additive nomenclature. The word 'hydrogen' is placed *separately* in forming analogous names in organic nomenclature, for example, dodecyl hydrogen sulfate, $C_{12}H_{25}OS(O)_2OH$. This difference between the two systems has the consequence that the important carbon containing ion HCO_3^- can be named equally correctly as 'hydrogen carbonate' and as 'hydrogencarbonate' (but not as bicarbonate).

3.3 Stereo Descriptors

The approximate geometry around the central atom is described using a polyhedral symbol placed in front of the name. The symbol is made up of italicised letter codes for the geometry and a number that indicates the coordination number. Frequently used polyhedral symbols are *OC*-6 (octahedral), *SP*-4 (square-planar), *T*-4 (tetrahedral), *SPY*-5 (square-pyramidal), and *TBPY*-5 (trigonalbipyramidal). More complete lists are available.

The relative positions of ligating groups around a central atom can be described using a configuration index that is determined in a particular way for each geometry, based on the Cahn-Ingold-Prelog priorities of the ligating groups, and it may change if the ligands change, even if the geometry remains the same. The absolute configuration can also be described. Generally, configuration indices are used only if there is more than one possibility and a particular stereoisomer is to be identified. The full stereo descriptors for the particular square-planar platinum complexes shown below are (*SP*-4-2) and (*SP*-4-1), for the *cis* and *trans* isomers, respectively. Alternatively, a range of traditional stereo descriptors may be used in particular situations. Thus, the isomers that are possible when a square-planar centre is coordinated by two ligating groups of one type and two of another are referred to as *cis*- (when the identical ligands are coordinated next to each other) or *trans*- (when they are coordinated opposite to each other).

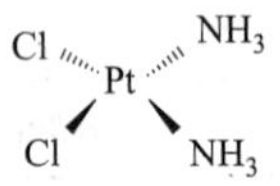

Cl Pt NH_3 Cl NH_3

cis-diamminedichloridoplatinum(II)

Cl Pt NH_3 H_3N Cl

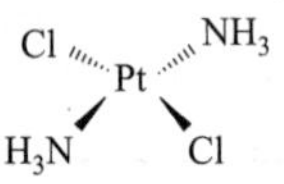

trans-diamminedichloridoplatinum(II)

Octahedral centres with four ligands of one kind and two of another can also be referred to as *cis*- (when the two identical ligands are coordinated next to each other) or *trans*- (when they are coordinated opposite each other). Octahedral centres with three of each of two kinds of ligands can be described as *fac*- (facial), when the three ligands of a particular kind are located at the corners of a face of the octahedron, or *mer*- (meridional), when they are not.

3.4 Summary

This document provides an outline of the essential nomenclature rules for producing names and formulae for inorganic compounds, coordination compounds, and organometallic compounds. The complementary document for nomenclature systems of organic chemistry will also be useful to the reader.

Names and formulae have only served half their role when they are created and used to describe or identify compounds, for example, in publications. Achieving their full role requires that the reader of a name or formula is able to interpret it successfully, for example, to produce a structural diagram. The present document is also intended to assist in the interpretation of names and formulae. Finally, we note that IUPAC has produced recommendations on the graphical representation of chemical structures and their stereochemical configurations.

References

[1] http://www.chem.qmul.ac.uk/iupac/.

[2] Hiorns R C, Boucher R J, Duhlev R, Hellwich K-H, Hodge P, Jenkins A D, Jones R G, Kahovec J, Moad G, Ober C K, Smith D W, Stepto R F T, Vairon J-P, Vohlídal J. Pure Appl. Chem., 2012, 84(10): 2167.

[3] Leigh G J. Principles of Chemical Nomenclature–A Guide to IUPAC Recommendations. Royal Society of Chemistry, Cambridge, U.K., 2011.

[4] Connelly N G, Damhus T, Hartshorn R M, Hutton A T. Nomenclature of Inorganic Chemistry–IUPAC Recommendations 2005. Royal Society of Chemistry, Cambridge, U.K., 2005.

[5] Favre H A, Powell W H. Nomenclature of Organic Chemistry–IUPAC Recommendations and Preferred Names 2013. Royal Society of Chemistry, Cambridge, U.K., 2014.

[6] Jones R G , Wilks E S, Metanomski W V, Kahovec J, Hess M. Compendium of Polymer Terminology and Nomenclature–IUPAC Recommendations 2008. Royal Society of Chemistry, Cambridge, U.K., 2009.

[7] Cahn R S, Ingold C, Prelog V. Angew. Chem., Int. Ed. Engl., 1996, 5: 385 and 511.

[8] Prelog V, Helmchen G. Angew. Chem., Int. Ed. Engl., 1982, 21: 567.

[9] Brecher J, Degtyarenko K N, Gottlieb H, Hartshorn R M, Moss G P, Murray-Rust P, Nyitrai J, Powell W, Smith A, Stein S, Taylor K, Town W, Williams A, Yerin A. Pure Appl. Chem., 2006, 78(10): 1897.

[10] Brecher J, Degtyarenko K N, Gottlieb H, Hartshorn R M, Hellwich K-H, Kahovec J, Moss G P, McNaught A, Nyitrai J, Powell W, Smith A, Taylor K, Town W, Williams A, Yerin A. Pure Appl. Chem., 2008, 80(2): 277.

Note: This chapter is quoted completely from the paper titled as *Brief Guide to the Nomenclature of Inorganic Chemistry* (IUPAC, *Pure Appl. Chem.* 2015, 87: 1039-1049), except for some necessary changes. (i. The numbers of the chapters, tables and charts etc. are changed to be suitable for this book; ii. Some pictures have been remade for reason of resolution; iii. The color of some words and pictures are changed to black, because this book will be printed in black and white.)

Key Words & Phrases

nomenclature *n.* 命名法
systematic name 系统命名法；系统名称
peroxide *n.* 过氧化物；过氧化氢
azide *n.* 叠氮化物
stoichiometric or compositional name 化学计量或组成名称
homoatomic *adj.* 同原子的
anion *n.* 阴离子
binary compound 二元化合物
element sequence 元素序列
electronegative *adj.* 电负性的
electropositive *adj.* 阳性的；带正电的；正电性的
heteropolyatomic *adj.* 杂多原子的
hydroxide *n.* 氢氧化物
cation *n.* 阳离子
locant *n.* 位次
hydrate *n.* 水合物
coordination *n.* 协作；协调；配合；配合物
mononuclear complex 单核配合物
ligand *n.* 配体
carbonyl *n.* 羰基
metal-metal bond 金属 - 金属键
charge number 电荷数
oxidation number 氧化数
thiocyanate *n.* 硫氰酸盐；硫氰酸酯
organometallic compound 金属有机化合物
parent hydride 母体氢化物
carbene *n.* 碳烯；卡宾
carbyne *n.* 碳炔；卡拜
contiguous *adj.* 相接的；相邻的
hapticity *n.* 扣数
ubiquitous *adj.* 无处不在的；十分普遍的
metallocene *n.* 茂金属
ferrocene *n.* 二茂铁
inorganic oxoacid 无机含氧酸
traditional name 传统名称
sulfuric acid 硫酸
sulfate *n.* 硫酸盐
nitric acid 硝酸
nitrate *n.* 硝酸盐

nitrous acid　亚硝酸
nitrite　*n.* 亚硝酸盐
phosphoric acid　磷酸
phosphate　*n.* 磷酸盐
arsenic acid　砷酸
arsenate　*n.* 砷酸盐
arsenous acid　亚砷酸
arsenite　*n.* 亚砷酸盐
silicic acid　硅酸
silicate　*n.* 硅酸盐
derivative　*n.* 衍生物
hydroxy　*n.* 羟基；氢氧根的
dodecyl　*n.* 十二烷基
stereo descriptor　立体描述符
polyhedral symbol　多面体符号
octahedral　*n.* 八面体
square-planar　平面正方形
tetrahedral　*adj.* 四面体的；有四面的
square-pyramidal　四角锥形
trigonal bipyramidal　三角双锥的
absolute configuration　绝对构型
stereoisomer　*n.* 立体异构体
cis and *trans* isomer　顺反异构体
meridional　*adj.* 子午线的；经向的

Chapter 4 Brief Guide to the Nomenclature of Organic Chemistry

4.1 Introduction

The universal adoption of an agreed nomenclature is a key tool for efficient communication in the chemical sciences, in industry and for regulations associated with import/export or health and safety. The International Union of Pure and Applied Chemistry (IUPAC) provides recommendations on many aspects of nomenclature. The basics of organic nomenclature are summarized here, and there are companion documents on the nomenclature of inorganic and polymer chemistry, with hyperlinks to original documents. An overall summary of chemical nomenclature can be found in Principles of Chemical Nomenclature. Comprehensive detail can be found in Nomenclature of Organic Chemistry, colloquially known as the Blue Book, and in the related publications for inorganic compounds (the Red Book), and polymers (the Purple Book).

It should be noted that many compounds may have non-systematic or semi-systematic names and IUPAC rules also allow for more than one systematic name in many cases. Some traditional names (*e.g.*, styrene, urea) are also used within systematic nomenclature. The new edition of the Blue Book incorporates a hierarchical set of criteria for choosing the single name which is to be preferred for regulatory purposes, the Preferred IUPAC Name, or PIN.

4.2 Substitutive Nomenclature

Substitutive nomenclature is the main method for naming organic chemical compounds. It is used mainly for compounds of carbon and elements of Groups 13-17. For naming purposes, a chemical compound is treated as a combination of a parent compound (Section 4.5) and characteristic (functional) groups, one of which is designated the principal characteristic group (Section 4.4). A systematic name is based on the name of the most senior parent compound (Section 4.6) in which the substitution of hydrogen atoms is represented by a suffix for the principal characteristic group(s), prefixes representing less senior characteristic groups and other substituent groups, and locants that specify their locations. Names created according to substitutive nomenclature may also include fragments named in accordance with other nomenclature types or operations. For example, addition and subtraction operations (Section 4.5.4) are performed mainly to define the hydrogenation state, while a replacement operation defines a replacement of (in most cases) carbon atoms with heteroatoms.

Components of systematic substitutive names

The most common components of a substitutive chemical name are illustrated with reference to the chemical structure shown in Table 4.1, along with its systematic name and the components of the name. Locants indicate the position of substituents or other structural features. They are generally placed before the part of the name that indicates the corresponding structural feature. Three kinds of enclosing mark are used, in the nesting order {[( )]}, when it is necessary to indicate which parts of a name belong together.

Table 4.1 Components of the substitutive name (4*S*,5*E*)-4,6-dichlorohept-5-en-2-one for

hept(a)	parent (heptane)	one	suffix for principal characteristic group
en(e)	unsaturation ending	chloro	substituent prefix
di	multiplicative prefix	*E*, *S*	stereo descriptors
2 4 5 6	locants	( )	enclosing marks

Multiplicative prefixes (Table 3.2) are used when more than one fragment of a particular kind is present in a structure. Which kind of multiplicative prefix is used depends on the complexity of the corresponding fragment–*e.g.* trichloro, but tris(chloromethyl).

4.3 Creation of Systematic Names

The formation of a systematic name requires several steps, to be taken (when they are applicable) in the following order:

(1) Determine the principal characteristic group to be cited as the suffix (see Section 4.4).

(2) Determine the senior parent amongst those structural components attached to a principal characteristic group (see Sections 4.5 and 4.6).

(3) Name the parent hydride and specify any unsaturation (Section 4.5).

(4) Combine the name of the parent hydride with the suffix for the principal characteristic group (Section 4.4).

(5) Identify the substituents and arrange the corresponding prefixes in alphabetical order.

(6) Insert multiplicative prefixes, without changing the already established order, and insert locants.

(7) Determine chirality centres and other stereogenic units, such as double bonds, and add stereodescriptors.

4.4 Characteristic Groups—Suffixes and Prefixes

The presence of a characteristic (or functional) group is denoted by a prefix or suffix attached to the parent name. The names of common characteristic groups are given in Table 4.2, in order of decreasing seniority. The most senior one, the principal characteristic group, is cited as the suffix, while all other groups are cited as prefixes. Note that, for nomenclature purposes, C—C multiple bonds are not considered to be characteristic groups (Section 4.5.4).

Table 4.2 Seniority order for characteristic groups

Class	Formula①	Suffix	Prefix
Carboxylates	—COO— —(C)OO—	carboxylate oate	carboxylato
Carboxylic acids	—COOH —(C)OOH	carboxylic acid oic acid	carboxy
Esters	—COOR —(C)OOR	(R) ...carboxylate② (R) ...oate②	(R)oxycarbonyl
Acid halides	—COX —(C)OX	carbonyl halide oyl halide	halocarbonyl
Amides	$—CONH_2$ $—(C)ONH_2$	carboxamide amide	carbamoyl
Nitriles	—C≡N —(C)≡N	carbonitrile nitrile	cyano
Aldehydes	—CHO —(C)HO	carbaldehyde al	formyl oxo
Ketones	=O	one	oxo
Alcohols	—OH	ol	hydroxy
Thiols	—SH	thiol	sulfanyl③
Amines	$—NH_2$	amine	amino
Imines	=NH	imine	imino

① Here —(C) indicates that the carbon atom is implied by the parent name.
② Here (R) means that the group R is expressed as a separate prefixed word.
③ Note: 'mercapto' is no longer acceptable (but is still used by CAS).

Depending on the number and arrangement of carbon-containing suffix groups, the carbon atom can be a part of the parent compound [*e.g.*—(C)OOH, 'oic acid'] or may be treated as an attachment to a parent compound [*e.g.*—COOH, 'carboxylic acid'].

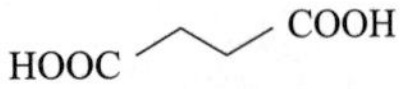

butanedioic acid

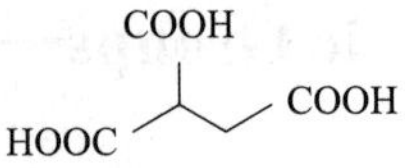

ethane-1,1,2-tricarboxylic acid

propanedinitrile

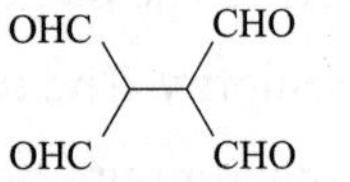

ethane-1,1,2,2-tetracarbaldehyde

Other characteristic groups on a parent compound are represented by appropriate prefixes cited in alphabetical order [here in blue (Note: In the original paper, different functional groups are distinguished by color), where R represents an alkyl or aryl group], including also ethers (—OR), (R)oxy; sulfides (—SR), (R)sulfanyl; —Br, bromo; —Cl, chloro; —F, fluoro; —I, iodo; and —NO_2, nitro.

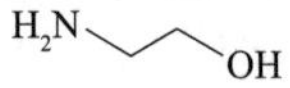

2-aminoethan-1-ol

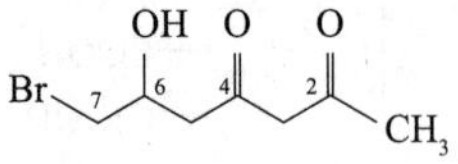

7-bromo-6-hydroxyheptane-2,4-dione

4.5 Parent Compounds, Parent Hydrides

Several types of parent compounds are used in substitutive nomenclature. Parent compounds without characteristic groups are called parent hydrides. These can be classified as either chains or rings, and may contain carbon atoms and/or heteroatoms. The ring parent compounds can be monocyclic, bridged polycyclic (rings sharing more than two atoms), fused polycyclic (rings sharing two neighbouring atoms), or spiro polycyclic (rings sharing only one atom). More complex parent compounds include bridged fused systems, ring assemblies, cyclophanes, and fullerenes. The atom numbering of a parent compounds is defined by the corresponding rules for each type of parent compound. Thereafter, the rules outlined in Section 4.7 are applied.

4.5.1 Acyclic parent hydrides

The names for saturated carbon chains (alkanes) are composed of the simple numerical term indicating the number of carbon atoms (Table 3.2, with the 'a' elided) together with an 'ane' ending (see Table 4.3), with the exception of the first four alkanes: methane, CH_4; ethane, CH_3CH_3; propane, $CH_3CH_2CH_3$; butane, $CH_3(CH_2)_2CH_3$.

Table 4.3 Names for some linear alkanes

$CH_3(CH_2)_3CH_3$	$CH_3(CH_2)_7CH_3$	$CH_3(CH_2)_{18}CH_3$
pentane	nonane	icosane
$CH_3(CH_2)_4CH_3$	$CH_3(CH_2)_{16}CH_3$	$CH_3(CH_2)_{20}CH_3$
hexane	octadecane	docosane

4.5.2 Monocyclic parent hydrides

The names of saturated carbon monocycles (cycloalkanes) are composed of the prefix 'cyclo' and the name of the corresponding alkane.

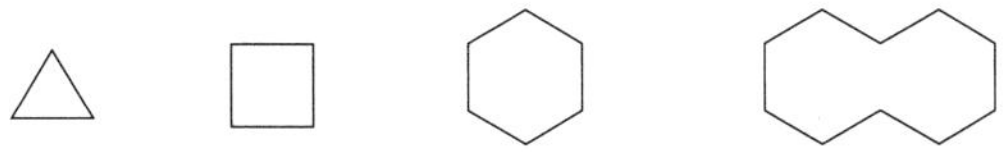

cyclopropane cyclobutane cyclohexane cyclodecane

A number of non-systematic names have been retained for common rings, for example benzene and the following heterocycles.

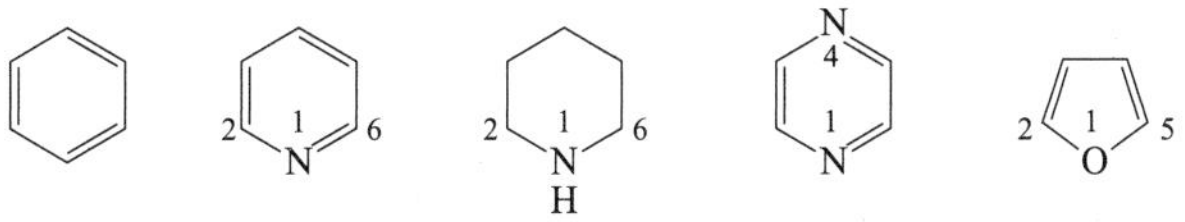

benzene pyridine piperidine pyrazine furan

Systematic names for monocycles that contain heteroatoms are constructed in accordance with either the Hantzsch-Widman (H-W) system (3- to 10-membered rings) or replacement nomenclature (larger rings). Both systems make use of the 'a' prefixes shown in Table 4.4, in which the seniority decreases from left to right across the first row and then the second row.

Table 4.4 Selected 'a' prefixes for H-W and replacement systems

O	oxa	S	thia	N	aza	P	phospha
As	arsa	Si	sila	Sn	stanna	B	bora

The H-W system combines the 'a' prefixes of Table 4.4 in decreasing order of seniority with endings, in the H-W system called stems, that indicate the size and saturation of the ring (Table 4.5). Appropriate locants are added to describe the location of the replacements in the ring and the 'a' is elided when followed by a vowel. If there are more than 10 atoms in the ring, replacement nomenclature is used, in which 'a' prefixes are again listed in decreasing order of seniority, with locants, before the parent name. The atom numbering is explained in Section 4.7.

Table 4.5 Stems in the Hantzsch-Widman system

Ring size	Unsaturated	Saturated
3	irine① /irene	iridine/iirane②
4	ete	etidine/etane②
5	ole	olidine/olane②
6	ine/ine/inine③	ane/inane/inane③
7	epine	epane

① For rings with only N heteroatom(s); ② For rings with/without N heteroatom(s); ③ For O, S / N, Si, Sn / P, As, B as the last cited heteroatom, respectively.

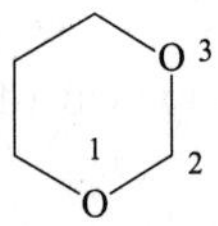

1,3-dioxane

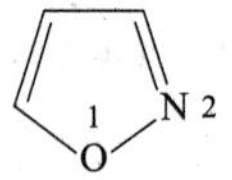

1,2-oxazole

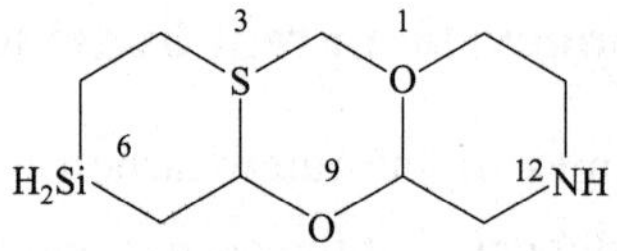

1,9-dioxa-3-thia-12-aza-
6-silacyclotetradecane

4.5.3 Polycyclic parent hydrides

The names of bridged polycyclic systems are based on the name of the alkane with the same number of carbon atoms, which is preceded by an indicator of the number of cycles present and a bridge descriptor that defines the sizes of the various rings; this descriptor gives the number of skeletal atoms in each of the bridges connecting the bridgeheads and is given by arabic numerals cited in descending numerical order, separated by full stops and enclosed in square brackets. Numbering starts at a bridgehead and goes around the rings in order (largest to smallest). Replacement nomenclature (see Section 4.5.2) is used to name the related heterocycles.

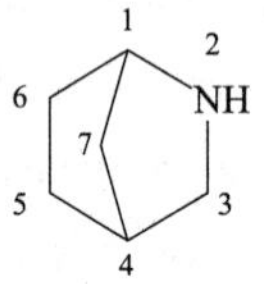

2-azabicyclo[2.2.1]heptane

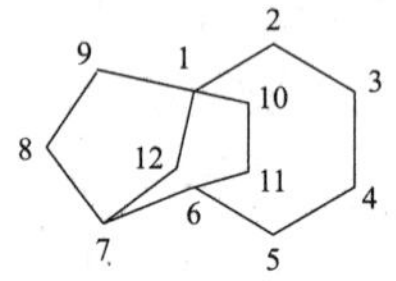

tricyclo[4.3.2.$1^{1,7}$]dodecane

The names of spiro polycyclic systems, in which there is a single atom in common to the rings, include the number of spiro junctions, a bridge descriptor, and the name of the alkane with the same number of carbon atoms. Again, the related heterocycles are named in accordance with replacement nomenclature (see Section 4.5.2).

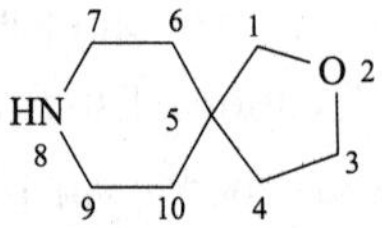

2-oxa-8-azaspiro[4.5]decane

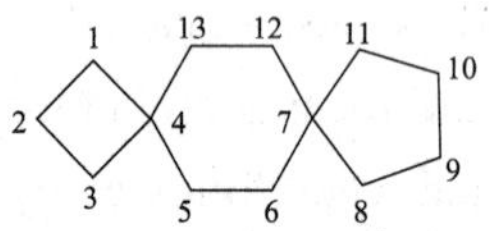

dispiro[3.2.4.7.2^4]tridecane

Fused polycycles are cyclic systems having one common bond for any pair of adjacent rings.

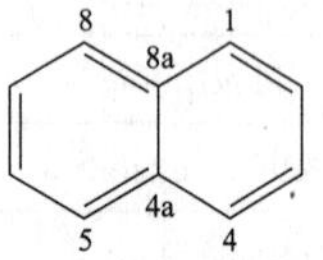

naphthalene

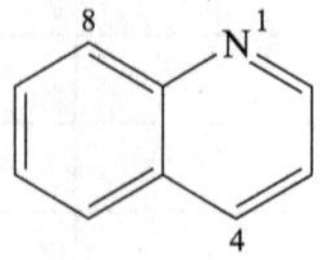

quinoline

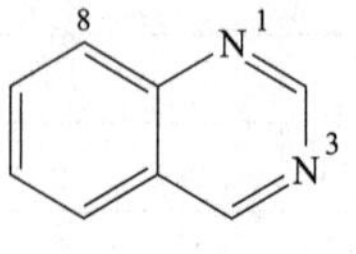

quinazoline

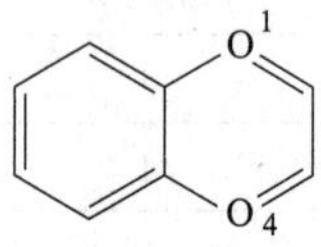

1,4-benzodioxine

In the systematic nomenclature of fused polycycles, the names for the components are combined and a fusion descriptor indicates how the components are connected. This process is beyond the scope of the current guide (see Ref. 6 for details).

furo[2,3-b]pyridine　　benzo[g]quinoline

4.5.4 Saturation and unsaturation

The degree of unsaturation of a compound in comparison to a saturated parent can be indicated by replacement of the 'ane' ending by 'ene' and 'yne' endings that define the presence of double and triple bonds, respectively, and addition of locants to define their locations.

buta-1,3-diene　　pent-1-en-4-yne　　cyclohexa-1,3-diene

The addition of hydrogen to unsaturated parent hydrides is represented by addition of hydro prefixes to indicate saturation of double bonds, again with locants to define where this occurs.

3,4-dihydropyridine　　1,2,3,4-tetrahydroisoquinoline

For some unsaturated parent hydrides, the saturated positions are specified using the indicated hydrogen convention.

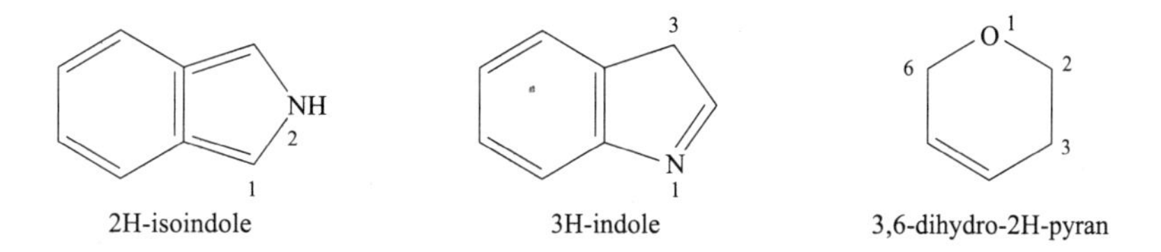

2H-isoindole　　3H-indole　　3,6-dihydro-2H-pyran

4.5.5 Substituent groups derived from parent hydrides

In cases where a group derived from a parent hydride is a substituent on another parent compound, the substituent name is created by addition of the suffixes 'yl' or 'ylidene' to the parent hydride name, with the corresponding locants indicating the position of the attachment. The attachment positions expressed by the suffixes 'yl' or 'ylidene' are senior to any characteristic group (see Section 4.4, Table 4.2).

butyl　　butan-2-ylidene　　pyrimidin-5-yl

4.5.6 Functional parents

A combination of a parent hydride with a functional group may form a functional parent named as single entity. Such names are used as systematic names only if they express the parent compound and the most senior characteristic group of the compound under consideration, *e.g.*, 4-chloroaniline, but 4-aminobenzoic acid (not 4-carboxyaniline or aniline-4-carboxylic acid).

NH_2 aniline　　OH phenol　　COOH benzoic acid　　H_3C O H acetaldehyde

4.6 Seniority of Parent Compounds

The systematic name is based on the name of the senior parent compound, which is chosen by applying the following criteria in the order described below and shown in Figure 4.1, until a decision is reached. For a complete set of criteria see Ref. 6 (section P-52).

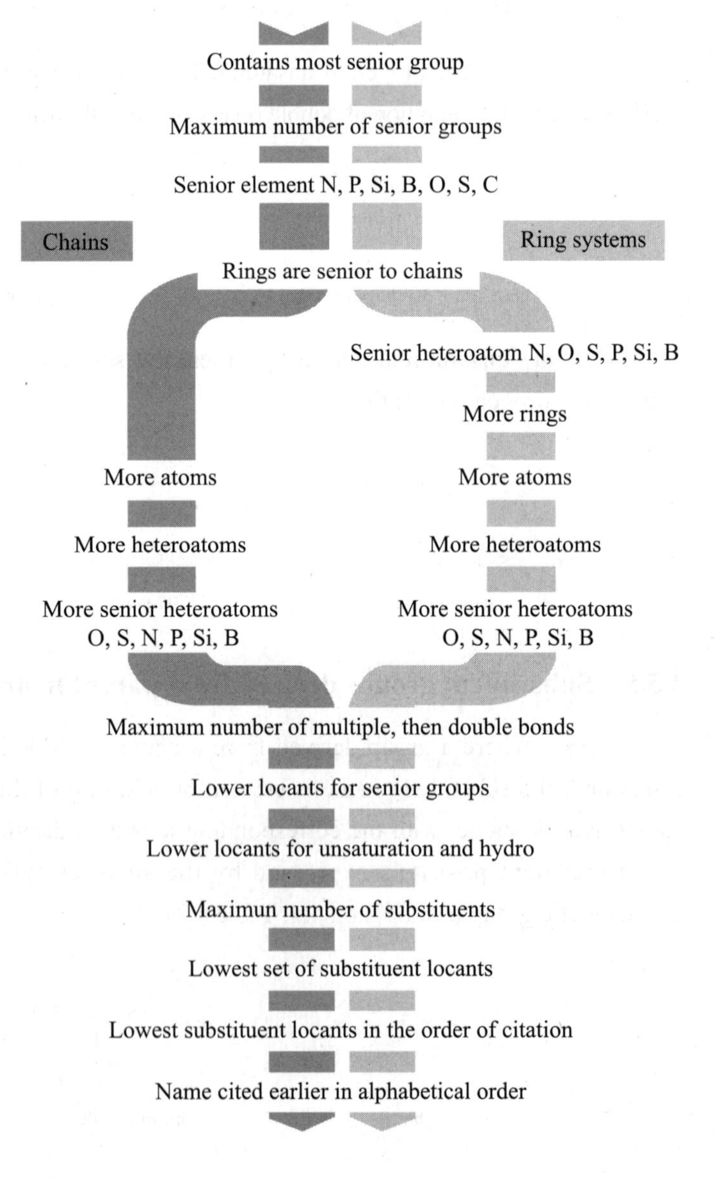

Figure 4.1 Criteria for choosing the senior parent compound

(1) Contains the principal characteristic group

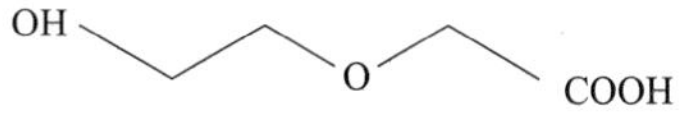

(2-hydroxyethoxy)acetic acid

acid is senior to alcohol

(2) Maximum number of principal characteristic groups

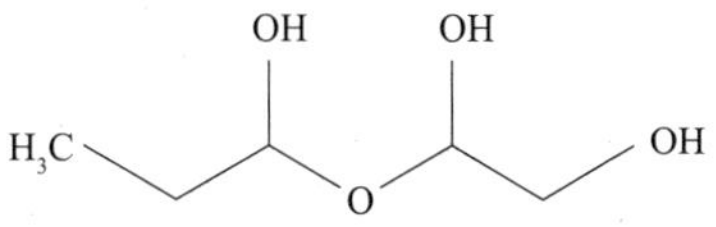

1-(1-hydroxypropoxy)ethane-1,2-diol

parent with two characteristic groups is senior

(3) Parent based on senior element (N, P, Si, B, O, S, C)

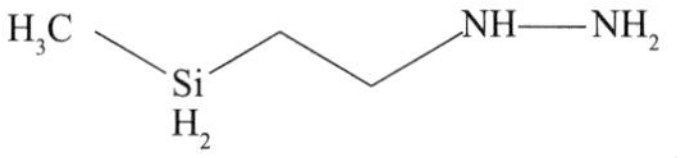

[2-(methylsilyl)ethyl]hydrazine

hydrazine is senior to silane(N senior to Si)

(4) Rings are senior to chains if composed of the same elements

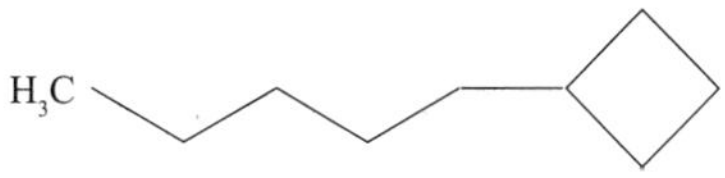

pentylcyclobutane

cyclobutane is senior to pentane

Note 1: After this criterion only rings or only chains remain for further choice.

Note 2: In earlier recommendations, seniority depended on number of atoms.

(5) Criteria for cyclic systems

① Contains most senior heteroatom in the order N, O, S, P, Si, B.

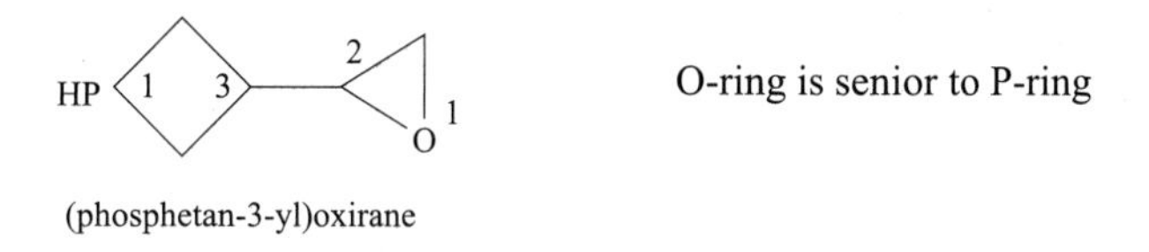

(phosphetan-3-yl)oxirane

O-ring is senior to P-ring

② Contains more rings

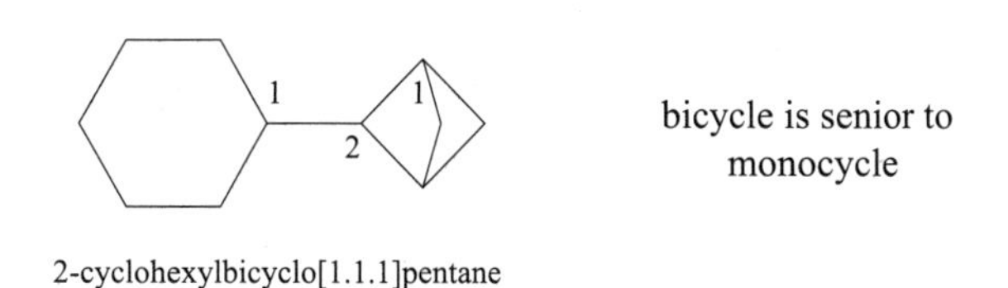

2-cyclohexylbicyclo[1.1.1]pentane

bicycle is senior to monocycle

③ Contains more atoms

cyclopentane is senior to cyclobutane

cyclobutylcyclopentane

④ Contains more heteroatoms

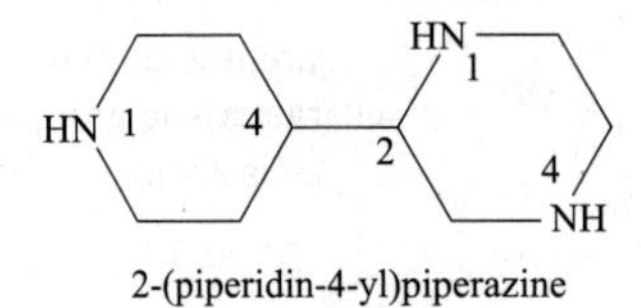

2-(piperidin-4-yl)piperazine

piperazine, having two heteroatoms, is senior to piperidine

⑤ Contains more senior heteroatoms

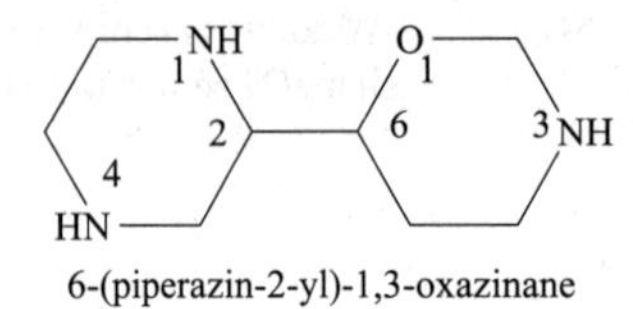

6-(piperazin-2-yl)-1,3-oxazinane

oxazinane, containing O and N, is senior to piperazine with two N atoms

(6) Criteria for chains　Contains more atoms

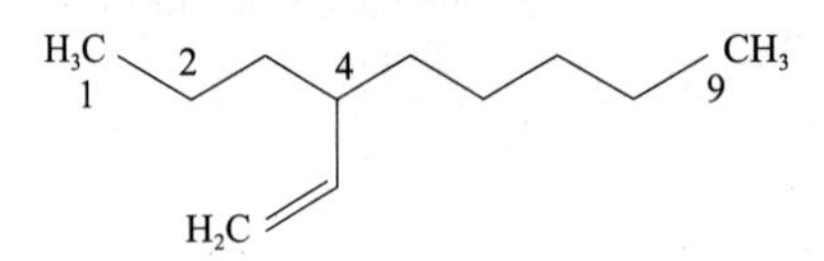

4-ethenylnonane

nine-atom chain is senior to eight-atom chain (even if it has fewer double bonds)

Note: In earlier recommendations unsaturation was senior to chain length

The following criteria are then applied to chains as well as rings:

(7) Contains more multiple, and then double bonds

3-ethylocta-1,4-diene

3-[(piperidin-2-yl)oxy]pyridine

(8) Having lower locants for principal characteristic groups

butan-1-ol is senior to butan-2-ol

1-(3-hydroxybutoxy)butan-1-ol

(9) Lower locants for unsaturation or hydro prefixes

3-[(but-2-en-2-yl)oxy]but-1-ene

5-[(3,4-dihydropyridin-3-yl)oxy]-2,5-dihydropyridine

(10) Maximum number of substituents

2-(1-bromo-2-hydroxyethoxy)-2,2-dichloroethan-1-ol

parent with three substituents is senior to parent with two substituents

(11) Lowest set of locants for all substituents

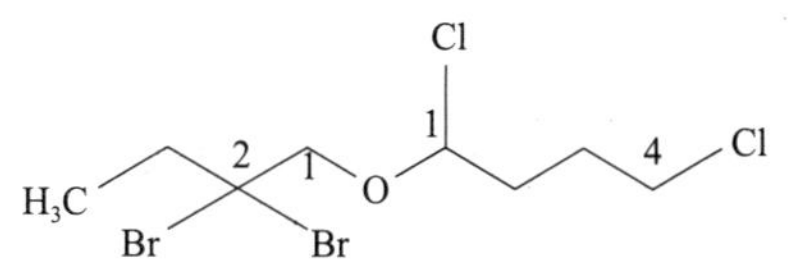

1,4-dichloro-1-(2,2-dibromobutoxy)butane
Note: Not 2,2-dibromo-1-(1,4-dichlorobutoxy)butane

all parent locants are arranged in increasing order and compared one by one: 1,1,4 is lower than 1,2,2

(12) Lowest substituent locants in order of citation

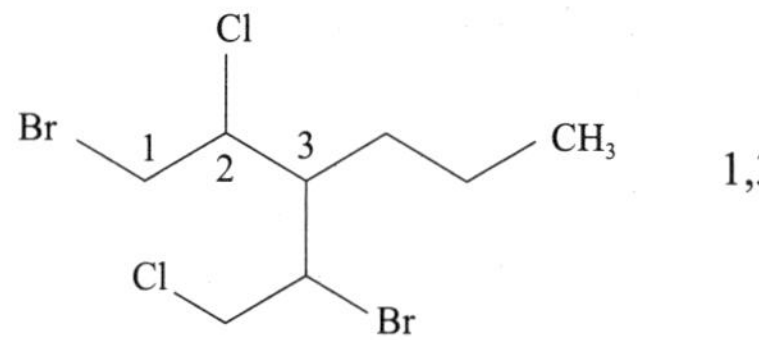

1,3,2 is lower than 2,3,1

1-bromo-3-(1-bromo-2-chloroethyl)-2-chlorohexane
Note: Not 2-bromo-3-(2-bromo-1-chloroethyl)-1-chlorohexane

(13) Name appearing earlier in alphabetical order

1-bromo-2-(2-chloroethoxy)ethane

Note: Not 1-(2-bromoethoxy)-2-chloroethane

4.7 Numbering of Parent Compounds

The numbering of the parent compound is determined by the compound class and then chosen by considering all possible sets of locants and successively applying the following criteria:

① Lowest locants for heteroatoms;
② Lowest locant(s) for indicated hydrogen;
③ Lowest locant(s) for principal characteristic group(s);
④ Lowest locants for 'ene' 'yne', and hydro prefixes;
⑤ Lowest locants as a set for all substituents cited by prefixes;
⑥ Lowest locants for substituents in the order of citation.

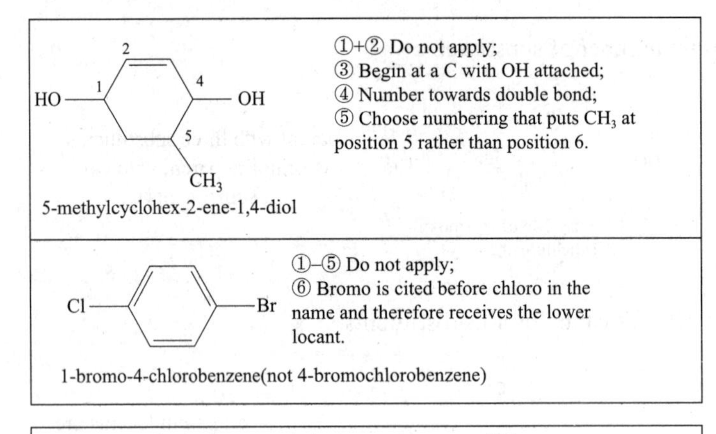

Correct numbering is extremely important, because a single incorrect locant makes it impossible for the reader of the name to work out the correct structure.

4.8 Functional Class Nomenclature

Functional class names (formerly radicofunctional names) are preferred for esters and acid halides. For other compound classes (*e.g.*, ethers, ketones, sulfoxides, and sulfones) functional class names are still in use, although substitutive names are preferred.

Functional class names consist of one or more substituent names, ordered alphabetically, and followed by the compound class name (separated with spaces as required). Thus $CH_3C(O)O—CH_3$ is named methyl acetate, $ClCH_2C(O)O—CH_3$ is methyl chloroacetate, $CH_3C(O)—Cl$ is acetyl chloride, $C_6H_5C(O)—Br$ is benzoyl bromide, and $(H_3C)_2SO_2$ is named dimethyl sulfone.

H_3C O CH_3 O — methyl propanoate

H_3C O CH_3 — diethyl ether or ethoxyethane

O CH_3 H_3C — ethyl methyl ketone or butan-2-one

4.9 Specifying Configuration of Stereoisomers

Stereoisomers are differentiated from each other by stereodescriptors cited in names and assigned in accordance with the Cahn-Ingold-Prelog (CIP) rules. The most common descriptors are those for the absolute configuration of tetrahedral stereogenic centres (R/S) and those for the configuration of double bonds (E/Z). Locants are added to define the locations of the stereogenic centres and the full set of descriptors is enclosed in parentheses.

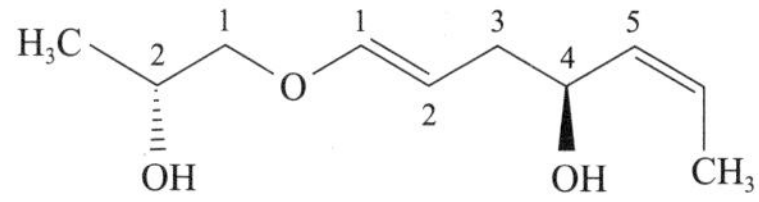

(1E,4S,5Z)-1-[(2R)-2-hydroxypropoxy]hepta-1,5-dien-4-ol

Other stereodescriptors (*e.g.*, *cis*/*trans*, M/P, C/A) are used in special cases. The non-italic descriptors α/β and D/L (small capitals) are commonly and only used for natural products, amino acids, and carbohydrates.

4.10 Chemical Abstracts Service (CAS) Names

CAS maintains a registry of chemical substances collected from publications. In the CAS system, compounds are named using methods similar to, but not identical with, those of IUPAC. The most prominent difference is the use of 'CA Index Names', which in the index are cited in a special inverted order that was devised for the creation of alphabetical indexes of chemical names. CAS also uses conjunctive nomenclature, in which parent compounds are combined to make a new, larger parent compound. In the example below, the conjunctive parent name is benzeneacetic acid (corresponding substitutive name: phenylacetic acid), while the substitutive name recommended by IUPAC for this example is based on the longer chain parent compound propanoic acid.

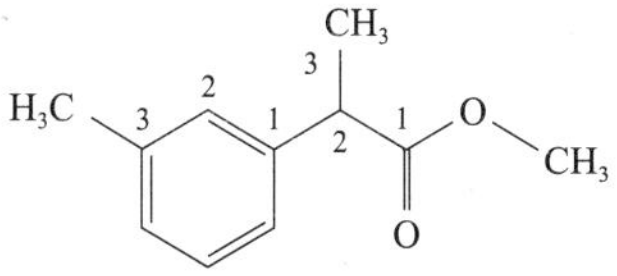

(1) 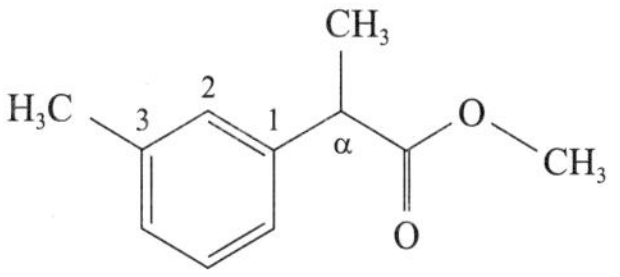 (2)

IUPAC name: methyl 2-(3-methylphenyl)propanoate(1)
CA name: methyl α,3-dimethylbenzeneacetate(2)
In index inverted to: benzeneacetic acid, α, 3-dimethyl-, methyl ester

Other differences include the position of locants and stereodescriptors, as well as some specific nomenclature procedures.

4.11 Graphical Representation

The structural formulae of organic-chemical compounds are usually drawn in accordance with the zig-zag convention as used widely above. In this convention, all carbon atoms (and their attached hydrogen atoms) attached to at least two other non-hydrogen atoms are represented by the intersection of two lines representing bonds. Hydrogen atoms attached to heteroatoms must not be omitted. In such graphical representations, each end of a line, each angle, and each intersection represents a carbon atom saturated with hydrogen. Special conventions are used to represent the configuration of stereogenic centres and double bonds.

References

[1] https://www.degruyter.com/view/j/pac
[2] https://www.qmul.ac.uk/sbcs/iupac/.
[3] Hartshorn R M, Hellwich K-H, Yerin A, Damhus T, Hutton A T. Pure Appl. Chem., 2015, 87: 1039.
[4] Hiorns R C, Boucher R J, Duhlev R, Hellwich K-H, Hodge P, Jenkins A D, Jones R G, Kahovec J, Moad G, Ober C K, Smith D W, Stepto R F T, Vairon J P, Vohlídal J. Pure Appl. Chem,, 2012, 84: 2167.
[5] Leigh G J. Principles of Chemical Nomenclature–A Guide to IUPAC Recommendations. 2011 Edition. Cambridge, UK: RSC Publishing, 2011.
[6] Favre H A, Powell W H. Nomenclature of Organic Chemistry–IUPAC Recommendations and Preferred Names 2013. Cambridge, UK: Royal Society of Chemistry, 2014.
[7] Connelly N G, Damhus T, Hartshorn R M, Hutton A T. Nomenclature of Inorganic Chemistry–IUPAC Recommendations 2005. Cambridge, UK: RSC Publishing, 2005.
[8] Jones R G, Kahovec J, Stepto R, Wilks E S, Hess M, Kitayama T, Metanomski W V. Compendium of Polymer Terminology and Nomenclature–IUPAC Recommendations 2008. Cambridge, UK: RSC Publishing, 2008.
[9] Cahn R S, Ingold C, Prelog V. Angew. Chem., 1966, 78: 413.
[10] Prelog V, Helmchen G. Angew. Chem., 1982, 94: 614.
[11] Chemical Abstracts Service, https://www.cas.org.
[12] Brecher J, Degtyarenko K N, Gottlieb H, Hartshorn R M, Hellwich K H, Kahovec J, Moss G P, McNaught A, Nyitrai J, Powell W, Smith A, Taylor K, Town W, Williams A, Yerin A. Pure Appl. Chem., 2008, 80: 277.
[13] Brecher J, Degtyarenko K N, Gottlieb H, Hartshorn R M, Moss G P, Murray-Rust P, Nyitrai J, Powell W, Smith A, Stein S, Taylor K, Town W, Williams A, Yerin A. Pure Appl. Chem., 2006, 78: 1897.

Note: This chapter is quoted completely from the paper titled as *Brief Guide to the Nomenclature of organic Chemistry* (IUPAC, *Pure Appl. Chem.* 2020, 92: 527-539; https://doi.org/10.1515/pac-2019-0104), except for some necessary changes. (i. The numbers of the chapters, tables and charts etc. are changed to be suitable for this book; ii. Some pictures have been remade for reason of resolution; iii. The color of some words and pictures are changed to black, because this book will be printed in black and white.)

Key Words & Phrases

styrene *n.* 苯乙烯
urea *n.* 尿素；脲
parent compound 母体化合物
characteristic (functional) group 特征基团；官能团
hydride *n.* 氢化物
chirality *n.* 手性
monocyclic *adj.* 单环的
bridged polycyclic 桥接多环
fused polycyclic 稠合多环
spiro polycyclic 螺多环
fullerene *n.* 富勒烯
heterocycle *n.* 杂环
bridgehead *n.* 桥头；桥塔
4-chloroaniline 4-氯苯胺
criterion *n.* 标准
substituent *n.* 取代基
functional class name 官能团类名
ester *n.* 酯
acid halide 酰基卤；酸性卤化物
ketone *n.* 酮
sulfoxide *n.* 亚砜
sulfone *n.* 砜
propanoic acid 丙酸
zig-zag *adj.* 锯齿状的

Chapter 5 A Brief Guide to Polymer Nomenclature

5.1 Introduction

The universal adoption of an agreed nomenclature has never been more important for the description of chemical structures in publishing and on-line searching. The International Union of Pure and Applied Chemistry (IUPAC) and Chemical Abstracts Service (CAS) make similar recommendations. The main points are shown here with hyperlinks to original documents. Further details can be found in the IUPAC Purple Book.

5.2 Basic Concepts

The terms polymer and macromolecule do not mean the same thing. A polymer is a substance composed of macromolecules. The latter usually have a range of molar masses (unit g/mol), the distributions of which are indicated by dispersity (*Đ*). It is defined as the ratio of the mass-average molar mass (M_{m}) to the number-average molar mass (M_{n}) *i.e.* $Đ=M_{m}/M_{n}$. Symbols for physical quantities or variables are in italic font but those representing units or labels are in roman font.

Polymer nomenclature usually applies to idealised representations; minor structural irregularities are ignored. A polymer can be named in one of two ways. Source-based nomenclature can be used when the monomer can be identified. Alternatively, more explicit structure-based nomenclature can be used when the polymer structure is proven. Where there is no confusion, some traditional names are also acceptable.

Whatever method is used, all polymer names have the prefix poly, followed by enclosing marks around the rest of the name. The marks are used in the order: {[( )]}. Locants indicate the position of structural features, *e.g.*, poly(4-chlorostyrene). If a source-based name is one word and has no locants, then the enclosing marks are not essential, but they should be used when there might be confusion, *e.g.*, poly(chlorostyrene) is a polymer whereas polychlorostyrene might be a small, multi-substituted molecule. End-groups are described with α- and ω-, *e.g.*, α-chloro-ω-hydroxy-polystyrene.

5.3 Source-based Nomenclature

5.3.1 Homopolymers

A homopolymer is named using the name of the real or assumed monomer (the 'source') from which it is derived, *e.g.*, poly(methyl methacrylate). Monomers can be named using IUPAC recommendations, or well-established traditional names. If ambiguity arise, class

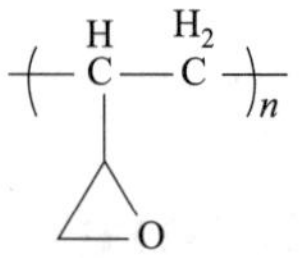

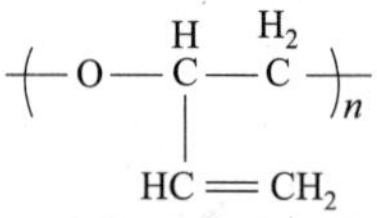

$\left[\underset{\text{(oxiranyl)}}{\overset{H}{C}} - \overset{H_2}{C} \right]_n$ $\left[O - \underset{HC=CH_2}{\overset{H}{C}} - \overset{H_2}{C} \right]_n$

names can be added. For example, the source-based name poly(vinyloxirane) could correspond to either of the structures shown below. To clarify, the polymer is named using the polymer class name followed by a colon and the name of the monomer, *i.e.*, class name:monomer name. Thus on the left and right, respectively, are polyalkylene:vinyloxirane and polyether:vinyloxirane.

5.3.2 Copolymers

The structure of a copolymer can be described using the most appropriate of the connectives shown in Table 5.1. These are written in italic font.

Table 5.1 Qualifiers for copolymers

Copolymer	Qualifier	Example
unspecified	co (C)	poly(styrene-co-isoprene)
statistical	stat (C)	poly[isoprene-stat-(methylmethacrylate)]
random	ran (C)	poly[(methylmethacrylate)-ran-(butylacrylate)]
alternating	alt (C)	poly[styrene-alt-(maleicanhydride)]
periodic	per (C)	poly[styrene-per-isoprene-per-(4-vinylpyridine)]
block	block (C)	poly(buta-1,3-diene)-block-poly(ethene-co-propene)
graft①	graft (C)	polystyrene-graft-poly(ethylene oxide)

① The first name is that of the main chain.

5.3.3 Non-linear polymers

Non-linear polymers and copolymers, and polymer assemblies are named using the italicized qualifiers in Table 5.2. The qualifier, such as branch, is used as a prefix (P) when naming a (co) polymer, or as a connective (C), *e.g.*, comb, between two polymer names.

Table 5.2 Qualifiers for non-linear (co)polymers and polymer assemblies

(Co)polymer	Qualifier	Example
blend	blend (C)	poly(3-hexylthiophene)-blendpolystyrene
comb	comb (C)	polystyrene-comb-polyisoprene
complex	compl (C)	poly(2,3-dihydrothieno[3,4-b][1,4]dioxine)-compl-poly (vinylbenzenesulfonic acid)①
cyclic	cyclo (P)	cyclo-polystyrene-graft-polyethylene
branch	branch (P)	branch-poly[(1,4-divinylbenzene)-stat-styrene]
network	net (C or P)	net-poly(phenol-co-formaldehyde)
interpenetrating network	ipn (C)	(net-polystyrene)-ipn-[netpoly(methyl acrylate)]
semi-interpenetrating network	sipn (C)	(net-polystyrene)-sipn-polyisoprene
star	star (P)	star-polyisoprene

① In accordance with IUPAC organic nomenclature, square brackets enclose locants that refer to the numbering of the components of the fused ring.

5.4 Structure-based Nomenclature

5.4.1 Regular single-strand organic polymers

In place of the monomer name used in source-based nomenclature, structure-based nomenclature uses that of the preferred constitutional repeating unit (CRU). It can be determined as follows: (i) a large enough part of the polymer chain is drawn to show the structural repetition, *e.g.*,

$$-CH(Br)-CH_2-O-CH(Br)-CH_2-O-CH(Br)-CH_2-O-$$

(ii) the smallest repeating portion is a CRU, so all such possibilities are identified. In this case:

$$-CH_2-O-CH(Br)- \qquad -O-CH(Br)-CH_2- \qquad -CH(Br)-CH_2-O-$$

$$-CH_2-CH(Br)-O- \qquad -CH(Br)-O-CH_2- \qquad -O-CH_2-CH(Br)-$$

(iii) the next step is to identify the subunits that make up each of these structures, *i.e.*, the largest divalent groups that can be named using IUPAC nomenclature of organic compounds such as the examples that are listed in Table 5.3;

Table 5.3 Representations of divalent groups in polymers

Name	Group①	Name	Group①
oxy	$-O-$	propylimino	$-N(CH_2CH_2CH_3)-$
sulfanediyl	$-S-$	hydrazine-1,2-diyl	$-\overset{1}{N}H-\overset{2}{N}H-$
sulfonyl	$-SO_2-$	phthaloyl	$C_6H_4(CO-)_2$ (1,2-)
diazenediyl	$-N{=}N-$	1,4-phenylene	$-C_6H_4-$ (1,4-)
imino	$-NH-$	cyclohexane-1,2-diyl	
carbonyl	$-C(=O)-$	butane-1,4-diyl	$-\overset{1}{C}H_2\overset{2}{C}H_2\overset{3}{C}H_2\overset{4}{C}H_2-$

(Continued)

Name	Group①	Name	Group①
oxalyl	$-\overset{\overset{O}{\|\|}}{C}-\overset{\overset{O}{\|\|}}{C}-$	1-bromoethane-1,2-diyl	$-\overset{1}{C}H(Br)-\overset{2}{C}H_2-$
silanediyl	$-SiH_2-$	1-oxopropane-1,3-diyl	$-\underset{1}{\overset{\overset{O}{\|\|}}{C}}-\overset{2}{C}H_2\overset{3}{C}H_2-$
ethane-1,2-diyl	$-\overset{1}{C}H_2-\overset{2}{C}H_2-$	ethene-1,2-diyl	$-\overset{1}{C}H=\overset{2}{C}H-$
methylene	$-CH_2-$	methylmethylene	$-CH(CH_3)-$

① To avoid ambiguity, wavy lines drawn perpendicular to the free bond, which are conventionally used to indicate free valences, are usually omitted from graphical representations in a polymer context.

(iv) using the shortest path from the most senior subunit to the next senior, the correct order of the subunits is determined using Figure 5.1;

(v) the preferred CRU is chosen as that with the lowest possible locant(s) for substituents.

In the above example, the oxy subunits in the CRUs are heteroatom chains. From Figure 5.1, oxy subunits are senior to the acyclic carbon chain subunits, the largest of which are bromo-substituted $-CH_2-CH_2-$ subunits. 1-Bromoethane-1,2-diyl is chosen in preference to 2-bromoethane-1,2-diyl as the former has a lower locant for the bromo-substituent. The preferred

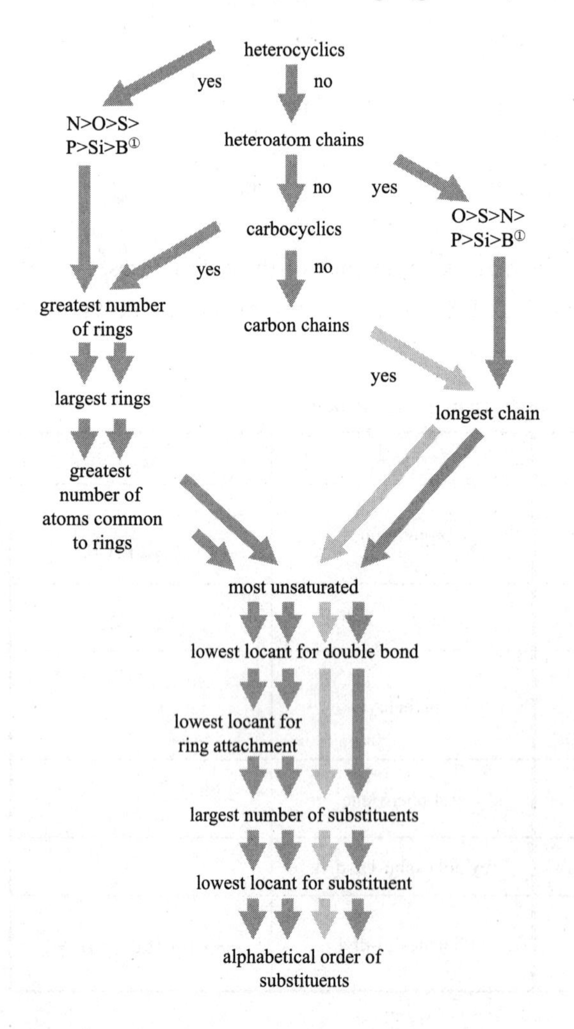

Figure 5.1 The order of subunit seniority (The senior subunit is at the top centre. Subunits of lower seniority are found by following the arrows. The type of subunit, being a heterocycle, a heteroatom chain, a carbocycle, or a carbon chain, determines the arrow to follow. ① Other heteroatoms may be placed in these orders as indicated by their positions in the periodic table)

CRU is therefore oxy(1-bromoethane-1,2-diyl) and the polymer is thus named poly[oxy(1-bromoethane-1,2-diyl)]. Please note the enclosing marks around the subunit carrying the substituent.

Polymers that are not made up of regular repetitions of a single CRU are called irregular polymers. For these, each constitutional unit (CU) is separated by a slash, *e.g.*, poly(but-1-ene-1,4-diyl/1-vinylethane-1,2-diyl).

5.4.2 Regular double-strand organic polymers

Double-strand polymers consist of uninterrupted chains of rings. In a spiro polymer, each ring has one atom in common with adjacent rings. In a ladder polymer, adjacent rings have two or more atoms in common. To identify the preferred CRU, the chain is broken so that the senior ring is retained with the maximum number of heteroatoms and the minimum number of free valences.

An example is . The preferred CRU is an acyclic subunit of 4 carbon atoms with 4 free valences, one at each atom, as shown below. It is oriented so that the lower left atom has the lowest number. The free-valence locants are written before the suffix, and they are cited clockwise from the lower left position as: lower-left, upper-left:upper-right, lower-right. This example is thus named poly(butane-1,4:3,2-tetrayyl). For more complex structures, the order of seniority again follows Figure 5.1.

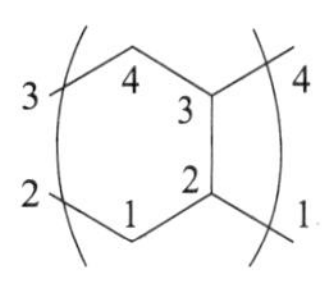

5.5 Nomenclature of Inorganic and Inorganic-Organic Polymers

Some regular single-strand inorganic polymers can be named like organic polymers using the rules given above, *e.g.*, $[O\text{-}Si(CH_3)_2]_n$ and $[Sn(CH_3)_2]_n$ are named poly[oxy(dimethylsilanediyl)] and poly(dimethylstannanediyl), respectively. Inorganic polymers can also be named in accordance with inorganic nomenclature, but it should be noted that the seniority of the elements is different to that in organic nomenclature. However, certain inorganic-organic polymers, for example those containing metallocene derivatives, are at present best named using organic nomenclature, *e.g.*, the polymer on the left can be named poly[(dimethylsilanediyl)ferrocene-1,1'-diyl].

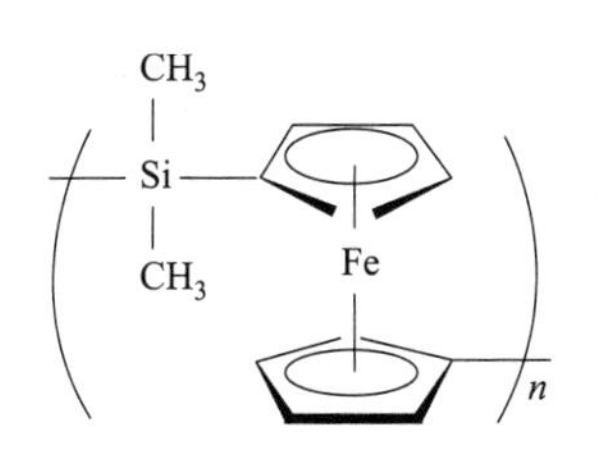

5.6 Traditional Names

When they fit into the general pattern of systematic nomenclature, some traditional and trivial names for polymers in common usage, such as polyethylene, polypropylene, and polystyrene, are retained.

5.7 Graphical Representations

The bonds between atoms can be omitted, but dashes should be drawn for chain-ends. The seniority of the subunits does not need to be followed. For single-strand (co)polymers, a dash is drawn through the enclosing marks, *e.g.*, poly[oxy(ethane-1,2-diyl)] shown below left. For irregular polymers, the CUs are separated by slashes, and the dashes are drawn inside the enclosing marks. End-groups are connected using additional dashes outside of the enclosing marks, *e.g.*, α-methyl-ω-hydroxy-poly[oxirane-co-(methyloxirane)], shown below right.

$$\left(\!-\mathrm{OCH_2}-\overset{\mathrm{H_2}}{\mathrm{C}}\!-\right)_n \qquad \mathrm{H_3C}-\left(-\mathrm{OCH_2}-\overset{\mathrm{H_2}}{\mathrm{C}}-/-\mathrm{OCH_2}-\overset{\mathrm{H_2}}{\mathrm{C}}-\right)_n-\mathrm{OH}$$

5.8 CA Index Names

CAS maintains a registry of substances. In the CAS system, the CRU is called a structural repeating unit (SRU). There are minor differences in the placements of locants, *e.g.*, poly(pyridine-3,5-diylthiophene-2,5-diyl) is poly(3,5-pyridinediyl-2,5-thiophenediyl) in the CAS registry, but otherwise polymers are named using similar methods to those of IUPAC.

References

[1] http://www.chem.qmul.ac.uk/iupac/.
[2] http://www.cas.org/.
[3] IUPAC. The "Purple Book", RSC Publishing, 2008.
[4] IUPAC. Pure Appl. Chem. 2009, 81, 351-353.
[5] IUPAC. Pure Appl. Chem. 1997, 69, 2511-2521.
[6] IUPAC. Pure Appl. Chem. 2001, 73, 1511-1519.
[7] IUPAC. Pure Appl. Chem. 1985, 57, 1427-1440.
[8] IUPAC. Pure Appl. Chem. 2002, 74, 1921-1956.
[9] IUPAC. Pure Appl. Chem. 1994, 66, 873-889.
[10] IUPAC. Pure Appl. Chem. 1993, 65, 1561-1580.
[11] IUPAC. Pure Appl. Chem. 1985, 57, 149-168.
[12] IUPAC. Pure Appl. Chem. 1994, 66, 2469-2482.
[13] IUPAC. Pure Appl. Chem. 2008, 80, 277-410.
[14] Macromolecules, 1968, 1: 193-198. The Committee on Nomenclature, American Chemical Society Division of Polymer Chemistry.
[15] Polym. Prepr. 2000, 41(1): 6a-11a.

Note: This chapter is quoted completely from the paper titled as *Brief Guide to Polymer Nomenclature* (IUPAC, *Pure Appl. Chem.*, 2012, 84: 2167-2169), except for some necessary changes. (i. The numbers of the chapters, tables and charts etc. are changed to be suitable for this book; ii. Some pictures have been remade for reason of resolution; iii. The color of some words and pictures are changed to black, because this book will be printed in black and white.)

Key Words & Phrases

polymer *n.* 聚合物
macromolecule *n.* 大分子
dispersity *n.* 色散度；弥散度
source-based nomenclature 基于来源的命名法
structure-based nomenclature 基于结构的命名法
end-group 端基
homopolymer *n.* 均聚物
monomer *n.* 单体
copolymer *n.* 共聚物
non-linear polymer 非线型聚合物
qualifier *n.* 修饰词
constitutional repeating unit 组成重复单元；重复结构单元
subunit *n.* 亚单元；亚基；亚单位
divalent group 二价基
oxy subunit 氧亚基
irregular polymer 非规整聚合物
spiro polymer 螺环聚合物
ladder polymer 梯形聚合物
free valence 自由价
trivial name 俗名

Chapter 6 International Nomenclature of Cosmetic Ingredients

6.1 What Is INCI?

INCI names (International Nomenclature Cosmetic Ingredient) are systematic names internationally recognized to identify cosmetic ingredients. They are developed by the International Nomenclature Committee (INC) and published by the Personal Care Products Council (PCPC) in the International Cosmetic Ingredient Dictionary and Handbook, available electronically as wINCI. INCI names often differ greatly from systematic chemical nomenclature or from more common trivial names and are a mixture of conventional scientific names, Latin and English words. INCI nomenclature conventions "are continually reviewed and modified when necessary to reflect changes in the industry, technology, and new ingredient developments".

Oversight for the INCI program is provided by PCPC as part of its mission to support the identification of the composition of personal care products, and publication of this information in a worldwide science-based dictionary. PCPC is committed to ensuring that the dictionary provides the world community with accurate, transparent, and harmonized nomenclature. By working closely with its international sister trade associations, and with other organizations around the world, PCPC strives to develop INCI names that accommodate differing labeling approaches described in national laws and regulations.

6.2 Why INCI?

There are many benefits to a uniform system of labeling names for cosmetic ingredients. Dermatologists and others in the medical community are ensured an orderly dissemination of scientific information, which helps to identify agents responsible for adverse reactions. Scientists are ensured that information from scientific and other technical publications will be referenced by a uniform name; and that multiple names for the same material will not lead to confusion, misidentification, or the loss of essential information. It also enables the cosmetic industry to track the safety and the regulatory status of ingredients efficiently on a global basis, enhancing its ability to market safe products in compliance with various national regulations. And finally, transparency is provided to consumers as ingredients are identified by a single labeling name regardless of the national origin of the product.

6.3 INCI Basics

The conventions used to determine INCI names for cosmetic ingredients are usually divided into three areas: general conventions, specific conventions (which are grouped primarily by chemical class), and miscellaneous conventions. These conventions are continually reviewed and modified when necessary to reflect changes in the industry, technology, and new ingredient developments. Every effort is made to ensure ingredients are named consistent with these principles. As new conventions are developed that give rise to INCI names that are different from those previously published, the older nomenclature is sometimes retained and considered to be "grandfathered". Grandfathered names are generally published for reference only.

There are 76 conventions at present, which is available online: https://www.personalcarecouncil.org/wp-content/uploads/2022/02/Conventions-2022_v1_rev.pdf. The INCI nomenclature system is fundamentally based on these core principles: the names should be science-based, systematic, informative, and unambiguous; reflect chemical composition; be of minimum length where possible; utilize existing food and drug terminology where appropriate; and be consistent with minimal change to avoid confusion and preserve understanding. Table 6.1 shows several common names and their corresponding INCI names.

Table 6.1 Several common names and their corresponding INCI names

Common name	INCI name
Purified water, deionized water, demineralized water, water, *etc.*	Aqua
Sodium coco sulfate	Sodium coco-sulfate
Sodium lauryl sulfate (from coconut oil)	Sodium lauryl sulfate
Sodium laureth sulfate (from coconut oil)	Sodium laureth sulfate
Cocamidopropyl betaine (from coconut oil)	Cocamidopropyl betaine
Decyl glucoside	Decyl glucoside①
Citric acid	Citric acid①
Paraben	Methylparaben
Cetyl alcohol	Cetyl alcohol
Vitamin E	Tocopherol
Beeswax	Beeswax①
Vegetable glycerin	Glycerin
Oat bran	*Avena sativa* (oat) bran
Shea butter	*Butyrospermum parkii* (shea butter)
Passion fruit juice	*Passiflora edulis* fruit juice
Red rose water	*Rosa damascena* flower water
Raspberry extract	*Rubus idaeus* (raspberry) fruit extract
Yucca herbal extract	*Yucca schidigera* stem extract
Aloe vera leaf gel	*Aloe barbadensis* leaf juice
Tea tree oil	*Melaleuca alternifolia* (tea tree) leaf oil
Peppermint leaf oil	*Mentha piperita* (peppermint) oil
Spearmint leaf oil	*Mentha viridis* (spearmint) leaf oil

（Continued）

Common name	INCI name
Wintergreen leaf oil	*Gaultheria procumbens* (wintergreen) leaf oil
Lavender oil	*Lavandula angustifolia* (lavender) oil
Cinnamon leaf oil	*Cinnamomum cassia* leaf oil
Lemon peel oil	*Citrus medica* limonum (lemon) peel oil
Valencia orange peel oil	*Citrus aurantium dulcis* (orange) peel oil
Pink grapefruit peel oil	*Citrus paradisi* (grapefruit) peel oil
Roman chamomile oil	*Anthemis nobilis* flower oil

① Some common names and INCI names are the same name.

6.4 INCI and CAS

According to CAS(chemical abstracts service), "CAS registry numbers are not dependent upon any system of chemical nomenclature. CAS numbers provide a common link between the various nomenclature terms used to describe substances and serve as an international resource for chemical substance identifiers used by scientists, industry and regulatory bodies" (https://www.cas.org/). Thus, CAS numbers can be interpreted to represent a link to numerous terms, including various INCI names and chemical names.

The relationship between a CAS number and an INCI name is not always one-to-one. In some cases, more than one INCI name may have the same CAS number, or more than one CAS number may apply to an INCI name. For example, the CAS number 1245638-61-2 has the CA index name of 2-propenoic acid, reaction products with pentaerythritol. This CAS number can accurately be associated with two INCI names: pentaerythrityl tetraacrylate and pentaerythrityl triacrylate. Alternatively, the INCI name, glucaric acid can be associated with two CAS numbers: 87-73-0 which has the CA index name of D-glucaric acid, and 25525-21-7, which has the CA index name of DL-glucaric acid. Both of these examples are accurate associations between CAS and INCI.

Scientists, regulatory professionals, and government agencies should keep in mind that while CAS numbers are very helpful for tracking substances and filing various registrations, multiple numbers might be suitable for substance identification. It is also important for ingredient suppliers and their business partners to understand that the association between a CAS number and an INCI name is interpretative and should not restrict, nor promote, the usage of an ingredient. Ultimately, the listing of a CAS number in the INCI data base does not imply any regulatory status for the said ingredient in any particular jurisdiction. Suppliers may use CAS numbers for product descriptions and registrations independent of their inclusion in the INCI data base.

6.5 INCI Labeling

In the U.S., under the Food, Drug, and Cosmetic Act and the Fair Packaging and Labeling Act, certain accurate information is a requirement to appear on labels of cosmetic products. In Canada, the regulatory guideline is the Cosmetic Regulations. Ingredient names must comply by law with EU requirements by using INCI names.

The cosmetic regulation laws are enforceable for important consumer safety. For example, the ingredients are listed on the ingredient declaration for the purchaser to reduce the risk of an allergic reaction to an ingredient the user has had an allergy to before. INCI names are mandated on the ingredient statement of every consumer personal care product. The INCI system allows the consumer to identify the ingredient content. In the U.S., true soaps (as defined by the FDA) are specifically exempted from INCI labeling requirements as cosmetics per FDA regulation.

6.6 What INCI Is Not

The designation of an INCI name for a cosmetic ingredient does not imply that the substance is safe for use as a cosmetic ingredient, nor does it indicate that its use as a cosmetic ingredient complies with the laws and regulations of the United States or any other country. INCI names do not imply standards or grades of purity. The assignment of an INCI name does not imply that the ingredient is "approved" "certified" or "endorsed" by the Council or any other organization or governmental body. Conversely, if an ingredient does not have an INCI name, it does not mean that the ingredient may not or should not be used in finished cosmetic and personal care products. The safety and fitness of use for an ingredient, along with regulatory considerations, are carefully evaluated by the manufacturer as part of the development process before the product is marketed.

6.7 Applying for an INCI Name

Anyone can submit an application for an INCI name through an electronic application process that is available online: https://inci.personalcarecouncil.org/. Details of the application include fields that relate to the composition of the ingredient and its manufacturing process. Applicants are also given the opportunity to suggest an INCI name. The application process does not require the submission of safety data because safety determinations are not within the purview of the Nomenclature Committee.

For more information about INCI nomenclature, you can visit the PCPC website: https://inci.personalcarecouncil.org/.

Reference

Nikitakis J, Sanzone J. Chapter 11 Nomenclature of Ingredients, Cosmetic Science and Technology[M]. Netherlands: Elsevier, 2017: 155-158.

Key Words & Phrases

dermatologist *n.* 皮肤科医生；皮肤科医师
transparency *n.* 透明度；透明；透明性；显而易见
pentaerythritol *n.* 季戊四醇
pentaerythrityl tetraacrylate 季戊四醇四丙烯酸酯
pentaerythrityl triacrylate 季戊四醇三丙烯酸酯
glucaric acid 葡萄糖二酸
sodium coco sulfate 椰油醇硫酸酯钠
sodium lauryl sulfate 十二烷基硫酸钠
sodium laureth sulfate 十二烷基醚硫酸钠
cocamidopropyl betaine 椰油酰胺丙基甜菜碱
decyl glucoside 癸基葡糖苷
citric acid 柠檬酸
paraben *n.* 尼泊金（苯甲酸酯类）
cetyl alcohol 十六醇
beeswax *n.* 蜂蜡
vegetable glycerin 蔬菜甘油；植物甘油
oat bran 燕麦麸
shea butter 乳木果油
passion fruit juice 西番莲汁；百香果汁
red rose water 红玫瑰水
raspberry extract 树莓精华；覆盆子提取物
Yucca herbal extract 丝兰草本提取物
Aloe vera leaf gel 芦荟叶凝胶
tea tree oil 茶树油
peppermint leaf oil 薄荷叶油
spearmint leaf oil 留兰香叶油
wintergreen leaf oil 冬青叶油
lavender oil 薰衣草油
cinnamon leaf oil 肉桂叶油
lemon peel oil 柠檬皮油
valencia orange peel oil 瓦伦西亚橘皮油
pink grapefruit peel oil 粉红葡萄柚皮油
roman chamomile oil 罗马洋甘菊油
methylparaben *n.* 羟苯甲酯；对羟基苯甲酸甲酯
tocopherol *n.* 生育酚；维生素E
partner *n.* 搭档；合伙人；同伴；舞伴；伙伴；配偶
interpretative *adj.* 解释的；作为说明的
jurisdiction *n.* 司法权；审判权；管辖权
allergic reaction 过敏反应
fitness *n.* 健身；健康；适合
purview *n.* 权限；（组织、活动等的）范围

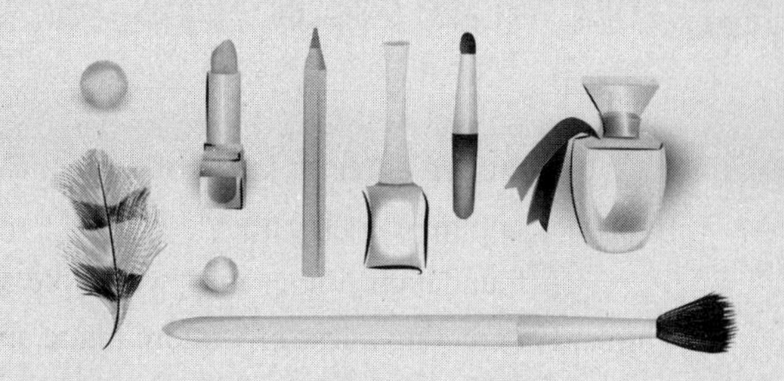

Unit 3

Introduction to Common Cosmetic Products

Though there are a large number of differing cosmetics used for a variety of different purposes, all cosmetics are typically intended to be applied externally. These products can be applied to the face (on the skin, lips, eyebrows and eyes), to the body (on the skin, in particular the hands and nails), and to the hair. These products may be intended for use as skincare, personal care or to alter the appearance, with the subset of cosmetics known as makeup primarily referring to products containing colour pigments intended for the purpose of altering the wearer's appearance; some manufacturers will distinguish only between "decorative" cosmetics intended to alter the appearance and "care" cosmetics designed for skincare and personal care. Therefore, cosmetics are usually classified into skincare, haircare, makeup, perfumes, toiletries, and others which includes feminine hygiene, baby care, and oral care in many countries.

In this unit, you will see:

① Decorative (or makeup) cosmetics;

② Skincare cosmetics;

③ Haircare cosmetics;

④ Perfumes.

Chapter 7 Decorative Cosmetics

7.1 Primers

Primers are used on the face before makeup is applied, creating a typically transparent, smooth layer over the top of the skin, allowing for makeup to be applied smoothly and evenly. Some primers may also be tinted, and this tint may match the wearer's skin tone, or may colour correct it, using greens, oranges and purples to even out the wearer's skin tone and correct redness, purple shadows or orange discolouration respectively.

There are different kinds of cosmetic primers such as foundation primer, eyelid primer, lip primer, and mascara primer.

A foundation primer may work like a moisturizer, or it may absorb oil with salicylic acid to aid in creating a less oily, more matte appearance. Some contain antioxidants such as vitamin A, C, and E, or other ingredients such as grape seed extract and green tea extract. There are water-based and silicon-based foundation primers. Ingredients may include cyclomethicone and dimethicone. Some primers do not contain preservative, oil or fragrance. Some may also have sun protection factor (SPF). Some foundation primers are tinted to even out or improve skin tone or color. Others give a pearlized finish to make the complexion more light reflective. There are also foundation primers which are mineral-based primers, which contain mica and silica.

Eyelid or eye shadow primers are similar, but made specifically for use near the eyes. An eyelid primer may help even the color of the lid and upper eye area, may reduce oiliness, may add shimmer. Eye primers aid in the smooth application of eye shadow, prevent it from accumulating in eyelid creases, and improve its longevity. Eye shadow primers are applied to the eyelid and lower eye area prior to the application of eye shadow. They even out the skin tone of the eyelids, hide eyelid veins, and smooth out the skin of the eyelids. Eye shadow primers help with the application of eye shadows. They intensify the color of the eye shadows and keep them from smearing or creasing by reducing the oiliness of the lids. Some eye shadows even state in the instruction sheet, that they are recommended for usage over the eye shadow primer. There is a real difference in the eye shadow color and time of wear when it is used over the primer on bare skin. The effect of eye shadow primers is not limited to eye shadows. They also work for eye liners and eye shadow bases.

Mascara primer is sometimes colorless. It usually thickens and/or lengthens the lashes before the application of mascara for a fuller finished look. It may also help keep mascara from smudging or flaking, and some claim to improve the health of the lashes.

Lip primers are intended to smooth the lips and help improve the application of lipstick or lip gloss, although exfoliating the lips is often recommended before applying. They also are intended to increase the longevity of lip color, and to prevent lipstick from "feathering", that is, smearing past the lip vermilion, and especially from migrating into any fine lines around the lips.

7.2 Concealers

A concealer or color corrector is a type of cosmetic that is used to mask dark circles, age spots, large pores, and other small blemishes visible on the skin. It is similar to foundation, but thicker and used to hide different pigments by blending the imperfection into the surrounding skin tone.

Both concealer and foundation are typically used to make skin appear more uniform in color. These two types of cosmetics differ in that concealers tend to be more heavily pigmented, though concealer and foundation are both available in a wide range of opacity. Concealer can be used alone or with foundation. It comes in different forms from liquid to powder. The first commercially available concealer was Max Factor's Erace, launched in 1954. Camouflage makeup is a much heavier pigmented form of concealer. It is used to cover serious skin discolorations such as birthmarks, scars and vitiligo.

Concealer is available in a variety of shades. When picking a concealer, people tend to choose one or two shades lighter than their skin tone to better hide their blemishes and dark circles under the eye. Some colors are intended to look like a natural skin tone, while others are meant to cancel out the color of a particular type of blemish. Concealers with yellow undertones are used to hide dark circles. Green and blue can counteract red patches on the skin, such as those caused by pimples, broken veins, or rosacea. A purple-tinted concealer can make sallow complexions look brighter.

There are different types of concealers. Each type of concealer is unique and has qualities that are better suited for different skin types.

(1) Liquid concealer

Liquid concealer is the most popular because it works for most skin types. This concealer also works well for acne spots because it will not bunch up around the acne or settle into the scars. The concealer does have to be set with powder or else it will crease into the fine lines of your skin. Liquid concealers can also have buildable coverage in different finishes to suit every need. A satin finish may look more natural on areas of drier skin. (Best suited for normal, combination, oily, or sensitive skin and skin that is prone to breakouts. Little risk of irritating pimples.)

(2) Stick concealer

Stick concealer is used for very specific purposes because it is thicker than liquid and contains a lot of coverage. This type of concealer is also very convenient because of the packaging and its ability to be used for touchups. This concealer works best on blemishes and small areas of discoloration. The formula is very creamy and blends easily into the skin allowing for flawless touchups throughout the day. Best suited for normal, dry, or sensitive skin.

(3) Cream concealer

This type of concealer is similar to the stick concealer because it offers heavy coverage. However, you need to apply it with a brush because it can look heavy due to the intense pigmentation. Best suited for normal, dry, or sensitive skin. Also, an option for concealing birthmarks.

(4) Pencil concealer

Manufactured in cream or wax forms, this type of concealer can be used to cover up small blemishes. It may also be used to define the shape of the eyebrows or line the inner lash line to brighten the appearance of the eyes.

(5) Cream to powder concealer

Available in a powder compact/concealer. Apply with a sponge for a powdery, matte finish.

7.3 Foundation

Foundation is a liquid, cream, or powder makeup applied to the face to create an even, uniform color to the complexion, cover flaws and, sometimes, to change the natural skin tone. Some foundations also function as a moisturizer, sunscreen, astringent or base layer for more complex cosmetics. Foundation applied to the body is generally referred to as "body painting" or "body makeup". Foundation provides a generally lower amount of coverage than concealer, and is sold in formulations that can provide sheer, matte, dewy or full coverage to the skin.

7.4 Rouge, Blush, or Blusher

Rouge, blush, or blusher is a liquid, cream or powder product applied to the centre of the cheeks with the intention of adding or enhancing their natural colour. Blushers are typically available in shades of pink or warm tan and brown, and may also be used to make the cheekbones appear more defined.

The ancient Egyptians were known for their creation of cosmetics, particularly their use of rouge. Ancient Egyptian pictographs show men and women wearing lip and cheek rouge. They blended fat with red ochre to create a stain that was red in color. Greek men and women eventually mimicked the look, using crushed mulberries, red beet juice, crushed strawberries, or red amaranth to create a paste. Those who wore makeup were viewed as wealthy and it symbolized status because cosmetics were costly. In China, rouge was used as early as the Shang Dynasty. It was made from the extracted juice of leaves from red and blue flowers. Some people added bovine pulp and pig pancreas to make the product denser. Women would wear the heavy rouge on their cheeks and lips. In Chinese culture, red symbolizes good luck and happiness to those who wear the color. In ancient Rome, men and women would create rouge using lead (Ⅱ , Ⅳ) (red lead) and cinnabar. The mixture was found to have caused cancer, dementia, and eventually death. In the 16th century in Europe, women and men would use white powder to lighten their faces. Commonly women would add heavy rouge to their cheeks in addition.

7.5 Bronzer

Bronzer is a powder, cream or liquid product that adds colour to the skin, typically in bronze or tan shades intended to give the skin a tanned appearance and enhance the colour of the face. Bronzer, like highlighter, may also contain substances providing a shimmer or glitter effect, and comes in either matte, semi-matte, satin, or shimmer finishes.

Bronzers are a temporary sunless tanning or bronzing option. Once applied, they create a tan that can easily be removed with soap and water. Like makeup, these products tint or stain a person's skin only until they are washed off. They are often used for "one-day" only tans, or to complement a DHA(docosahexaenoic acid)-based sunless tan. Many formulations are available, and some have limited sweat or light water resistance. If applied under clothing, or where fabric and skin edges meet, most will create some light but visible rub-off. Dark clothing prevents the rub-off from being noticeable. While these products are much safer than tanning beds, the color produced can sometimes look orangey and splotchy if applied incorrectly. A recent trend is that of lotions or moisturizers containing a gradual tanning agent. A slight increase in color is usually observable after the first use, but color will continue to darken the more frequently the product is used. Just as with the term "sunless tanner", the term "bronzer" is likewise not defined by law, or by regulations enforced by the FDA. What is defined and regulated is the color additive DHA, or dihydroxyacetone.

Air brush tanning is a spray on tan performed by a professional. An air brush tan can last five to ten days and will fade when the skin is washed. It is used for special occasions or to get a quick dark tan. At-home airbrush tanning kits and aerosol mists are also available.

Tanners usually contain a sunscreen. However, when avobenzone is irradiated with UVA light, it generates a triplet excited state in the keto form which can either cause the avobenzone to degrade or transfer energy to biological targets and cause deleterious effects.

7.6 Highlighter

Highlighter is a type of cosmetic product that reflects light. Often used for contouring, it can be applied to the face or other parts of the body to brighten the skin on a given area, create the perception of depth and angles. The product can come in a variety of forms, including powder, liquid, cream, gloss, solid stick and jelly. Highlighters became a significant tool among theater and film actors shooting or performing indoors, where natural light was not available to provide definition of facial features like cheekbones, nose, and jawline. Highlighter also offers the possibility of heightening or diminishing a given feature to suit the character portrayed, as well as aesthetic trends.

7.7 Eyebrow Pencils, Creams, Waxes, Gels, and Powders

Eyebrow pencils, creams, waxes, gels, and powders are used to color, fill in, and define the brows. Eyebrow tinting treatments are also used to dye the eyebrow hairs a darker colour, either temporarily or permanently, without staining and colouring the skin underneath the eyebrows.

7.8 Eyeshadow

Eyeshadow is a powder, cream or liquid pigmented product used to draw attention to, accentuate and change the shape of the area around the eyes, on the eyelid and the space below the eyebrows. Eyeshadow is typically applied using an eyeshadow brush, with generally small and rounded bristles, though liquid and cream formulations may also be applied with the fingers. Eyeshadow is available in almost every colour, as well as being sold in a number of different finishes, ranging from matte finishes with sheer coverage to glossy, shimmery, glittery and highly pigmented finishes. Many different colours and finishes of eyeshadow may be combined in one look and blended together to achieve different effects.

Civilizations around the world use eyeshadow predominantly on females but also occasionally on males. In western society, it is seen as a feminine cosmetic, even when used by men. In Gothic fashion, black or similarly dark-colored eye shadow and other types of eye makeup are popular among both sexes. In India, eye liner, called Kohl, played a prominent role in various dance forms and ceremonies such as weddings.

Many people use eyeshadow simply to improve their appearance, but it is also commonly used in theatre and other plays, to create a memorable look, with bright, bold colors. Depending on skin tone and experience, the effect of eyeshadow usually brings out glamour and gains attention. The use of eyeshadow attempts to replicate the natural eyeshadow that some women exhibit due to a natural contrasting pigmentation on their eyelids. Natural eyeshadow can range anywhere from a glossy shine to one's eyelids, to a pinkish tone, or even a silver look.

7.9 Eyeliner

Eyeliner is used to enhance and elongate the apparent size or depth of the eye; though eyeliner is commonly black, it can come in many different colours, including brown, white and blue. Eyeliner can come in the form of a pencil, a gel or a liquid.

Eyeliner was first used in ancient India, ancient Egypt and Mesopotamia as a dark black line around the eyes. In the late twentieth and early twenty-first century, heavy eyeliner use has been associated with Goth fashion and Punk fashion. Eyeliner of varying degrees of thickness has also become associated with the emo subculture and various alternative lifestyles. Guyliner is also a special style the emo subculture tends to use after being popularized from Pete Wentz, bassist of the pop-punk band Fall Out Boy. Eyeliner is commonly used in a daily makeup routine to define the eye or create the look of a wider or smaller eye. Eyeliner can be used as a tool to create various looks as well as highlighting different features of the eyes. Eyeliner can be placed in various parts of the eye to create different looks with a winged eyeliner or a tight lined at the waterline. Eyeliner can be drawn above upper lashes or below lower lashes or both, even on the waterlines of the eyes. Its primary purpose is to make the lashes look lush, but it also draws attention to the eye and can enhance or even change the eye's shape. Eyeliner is available in a wide range of hues, from the common black, brown and grey to more adventurous shades such as bright primary colors, pastels, frosty silvers and golds, white and even glitter-flecked colors. Eyeliner can also be used for showing depression in photographs, such as the famous "Bleeding Mascara".

7.10 False Eyelashes

False eyelashes are used to extend, exaggerate and add volume to the eyelashes. Consisting generally of a small strip to which hair–either human, mink or synthetic–is attached, false eyelashes are typically applied to the lash line using glue, which can come in latex and latex free varieties; magnetic false eyelashes, which attach to the eyelid after magnetic eyeliner is applied, are also available. Designs vary in length and colour, with rhinestones, gems, feathers and lace available as false eyelash designs. False eyelashes are not permanent, and can be easily taken off with the fingers. Eyelash extensions are a more permanent way to achieve this look. Each set lasts for two to three weeks, then the set can be filled, similar to the maintenance of acrylic nails. To apply to extensions the certified lash artist would start by taping down the bottom eyelashes. The lash artist would then use two tweezers, one to isolate the natural eyelash and one to apply the false eyelash. An individual false eyelash, or lash fan, is applied to one natural eyelash using a lash glue specific for this process. The eyelashes should not be stuck together. The length and thickness of the false lash should not be to heavy for the natural eyelash. If this process is done correctly no harm will be done to the natural eyelashes.

7.11 Mascara

Mascara is used to darken, lengthen, thicken, or enhance the eyelashes through the use of a typically thick, cream consistency product applied with a spiral bristle mascara brush. Mascara is commonly black, brown or clear, though a number of different colours, some containing glitter, are available. Mascara is typically advertised and sold in a number of different formulations that advertise qualities such as waterproofing, volume enhancement, length enhancement and curl enhancement, and may be used in combination with an eyelash curler to enhance the natural curl of the eyelashes.

7.12 Lip Products

Lip products include lipstick, lip gloss, lip liner and lip balms. Lip products commonly add colour and texture to the lips, as well as serving to moisturise the lips and define their external edges. Products adding colour and texture to the lips, such as lipsticks and lip glosses, often come in a wide range of colours, as well as a number of different finishes, such as matte finishes and satin or glossy finishes. Other styles of lip colouration products such as lip stains temporarily saturate the lips with a dye, and typically do not alter the texture of the lips. Both lip colour products and lip liners may be waterproof, and may be applied directly to the lips, with a brush, or with the fingers. Lip balms, though designed to moisturise and protect the lips (such as through the addition of UV protection) may also tint the lips.

7.13 Face Powder, Setting Powder, or Setting Sprays

Face powder, setting powder, or setting sprays are used to 'set' foundation or concealer, giving it a matte or consistent finish whilst also concealing small flaws or blemishes. Both powders and setting sprays claim to keep makeup from absorbing into the skin or melting off.

Whilst setting sprays are generally not tinted, setting powder and face powder can come in translucent or tinted varieties, and can be used to bake foundation in order for it to stay longer on the face. Tinted face powders may also be worn alone without foundation or concealer to give an extremely sheer coverage base.

7.14 Nail Polish

Nail polish (also known as nail varnish or nail enamel) is a liquid used to colour the fingernails and toenails. Transparent, colorless nail polishes may be used to strengthen nails or be used as a top or base coat to protect the nail or nail polish. Nail polish, like eyeshadow, is available in almost every colour and a number of different finishes, including matte, shimmer, glossy and crackle finishes.

Nail polish has three main types. The first is the basecoat, which serves the purpose of creating a smooth and uniform layer upon which the pigmented nail polish may be applied. The second is the pigmented nail polish, itself. The third is the aftercoating, which is applied on top of the pigmented nail polish to provide fortification against chipping as well as an added sheen. Nail polish often contains the following components: a film former, a plasticizer, a thermoplastic resin, a solvent extender, pigment, and possibly a suspending agent. The film former is often composed of nitrocellulose. The plasticizer, often dibutyl phthalate, serves to enhance adhesion and provides flexibility. The thermoplastic resin, often toluene sulfonamide-formaldehyde, improves adhesion, hardening, and gloss. The solvent extenders allow the components of the nail varnish to remain in a liquid form, and compounds used are usually ethyl acetate, isopropyl alcohol, butyl acetate, or toluene. Finally, the pigment component may be highly variable and may include elements such as iron oxides, color lakes, and mica. The pigment may be organic or inorganic, although inorganic compounds must have low heavy-metal content.

Nail polish originated in China and dates back to 3000 BC. Around 600 BC, during the Zhou Dynasty, the royal house preferred the colors gold and silver. However, red and black eventually replaced these metallic colors as royal favorites. During the Ming Dynasty, nail polish was often made from a mixture that included beeswax, egg whites, gelatin, vegetable dyes, and gum arabic.

Reference

Arora, H., Tosti, A. Safety and Efficacy of Nail Products. Cosmetics, 2017, 4: 24.

Key Words & Phrases

pigment *n.* 颜料；色素
decorative cosmetic 装饰性化妆品
hygiene *n.* 卫生
primer *n.* 底漆；底层涂料；打底妆
moisturizer *n.* 润肤霜；润肤膏
salicylic acid 水杨酸；邻羟基苯甲酸
antioxidant *n.* 抗氧化剂；抗氧化物
cyclomethicone *n.* 环聚二甲基硅氧烷；环甲硅脂
dimethicone *n.* 聚二甲基硅氧烷；二甲基硅油
preservative *n.* 防腐剂；保鲜剂
fragrance *n.* 香味；香气；香水；芳香
sun protection factor (SPF) 防晒系数

mica *n.* 云母
shimmer *n.* 微光；闪烁的光
dark circle 黑眼圈；黑圈
age spot 老年斑
large pore 大孔；粗大毛孔
camouflage makeup 迷彩妆
rosacea *n.* 酒渣鼻；红鼻头
sallow complexion 面色萎黄
acne *n.* 痤疮；粉刺
touchup *n.* 润色；修补画面
foundation *n.*（化妆打底用的）粉底霜；地基；基础
astringent *n.*（用于化妆品或药物中的）收敛剂；止血剂
blusher *n.* 胭脂；脸红的人
mulberry *n.* 桑树；桑葚；深紫红色
red beet juice 红甜菜汁
amaranth *n.* 紫红色；苋属植物；苋菜
pig pancreas 猪胰腺
cinnabar *n.* 辰砂；朱砂（可用作颜料）
bronzer *n.* 古铜色化妆品
formulation *n.*（药品、化妆品等的）配方；剂型；配方产品
lotion *n.* 洗剂；化妆水；水粉剂
dihydroxyacetone *n.* 二羟基丙酮
avobenzone *n.* 阿伏苯宗；亚佛苯酮
highlighter *n.* 荧光笔；高光色
eyeliner *n.* 眼线笔，眼线膏
false eyelash 假睫毛
latex *n.* 乳胶
rhinestone *n.* 水钻；莱茵石（用于仿钻石首饰）
lace *n.* 网眼织物；花边；蕾丝
acrylic nail 水晶指甲
lip gloss 唇彩；亮唇膏
lip liner 唇线笔；唇线
lip balm 润唇膏；护唇膏
basecoat *n.* 底漆；最下面的一层
plasticizer *n.* 增塑剂；塑化剂
thermoplastic resin 热塑性树脂
suspending agent 助悬剂；悬浮剂
nitrocellulose *n.* 硝化纤维素；硝酸纤维素
dibutyl phthalate 邻苯二甲酸二丁酯
toluene *n.* 甲苯
sulfonamide *n.* 磺胺
color lake 色淀

Chapter 8 Skincare Cosmetics

8.1 Cleansers

A cleanser is a facial care product that is used to remove makeup, skin care product residue, microbes, dead skin cells, oils, sweat, dirt and other types of daily pollutants from the face. These washing aids help prevent filth-accumulation, infections, pores clogs, irritation and cosmetic issues like dullness from dead skin buildup & excessive skin shine from sebum buildup. This can also aid in preventing or treating certain skin conditions such as acne. Cleansing is the first step in a skin care regimen and can be used in addition of a toner and moisturizer, following cleansing. Sometimes "double cleansing" before moving on to any other skincare product is encouraged to ensure the full dissolution & removal of residues that might be more resistant to cleansing, such as waterproof makeup, water-resistant sunscreen, the excess sebum of oily skin-type individuals and air pollution particles. Double cleansing usually involves applying a lipid-soluble cleanser (*e.g.*, cleansing balm, cleansing oil, micellar cleansing water, or others) to dry skin and massaging it around the face for a length of time, then the area may or may not be splashed with water. Any type of aqueous cleanser is then emulsified with water and used as the main cleanser that removes the first cleanser and further cleans the skin. Then the face is finally thoroughly rinsed with water until no filth or product residue remains.

Using a cleanser designated for the facial skin to remove dirt is considered to be a better alternative to bar soap or another form of skin cleanser not specifically formulated for the face for the following reasons:

(1) Bar soap has an alkaline pH (in the area of 9 to 10), and the pH of a healthy skin surface is around 4.7 on average. This means that soap can change the balance present in the skin to favor the overgrowth of some types of bacteria, increasing acne. In order to maintain a healthy pH balance and skin health, your skin must sit on the proper pH level; some individuals who use bar soap choose to use pH-balancing toners after cleaning in attempts to compensate for the alkalinity of their soaps.

(2) Bar cleansers have thickeners that allow them to assume a bar shape. These thickeners can clog pores, which may lead to pimples in susceptible individuals.

(3) Using bar soap on the face can remove natural oils from the skin that form a barrier against water loss. This causes the sebaceous glands to subsequently overproduce oil, a condition known as reactive seborrhoea, which will lead to clogged pores. In order to prevent drying out the skin, many cleansers incorporate moisturizers.

Facial cleansers include the following: balm cleansers, bar cleansers, clay cleansers, cold cream cleansers, creamy cleansers, exfoliant scrub cleansers, foam/foaming cleansers, gel/

jelly cleansers, lotion cleansers, micellar cleansers, milky cleansers, oil cleansers, powder cleansers, treatment/medicated cleansers (*Aloe vera*, benzoyl peroxide, carboxylic acids, charcoal, colloidal oatmeal, honey, sulphur, vitamin C, lighteners) and tool cleansers (cotton rounds, konjac sponges, microfiber cloths, mitts, silicone brushes, spinning brushes, sponges, towelettes/wipes).

Cleansers that have active ingredients are more suitable for oily skins to prevent breakouts. But they may overdry and irritate dry skin, this may make the skin appear and feel worse. Dehydrated skin may require a creamy lotion-type cleanser. These are normally too gentle to be effective on oily or even normal skin, but dry skin requires much less cleansing power. It may be a good idea to select a cleanser that is alcohol-free for use on dry, sensitive, or dehydrated skin.

Some cleansers may incorporate fragrance or essential oils. However, for some people, these cleansers may irritate the skin and often provoke allergic responses. People with such sensitivity should find cleansers that are pH-balanced/cosmetic balanced, contain fewer irritants, suit many variating skin types, and do not make the skin feel dehydrated directly after cleansing. Tight, uncomfortable skin is often dehydrated and may appear shiny after cleansing, even when no sebum is present. This is due to the tightening and 'stripping' effect some cleaners can have on the skin. One should discontinue use of a cleanser that upsets the balance of the skin; cleansers should work with the skin not against it. Finding the right cleanser can involve some trial-and-error.

8.2 Toners

In cosmetics, skin toner or simply toner refers to a lotion, tonic or wash designed to cleanse the skin and shrink the appearance of pores, usually used on the face. It also moisturizes, protects and refreshes the skin. Toners can be applied to the skin in different ways: on a cotton round (the most frequently used method), spraying onto the face, or by applying a tonic gauze facial mask (a piece of gauze is covered with toner and left on the face for a few minutes). Some toners may cause some irritation to the skin upon their initial use. Users often apply serum and moisturizer after the toner has dried.

Toners usually contain water, citric acid, herbal extracts and other ingredients. Witch hazel is still commonly used in toners to tighten the pores and refresh the skin. Alcohol is used less often as it is drying and can be irritating to the skin. It may still be found in toners specially for those with oily skin. Some toners contain active ingredients and target particular skin types, such as tea tree oil, salicylic acid, or glycolic acid.

There are different kinds of toners:

(1) Skin bracers or fresheners are the mildest form of toners, which contain water and a humectant such as glycerine, and little if any alcohol (0-10%). Humectants help to keep the moisture in the upper layers of the epidermis by preventing it from evaporating. A popular example of this is rosewater. These toners are the gentlest to the skin, and are most suitable for use on dry, dehydrated, sensitive and normal skins. It may give a burning sensation to sensitive skin.

(2) Skin tonics are slightly stronger and contain a small quantity of alcohol (up to 20%), water and a humectant ingredient. Orange flower water is an example of a skin tonic. Skin tonics are suitable for use on normal, combination, and oily skin.

(3) Acid Toners are a strong form of toner that typically contains alpha hydroxy acid and/or beta hydroxy acid. Acid toners are formulated with the intent of chemically exfoliating the skin. Glycolic, Lactic, and Mandelic acids are the most commonly used alpha hydroxy acids, best suited to exfoliate the surface of the skin. Salicylic acid is the most commonly used beta hydroxy acid best for exfoliating into the deeper layers of the skin.

(4) Astringents are the strongest form of toner and contain a high proportion of alcohol (20%-60%), antiseptic ingredients, water, and a humectant ingredient. These can be irritating and damaging to the skin as they can remove excess protective lipids as well as denature proteins in the skin when a high percentage of alcohol is used.

8.3 Hyperpigmentation Treatment

Hyperpigmentation can be caused by sun damage, inflammation, or other skin injuries, including those related to acne vulgaris. People with darker skin tones are more prone to hyperpigmentation, especially with excess sun exposure. Kojic acid soap, cream or powder and arbutin (β-D-glucopyranoside derivative of hydroquinone) serum or cream help to get rid of hyperpigmentation spots of the skin.

8.4 Facial Masks

A facial is a family of skin care treatments for the face, including steam, exfoliation (physical and chemical), extraction, creams, lotions, facial masks, peels, and massage. They are normally performed in beauty salons, but are also a common spa treatment. They are used for general skin health as well as for specific skin conditions. Facial masks are treatments applied to the skin and then removed. Typically, they are applied to a dry, cleansed face, avoiding the eyes and lips.

There are different kinds of masks (*e.g.*, clay, cactus, cucumber, *etc.*) for different purposes: deep-cleansing, by penetrating the pores; healing acne scars or hyperpigmentation; brightening, for a gradual illumination of the skin tone. Facial masks also help with anti-ageing, acne, crow's feet, under eye bags, sagging eyelids, dark circles, puffiness and more. Some masks are designed to dry or solidify on the face, almost like plaster; others just remain wet. The perceived effects of a facial mask treatment include revitalizing, healing, or refreshing; and, may yield temporary benefits (depending on environmental, dietary, and other skincare factors). There is little to no objective evidence that there are any long-term benefits to the various available facial treatments.

Masks are removed by either rinsing the face with water, wiping off with a damp cloth, or peeling off of the face. Duration for wearing a mask varies with the type of mask, and manufacturer's usage instructions. The time can range from a few minutes to overnight. Those with sensitive skin are advised to first test out the mask on a small portion of the skin, in order to check for any irritations. Some facial masks are not suited to frequent use. A glycolic mask should not be used more frequently than once a month without the risk of burning the skin.

Masks can be found anywhere from drugstores to department stores and can vary in consistency and form. Setting masks include: clay, which is a thicker consistency, and will draw out impurities (and sometimes, natural oils, too) from the pores; a cream, which stays damp to hydrate the skin; sheet-style, in which a paper mask is dampened with liquid to tone and moisturize the skin; and lastly, a hybrid/clay and cream form that includes small beads for removing dead surface skin cells. Non-setting facial masks include warm oil and paraffin wax masks. These different forms are made to suit different skin types (*e.g.*, oily or dry), and different skincare goals or needs (*e.g.*, moisturizing, cleansing, exfoliating). Clay and mud masks suit oily and some "combination" skin types, while cream-based masks tend to suit dry and sensitive skin types. There are also peel-off masks which are great for when you need to remove thin layers of dead skin cells and dirt.

8.5 Moisturizers

Moisturizers, or emollients, are creams or lotions which are usually used for protecting, moisturizing, and lubricating the skin. These functions are normally performed by sebum produced by healthy skin.

Moisturizers may contain essential oils, herbal extracts, or chemicals to assist with oil control or reducing irritation. Night creams are typically more hydrating than day creams, but may be too thick or heavy to wear during the day. Tinted moisturizers contain a small amount of foundation, which can provide light coverage for minor blemishes or to even out skin tones. They are usually applied with the fingertips or a cotton pad to the entire face, avoiding the lips and area around the eyes. Eyes require a different kind of moisturizer compared with the rest of the face. The skin around the eyes is extremely thin and sensitive, and is often the first area to show signs of aging. Eye creams are typically very light lotions or gels, and are usually very gentle; some may contain ingredients such as caffeine or vitamin K to reduce puffiness and dark circles under the eyes. Eye creams or gels should be applied over the entire eye area with a finger, using a patting motion. Finding a moisturizer with SPF is beneficial to prevent aging and wrinkles.

There are many different types of moisturizers. Petrolatum is one of the most effective moisturizers, although it can be unpopular due to its oily consistency. Other popular moisturizers are cetearyl alcohol, cocoa butter, isopropyl myristate, isopropyl palmitate, lanolin, liquid paraffin, polyethylene glycol, shea butter, silicone oil, stearic acid, stearyl alcohol and castor oil, as well as other oils.

8.6 Barrier Cream

A barrier cream is a cosmetic used to place a physical barrier between the skin and contaminants that may irritate the skin (contact dermatitis or occupational dermatitis). Three classes of barrier creams are used: water repellent creams, water-soluble creams, and creams designed for special applications. Barrier creams may contain substances such as zinc oxide, talc or kaolin to layer over the skin. For hand care they are designed to protect against the harm from detergents and other irritants.

The efficacy of barrier creams is controversial. They have not been demonstrated to be useful in preventing hand eczema. A 2018 Cochrane review concluded that the use of moisturizers alone or in combination with barrier creams may result in important protective effects for the prevention of Occupational Irritant Hand Dermatitis (OIHD). They are a poor substitute for protective clothing for workers. Gloves provide a greater protection than barrier creams. However, they are reasonably effective for the protection of face against some airborne substances.

Some evidence suggests that improper use of barrier cream could cause a harmful rather than a beneficial effect. Skin that has been moisturized by barrier cream may be more susceptible to irritation by sodium lauryl sulfate, which can permeate hydrated skin more easily because of its hydrophilia. Barrier creams that contain petroleum jelly or certain oils may cause rubber or latex gloves to deteriorate.

8.7 Sunscreens

Sunscreen, also known as sunblock or suntan lotion, is a photoprotective topical product for the skin that absorbs or reflects some of the sun's ultraviolet (UV) radiation and thus helps protect against sunburn and most importantly prevent skin cancer. Sunscreens come as lotions, sprays, gels, foams (such as an expanded foam lotion or whipped lotion), sticks, powders and other topical products. Sunscreens are common supplements to clothing, particularly sunglasses, sunhats and special sun protective clothing, and other forms of photoprotection (*e.g.*, umbrellas, *etc.*).

Solar radiation reaching the terrestrial surface comprises ultraviolet (UV), visible light, and infrared (IR) rays. The spectra of all electromagnetic radiation range from 100 nm to 1 mm, in which UV radiation has the shortest wavelength (200-400 nm) compared to visible light (400-740 nm) and IR (760-1000000nm). UV radiation constitutes about 10% of the total light output of the sun. The broad spectrum of UV radiation is subdivided into three recommended ranges (UVA, UVB, and UVC). Therein, UVA has the longest wavelength (320-400nm) but the least energy photon, while UVB wavelength is in the middle span (280-320nm) and UVC has the shortest wavelength (100-280nm) but the highest energy. It has been reported that moderate sun exposure offers a number of beneficial effects, including production vitamin D, antimicrobial activity, and improved cardiovascular health. However, long-term exposure to UV rays is considered to be a potential risk of skin cancer and acute and chronic eye injuries.

UV-induced skin damage is one of the most common concerns in the world. Certainly, UVA is a risk of skin aging, dryness, dermatological photosensitivity, and skin cancer. It damages

DNA through the generation of reactive oxygen species (ROS), which causes oxidative DNA base modifications and DNA strand breaks, resulting in mutation formation in mammalian cells. On the other hand, UVB can directly damage DNA through the formation of pyrimidine dimer and then cause apoptosis or DNA replication errors, leading to mutation and cancer. Although UVC is the shortest and most energetic wavelength, it is the most dangerous type of UV ray because it can cause various adverse effects (*e.g.*, mutagenic and carcinogenic). However, UVC rays do not penetrate through the atmosphere layer.

It has been proved that photoprotectors, especially sunscreen, play a critical role in reducing the incidence of human skin disorders (pigment symptoms and skin aging) induced by UV rays. Sunscreen was first commercialized in the United States in 1928 and has been expanded worldwide as an integral part of the photoprotection strategy. It has been found to prevent and minimize the negative effects of UV light based on its ability to absorb, reflect, and scatter solar rays. Over the decades of development, sunscreens have been improved step-by-step, accompanying the innovation of photoprotective agents. Certainly, recent sunscreens are found to not only address UV effects, but also protect the skin from other risks (*e.g.*, IR, blue light, and pollution). Indeed, while UV radiation is most commonly implicated in skin disorder development, it is crucial to note the potential role of these considerable harmful factors. It has been suggested that these factors can worsen disorders of dyspigmentation, accelerating aging, and eliciting genetic mutations.

Furthermore, the photoprotective efficiency of sunscreen is determined through sun protection factor (SPF) and the protection grade of UVA (PA) values. According to Food and Drug Administration (FDA) regulations, commercial products must be labeled with SPF values that indicate how long they will protect the user from UV radiation and must show the effectiveness of protection. Certainly, the SPF values are generally in the range of 6-10, 15-25, 30-50, and 50+, corresponding to low, medium, high, and very high protection, respectively. Nevertheless, there are some fundamental misunderstandings of the SPF. Some argument is that an SPF 15 sunscreen can absorb 93% of the erythemogenic UV radiations, while an SPF 30 product can block 96%, which is just over 3% more. The argument may be correct when evaluating sun protection capacity, but is not sufficient in assessing the amount of UV radiation entering the skin. In other words, half as much UV radiation will penetrate into the skin when applying an SPF 30 product compared to an SPF 15 product. This is also illustrated by comparing SPF 10 with SPF 50 sunscreen. Ten and two photons transmit through sunscreen film and enter the skin when applying SPF 10 and SPF 50 products, respectively, as a difference factor of five it is expected. On the other hand, in 1996, the Japan Cosmetic Industry Association (JCIA) developed an *in vivo* persistent pigment darkening (PPD) method to evaluate UVA efficacy of sunscreen. Sunscreens are labeled with PA+, PA++, PA+++, and PA++++, corresponding to the level of protection grade of UVA (PA) obtained from the PPD test. Sunscreens labeled as PA+ express low protection, mainly contributed by between two and four UVA filters. Sunscreens containing four to eight sunscreen agents show moderate levels of UVA blocking and are labeled as PA++. In contrast, the PA+++ and PA++++ symbols represent products that are composed of more than eight UVA filters and provide a high sunscreen efficacy.

Reference

Ngoc L T N, Tran V V, Moon J-Y, Chae M, Park D, Lee Y-C. Recent Trends of Sunscreen Cosmetic: An Update Review. Cosmetics, 2019, 6: 64.

Key Words & Phrases

cleanser *n.* 洁肤液；洁肤霜；清洁剂
sebum *n.* 皮脂
cleansing balm 卸妆膏；洁面膏
compensate *v.* 补偿；弥补
exfoliant scrub cleanser 去角质磨砂洁面乳
Aloe vera 芦荟汁（用于生产护肤霜等）；芦荟
benzoyl peroxide 过氧化苯甲酰
carboxylic acid 羧酸
colloidal oatmeal 胶态燕麦；胶原燕麦；燕麦凝胶
lightener *n.* 美白剂
konjac sponge 魔芋海绵
microfiber cloth 超细纤维布
mitt *n.* 接球手套；棒球手套；手
silicone brush 硅胶刷
spinning brush 旋转刷
essential oil 精油
toner *n.* 爽肤水；呈色剂；色粉
tonic *n.* 补品；补药；护发液；护肤液
facial mask 面膜
witch hazel 金缕梅酊剂（用于治疗皮肤创伤）；金缕梅
glycolic acid 乙醇酸
humectant *n.* 保湿剂
antiseptic *n.* 防腐剂；抗菌防腐药
hyperpigmentation *n.* 色素沉着过度
kojic acid 曲酸
arbutin *n.* 熊果苷
β-D-glucopyranoside *n.* β-D-吡喃葡萄糖苷
hydroquinone *n.* 氢醌；对苯二酚
cactus *n.* 仙人掌；仙人掌科植物
crow’s feet 鱼尾纹；眼角的鱼尾纹
sagging eyelid 眼睑或眼皮下垂
puffiness *n.* 浮肿；自夸；虚胖；膨胀；肿胀
non-setting facial mask 不定型面膜
emollient *n.* 润肤剂；润肤霜
night cream 晚霜
day cream 日霜
caffeine *n.* 咖啡碱
petrolatum *n.* 矿脂
cetearyl alcohol 鲸蜡醇；十六硬脂酸酯；棕榈醇
cocoa butter 可可脂
isopropyl myristate 肉豆蔻酸异丙酯；十四酸异丙酯
isopropyl palmitate 十六酸异丙酯；棕榈酸异丙酯
lanolin *n.* 羊毛脂
liquid paraffin 液体石蜡
polyethylene glycol 聚乙二醇
silicone oil 硅油
stearic acid 硬脂酸；十八烷酸
stearyl alcohol 十八烷醇
castor oil 蓖麻子油
barrier cream 护肤霜；隔离霜
eczema *n.* 湿疹
hydrophilia *n.* 亲水性；吸水性
petroleum jelly 蜡膏
latex glove 乳胶手套
sunblock/suntan lotion 防晒霜；防晒油
cardiovascular *adj.* 心血管的
reactive oxygen species 活性氧物种；活性氧类
mutation *n.* 突变；（生物物种的）变异
mammalian *n.* 哺乳类动物
pyrimidine dimer 嘧啶二聚体
apoptosis *n.* 凋亡；细胞凋亡
mutagenic *adj.* 诱变的
carcinogenic *adj.* 致癌的
pigment symptom 色素症状
erythemogenic *adj.* 引起红斑的
persistent pigment darkening 持续性色素沉着

Chapter 9 Hair Care Cosmetics

Hair care is an overall term for hygiene and cosmetology involving the hair which grows from the human scalp and other body hair. Hair care products are a category of cosmetics which are used to improve the appearance of hair. Hair care services are offered in salons, barbershops and day spas, and products are available commercially for home use. Nowadays, hair care and style play a very important role in people's lives, both for men and women, so knowledge of hair products, mode of action, efficacy, ingredients and hair procedures has become more relevant in dermatologists' medical practice. The amount of money spent to enhance the hair beauty is an indication of how much attention is given today to the hair appearance. On the other hand, these data are emphasized in patients suffering from hair disease.

Hair cosmetics are also an important tool for increasing patient's adhesion to scalp treatments, according to the diversity of hair types and ethnicity.

Trueb described them as "preparations intended for placing in contact with the hair and scalp, with the purpose of cleansing, promoting attractiveness, altering appearance, and/or protecting them in order to maintain them in good condition".

Hair cosmetics can be distinguished into two main categories:

(1) Cosmetics with temporary effect on the hair, for example shampoos, conditioners, sprays, and temporary colors; and

(2) Cosmetics that produce permanent effect on the hair shaft, such as permanent waves, relaxers, bleaches and permanent colors.

9.1 Shampoos

Shampoos are the most commonly prescribed treatment for hair and scalp. In the past, soap was the only available cleanser for the hair, while the introduction of the first non-alkaline shampoo dated back to 1933.

The modern society requires many features in a formulation: a shampoo has primarily to clean the scalp and the hair, but also has to be cosmetically pleasing, not dry out the hair, not irritate the skin, improve hair beautification, and be less expensive.

A shampoo consists of 10-30 ingredients, classified according to their different activities. These ingredients include: (1) cleansing agents, called surfactants, that remove sebum or skin scale; (2) conditioning agents that give softness to the hair; (3) active ingredients, for treating specific diseases like dandruff; and (4) additives, such as preservatives, that contribute to the stability and comfort of the product. Every ingredient is tested, officially approved and declared on the label.

Thanks to surfactant, a shampoo can remove dirty particles and sebum from the scalp and from the hair shafts, but this cleansing activity has to be mild in order to avoid complete removal of the natural oils and sebum from the skin. Modern shampoos are sensitive on skin or mucous membranes, and the development of contact dermatitis is rare, due to the short contact time of the shampoo with the scalp. Typical ingredients of shampoos are foaming agents and detergent agents, which prevent the formation of insoluble soap on hair and scalp. Surfactants are the most important cleaning substances of shampoos, thanks to their molecular structure. In particular, one end of the molecule has a negative electrical charge and is soluble in water, so it does not blend with oils. The other end is soluble in oil and grease and does not blend with water. These molecules surround the fragments of grease localized on hair and scalp: the oil soluble parts go into the grease and the water-soluble parts remain located outward, forming a hydrophilic mass that is totally negatively charged. The surface of the hair shaft has negative charges, which tend to force the mass negative charges apart, resulting in a reduction of tension between water and grease on the surface, and production of foam, which incorporates the dirt, so that it cannot be re-deposited on the hair and scalp.

Conditioning molecules contained in shampoos combine the cleansing action with the function to impart manageability, gloss and antistatic properties of the hair. This conditioning shampoos are composed by fatty substances, like vegetable oils, wax, lecithin (a lanolin derivative), collagen, animal proteins, quaternium ammonium compounds or cationic polymers (which are the main component of the “2 in 1 shampoo”) and silicone, that reduce the friction obtained by combing the hair, and then the risk of hair shaft damage.

Among other substances we can mentioned: glycerin, polyvinylpyrrolidone, propylene glycol and stearalkonium chloride. In particular, protein-derived substances have the property to be attractive by the keratin (“substantivity”), with the result of temporarily adhere the cortex fragments together. Additives of shampoos include ingredients that are able to control the thickness of the formulation like citrate and lactate.

Another key consideration is shampoos pH: many of them have an alkaline pH, which causes hair shaft swelling, predisposing the hair to the damage. A neutral pH shampoo is the best choice for chemically treated hair from either permanent dyeing or permanent waving.

Other additives include UV absorbers, like benzophenone derivatives and antioxidants, like ascorbic acid and α-tocopherol. Preservatives of shampoos prevent bacterial contamination, and include sodium benzoate, parabens, ethylenediaminetetraacetic acid (EDTA), DMDM hydantoin and tetrasodium EDTA.

Moreover, most shampoos contain other ingredients like colors, perfumes, pearlised agents, moisturizers such as natural oils and fatty acid esters, and humectants, such as glycerin and sorbitol.

“Medicated” shampoos contain active ingredients selected to treat medical conditions, mainly seborrheic dermatitis or psoriasis. Active ingredients include ketoconazole, ciclopirox olamine, zinc pyrithione, piroctone olamine, tar derivates, salicylic acid, selenium sulfide, poly-vinyl-pyrrolidone, iodine complex, menthol and colloidal sulfur. The combination of cosmetic technology and medical therapy allows the benefits of cosmetic products together with the efficacy of medical agents.

Adverse reactions to shampoos are rare. The most common one is accidental contact with the mucous membranes, such as the nose and eye.

If patch testing to a shampoo is required, the shampoos should be diluted to form a 1%-2% aqueous solution for closed patch testing and a 5% aqueous solution for open patch testing. False positive reactions due to irritation can occur. Patch testing individual ingredients separately is advisable.

9.2 Hair Conditioners

Most of the shampoos contain conditioning agents, but consumers often apply a conditioner after shampooing, to minimize hair frizzing, increasing manageability and gloss of the hair. A conditioner can therefore be used to reduce a chemical or mechanical trauma of the hair, such as permanent dyes, bleaching, and excessive brushing.

As already mentioned, the hair shaft has negative electrical charges, while conditioning agents have positive charges: conditioners act by neutralizing the electrical negative charge of the hair fiber by addicting positive charges and by greasing the cuticle, with the final result of reduce fiber hydrophilicity. Conditioners also help the raised cuticles to lie down against the hair surface, and in this way preserve hair color and enhance its shyness and smoothness too.

Common ingredients of conditioners include mainly cationic surfactants, like cetyltrimethylammonium chloride, propyltrimonium, and stearamidopropyl dimethylamine.

Polymers, like mono and polypeptides such as hydrolyzed proteins (amino acids) derived from collagen and polyvinylpyrrolidone (PVP), fill the hair shaft defects, creating a smooth surface to increase shine while eliminating static electricity thanks to its cationic nature.

Emollients/oily compounds are composed of natural or synthetic oils, but also esters and waxes. Jojoba oil, olive oil, and grape seed oil are some of these natural oils.

The most frequently used synthetic and active oils are silicones (*e.g.*, dimethicone, dimethiconol, amodimethicone and cyclomethicone). Dimethicone is the most widely used silicone in hair care cosmetics, with the effect of protecting the hair shaft from abrasive actions.

Other conditioners components are: fatty alcohols (*e.g.*, cetyl alcohol and stearyl alcohol), waxes (*e.g.*, carnauba wax and paraffin wax), gums (*e.g.*, guar gum), salt (sodium chloride) and emulsifiers like ethoxylated fatty alcohols (*e.g.*, polysorbate-80).

In daily practice, hair conditioners can be used in different ways.

Instant conditioners are applied immediately after shampooing and rinsed off after few minutes, while deep conditioners, formulated usually in creams, can remain on hair for 20-30 min with an increase of their penetration.

Blow drying lotions are another formulation that can be left on the hair because are oil-free. They are very useful for people with fine hair or excessive scalp sebum. A particular type of conditioner, called hair glaze or hair thickeners conditioners, increase hair diameter by coating the hair shaft. Similar to conditioners, they usually contain proteins.

9.3 Hair Sprays

Hair sprays are the simplest of styling aids. These products are applied to dry hair and hold it in place by a combination of coating each hair with a thin deposit of stiff polymer, and 'gluing' hairs together at points where they cross. Such products evaporate rapidly and the effect is 'instant', *i.e.*, the product is applied to the finished hairstyle to hold it in place. Hair sprays were revolutionised by the introduction of the aerosol with its concomitant huge increase in product performance. The pump-spray revolution has been altogether quieter, though no less effective, with concern over ozone depletion adding to already increasing sales. It is now possible to achieve a very fine spray, a choice of spray angles and different delivery quantities per stroke by careful selection of the most suitable pump. Further legislation to control the emission of volatile organic compounds (VOCs) may ultimately favour pump sprays at the expense of aerosols, although water-based aerosol sprays using lower levels of VOCs are being developed.

9.4 Permanent Waves

The topical application of cosmetics in lotions or sprays may give hair a temporary different "style" that is lost after the first shampooing. A modification of hair internal chemical structure with a specific chemical process is needed to change in a permanent way the shape of the hair.

A permanent wave is a process that makes curly the hair, using chemicals that break and reform the strong disulfide bonds of the hair shaft. First the hair is washed and wrapped around the curler (the amplitude of which will determinate the thickness of the curl), then waving lotion in alkaline base is then applied (the most commonly used is ammonium thioglycolate): it raises the scales of the cuticle and, reaching inside the cortex, breaks the disulfide bonds that give the natural shape of the hair. After a fixed time, the waving lotion is removed with the rinse and a neutralizer is applied, which contains oxidizing agents, like hydrogen peroxide, that rebuild the disulfide bonds in a different order, stabilizing the new shape of the hair. Hair waving is for this reason the most dangerous cosmetic procedures for the hair and should be monitored carefully.

Trichorrhexis nodosa is frequent in bleached and permed hair. The damage to the cuticle cells makes them detach and leads to exposure to the hair shaft cortex. The hair shaft shows small whitish knots visible with the unaided eye. The hair is fragile and short due to breakage. Microscopic observation of the hair shaft shows one or more knots that easily break off producing two brush-like tips.

9.5 Hair Dyes

In modern society, people want to change the hair color according to the fashion trends of the moment, and thanks to developments in the field of dyestuffs, now this desire can be obtained in a few minutes. Nowadays, hair dyes are utilized by both men and women to modify natural hair color or restore pigmentation once graying has set in.

Modern hair dyes can be classified into three categories: temporary, semi-temporary and permanent. The main difference between them is their capacity to reach the hair cortex, obtained by an alkaline pH. In many of these products, ammonia or ethanolamine are added to increase the pH.

Temporary colors are applied in the form of shampoos, lotions or foam and change the hair color for a relatively short time, normally until the first wash. They are generally used to add color highlights, remove yellowish hues from white hair, or cover up few graying hairs.

Ingredients of temporary colors are products with a high molecular mass, which simply deposit on the cuticle of the hair and are then removed easily. The dye is bonded to a cationic polymer to reduce its solubility and increase its affinity to hair. The resulting complex is dispersed in a base using surfactants to make the final product.

One example is henna, obtained from the plant *Lawsonia alba*, which is used on skin and nails, too. In the literature, few cases of allergic contact dermatitis to henna have been described.

Semi-permanent colors are composed by ingredients like nitroaromatic amines or aromatic dyes (of low molecular mass), which do not bind to the hair protein because they do not oxidize: they are water soluble, so they can be washed out after usually 6-10 wash cycles or earlier. They do not contain ammonia or ethanolamine, but hydrogen peroxide or resorcinol can be present. The main purpose of semi-permanent colors is to color white hair or to give more tone to the natural color, but they cannot lighten hair. One important characteristic is low molecular mass, which enables these products to diffuse into the middle layers of the cuticle without binding firmly to the hair protein itself. Consequently, the application of semi-permanent colors does not require any prior hair modification, required instead by permanent hair color.

Permanent hair dyeing is a process that permanently changes the color of the hair, through a chemical reaction of oxidation, which allows the molecules of color to penetrate into the hair. Permanent colors allow darkening or brightening hair color, and are resistant to any external factor, including washings, for 4-6 weeks. They offer many ranges of hues and can cover the highest gray-hair numbers.

The principle consists in the penetration of the coloring molecules into pores of hair shaft (previously increased in size, by hydration and alkalinization). At this point, the molecules are oxidized and assume color, which is transmitted to the keratin of the cuticle and the cortex, mimicking natural melanin granules.

During these reactions, some intermediates are formed, usually derivates of *p*-phenylenediamine (PPD), responsible for dark brown and black shades, or of *p*-aminophenol, responsible for blonde colors. Some dyes contain also resorcinol, but many companies have eliminated it for security reasons. PPD is an hapten (an incomplete antibody stimulating substance), which can combine with a protein molecule in the skin, leading to sensitization, which in future reapplication will lead to an allergic reaction. Allergic skin reactions to PPD may have different severity and sometime include the mucosae. For these reasons, a skin allergy test, made at least 48 hours before coloring, is mandatory before applying the coloring agent on the scalp in people with suspected PPD sensitization.

As mentioned, in dyes active ingredients are also present, like hydrogen peroxide, which bleaches out the original color and, oxidizing the molecules of the dye, activate the colorant reactions, and ammonia, which is the alkalizer. Other ingredients are: chelating agents, which maintain chemically stable peroxide and its activity, deactivating any catalytic metals in the product; and solvents, which help to improve viscosity and solubilization of the dye; surfactants, which remove oil and sebum from the hair surface.

Dyes can interfere with some dermoscopic diagnosis of hair diseases: for example, permanent dye can penetrate hair follicle, giving a "black dot" appearance, distinctive for alopecia areata; semi-permanent dye can deposit on the scalp, mimicking sun exposure or excoriated lesions.

Long-term contact of dyes with the scalp can determine a contact dermatitis which can cause a telogen effluvium, but in severe cases can induce facial edema.

As mentioned, PPD is an extremely potent contact sensitizer, allowed at a concentration up of 6% in hair dyes. Other frequent reported hair dye allergens are toluene-2,5-diamine, allowed for use at a concentration up to 10%, *p*-aminophenol and *m*-aminophenol, the last at up to 2%.

A study analyzed chemically the actual content of PPD and selected derivatives in hair dye products causing allergic contact dermatitis in nine patients: more specifically, chemical analysis of the randomly collected products identified concentrations of the oxidative hair dyes similar to the concentrations in the products used by the patients with hair dye dermatitis. This study shows that patients can develop hair dye dermatitis from concentrations lower than that permitted in the European Union (EU) Cosmetic Directive.

In another interesting study, authors provided a ranking concerning contact allergy risk, analyzed the predictions for sensitization potential of 229 hair dyes substances: as result, about the 75% of all hair dyes were predicted to be strong/moderate sensitizers, even stronger than PPD, and many of these are not currently used for diagnosing contact allergy to hair dyes. PPD is resulted more potent than toluene-2,5-diamine, while resorcinol seems to be less potent than PPD. Only 22% of hair dye substances were predicted to be weak sensitizers and 3% were predicted to be extremely weak or non-sensitizing.

9.6 Bleaching Agents

These products provide permanent hair lightening without the addition of a coloring tint, thanks to substances like hydrogen peroxide, ammonia and persulfate salts. The process requires time and is realized by a series of color stages. As mentioned, hydrogen peroxide bleaches out the original color of the hair. The oxidizing agent, in alkaline solution, penetrates into the cortex of the hair, after raising the scales of the cuticle, and chemically modifies the melanin: as result, melanin granules are dissolved, leaving tiny gaps in the cortex. For this reasons, repeated bleaching treatments can cause an increased porosity of hair and permanently raised scales, leading to hair weathering.

Persulfates can trigger immunomediate reactions, such as rhinitis, asthma, contact urticarial and even anaphylaxis, but the exact immune mechanism is still unknown.

9.7 Hair Waxes

Hair waxes are a kind of thick hairstyling products containing wax, used to assist with holding the hair. In contrast with hair gels, most of which contain alcohol, hair waxes remain pliable and have less chance of drying out. Consequently, hair waxes are currently experiencing an increase in popularity, often under names such as pomade, putty, glue, or styling paste. The texture, consistency, and purpose of these products vary widely and each has a different purported purpose depending on the manufacturer. Traditionally, pomades are a type of hair wax that also adds shine to one's hair.

Hair waxes have been used for many years and a waxy soap-like substance was invented by the ancient Gauls as a hair styling agent and was not used as a cleaning agent until many years later.

9.8 Relaxers

The mechanisms of relaxing are the breaking of hair disulfide bonds, re-forming of hair shape and re-making of the linkages. The shape of the hair changes from curly to straight. Multiple procedures or prolonged time of exposure to the relaxer may be needed to straighten tightly curled hair, such as that of African-Americans, producing hair weathering and breakage.

9.9 Hair Straightening

In the past, hair straightener procedure consisted on applying petrolatum-based oils on the hair combined with hot irons or hot combs pressed to the hair, but the effect was only temporary.

Nowadays, hair straighteners are called chemical relaxers, and the effect of hair straightening is permanent. Thanks to the high pH (9.0-14.0), the solution inflate the hair, opening cuticle scales, which allows the alkaline agent (OH^-) to penetrate into the hair fibers up to the endocuticle. The straightening product reacts with keratin, breaking and rearranging the disulfide bridges, which making the spiral keratin molecule soft and stretched. This procedure needs to be repeated every 12 weeks or longer.

Alkaline straighteners contain 1%-10% sodium hydroxide (lye-relaxer), lithium hydroxide, calcium hydroxide or a combination of these ingredients such as guanidine carbonate and calcium hydroxide (no-lye relaxers). Ammonium thioglycolate is another "no-lye" relaxers, which selectivity weakens the hair cysteine bonds: it is then oxidized by hydrogen peroxide and, applying a hot iron during the process, the hair straightening can be obtained. Thioglycolate and hydroxides are not mutagenic or carcinogenic. The main adverse effects of hair straightening are scalp burns and hair breakage. In particular, in the study of Shetty *et al.*, the most common adverse effects reported after chemical hair straightening were: frizzy hair in 67%, dandruff in 61%, hair loss in 47%, thinning and weakening of hair in 40%, greying of hair in 22%, and split ends in only 17%.

A keratin treatment containing formaldehyde, a well-known carcinogen has been performed in Brazil with success: soon the health vigilance organizations in Brazil prohibited the use of any product containing formaldehyde in concentrations above 0.2% for cosmetics. Afterward, the formaldehyde was replaced by a potentially more mutagenic and neurotoxic product, glutaraldehyde, which belongs to the same aldehyde group, sold as a homemade treatment. The ingredients used today do not contain any of these two substances, but are based on formaldehyde-releasers such as methylene glycol or glyoxylic acid. Acute contact dermatitis may occur as adverse effect in previously sensitized patients.

References

[1] Trueb R M. Dermocosmetic Aspects of Hair and Scalp. Journal of Investigative Dermatology Symposium Proceedings, 2005, 10: 289-292.
[2] Aurora A, Bianca P. Essential of Hair Care Cosmetics. Cosmetics, 2016, 3: 34.

Key Words & Phrases

hair care 护发
ethnicity *n.* 种族渊源；种族特点
conditioner *n.* 护发剂；护发素；（洗衣后用的）柔顺剂
permanent wave 卷发；烫发
relaxer *n.* 直发膏；顺发剂
permanent color 永久性染色；固定的颜色
dandruff *n.* 头皮屑
surfactant *n.* 表面活性剂
detergent agent 洗洁精
blend *v.*（和某物）混合；融合；掺和；调制
hydrophilic *adj.* 亲水（性）的
antistatic *adj.* 抗静电的
lecithin *n.* 卵磷脂；磷脂酰胆碱
collagen *n.* 胶原蛋白；胶原
quaternium ammonium compound 季铵化合物
cationic polymer 阳离子聚合物
friction *n.* 摩擦；摩擦力；争执；分歧；不和
polyvinylpyrrolidone *n.* 聚乙烯吡咯烷酮
keratin *n.* 角蛋白
cortex fragment 皮质片段；皮层片段
citrate *n.* 柠檬酸盐
lactate *n.* 乳酸盐
permanent dyeing 永久性染色
ascorbic acid 抗坏血酸；维生素C
ethylenediaminetetraacetic acid 乙二胺四乙酸
DMDM hydantoin 二羟甲基二甲基乙内酰脲
pearlised agent 珠光剂
seborrheic dermatitis 脂溢性皮炎
psoriasis *n.* 牛皮癣；银屑病
ketoconazole *n.* 酮康唑
ciclopirox olamine 环吡司；环吡酮胺
zinc pyrithione 吡啶硫酮锌
piroctone olamine 吡啶酮乙醇胺盐
selenium sulfide 硫化硒；二硫化硒
menthol *n.* 薄荷醇
colloidal sulfur 胶体硫黄；硫黄胶；胶态硫
hair conditioner 护发素
conditioning agent 调节剂
hair frizzing 头发卷曲
trauma *n.* 创伤；损伤；痛苦经历；挫折
cuticle *n.* 角质层
hydrophilicity *n.* 亲水性
cetyltrimethylammonium chloride 十六烷基三甲基氯化铵
propyltrimonium *n.* 丙基三甲基铵
stearamidopropyl dimethylamine 硬脂酰胺丙基二甲胺
polypeptide *n.* 多肽
jojoba oil 荷荷巴油；霍霍巴油
dimethiconol *n.* 聚二甲基硅氧烷醇
amodimethicone *n.* 氨端聚二甲基硅氧烷

cyclomethicone *n.* 环甲硅脂；环聚二甲基硅氧烷
abrasive action 碾削作用
carnauba wax 巴西棕榈蜡；棕榈蜡
paraffin wax 石蜡
guar gum 瓜尔胶；古尔胶
ethoxylated fatty alcohol 乙氧基脂肪醇
polysorbate-80 吐温类乳化剂-80
penetration *n.* 穿透；渗透
glaze *n.* 釉；瓷釉
hair spray 发胶
stiff *adj.* 不易弯曲（或活动）的；硬的；挺的；僵硬的
aerosol *n.* 气溶胶；气雾剂
ammonium thioglycolate 巯基乙酸铵；硫代乙醇酸铵
trichorrhexis nodosa 结节性脆发病
permed hair 烫发
cortex *n.* 皮层；皮质
henna *n.* 散沫花染剂（棕红色，尤用于染发）
Lawsonia alba 散沫花；指甲花
nitroaromatic amine 硝基芳香胺
aromatic dye 芳香染料
resorcinol *n.* 间苯二酚
melanin granule 黑色素颗粒
p-phenylene diamine 对苯二胺
p-aminophenol *n.* 对氨基苯酚
mucosa *n.* 黏膜
chelating agent 螯合剂；络合剂
dermoscopic *n.* 皮肤镜
hair follicle 毛囊
alopecia areata 斑秃；局限性脱发
telogen effluvium 休止期脱发；休止期落发；静止期脱发
edema *n.* 浮肿；水肿
bleaching agent 漂白剂；增白剂
persulfate salt 过硫酸盐
trigger *v.* 触发；引起；发动；开动；起动
immunomediate *adj.* 免疫介导的
rhinitis *n.* 鼻炎（感染或过敏引起）
asthma *n.* 气喘；哮喘
contact urticarial 接触性荨麻疹
anaphylaxis *n.* 过敏反应
hair wax 发蜡
pomade *n.* 发油
putty *n.* 腻子
endocuticle *n.* 内表皮；内角皮
guanidine carbonate 碳酸胍；胍碳酸盐
frizzy hair 卷发
carcinogen *n.* 致癌物
glutaraldehyde *n.* 戊二醛
formaldehyde-releaser 甲醛释放剂
methylene glycol 亚甲二醇
glyoxylic acid 乙醛酸

Chapter 10 Perfumes

Perfume is a mixture of fragrant essential oils or aroma compounds (fragrances), fixatives and solvents, usually in liquid form, used to give the human body, animals, food, objects, and living-spaces an agreeable scent.

10.1 History

The word perfume derives from the Latin "perfumare", meaning "to smoke through". Perfumery, as the art of making perfumes, began in ancient Mesopotamia, Egypt, the Indus valley civilization and possibly ancient China. It was further refined by the Romans and the Muslims.

On the Indian subcontinent, perfume and perfumery existed in the Indus civilization (3300 BC-1300 BC).

In the 9th century the Arab chemist Al-Kindi (Alkindus) wrote the book of the chemistry of perfume and distillations, which contained more than a hundred recipes for fragrant oils, salves, aromatic waters, and substitutes or imitations of costly drugs. The book also described 107 methods and recipes for perfume-making and perfume-making equipment, such as the alembic.

The Persian chemist Ibn Sina (also known as Avicenna) introduced the process of extracting oils from flowers by means of distillation, the procedure most commonly used today. He first experimented with the rose. Until his discovery, liquid perfumes consisted of mixtures of oil and crushed herbs or petals, which made a strong blend. Rose water was more delicate, and immediately became popular. Both the raw ingredients and the distillation technology significantly influenced western perfumery and scientific developments, particularly chemistry.

Between the 16th and 17th centuries, perfumes were used primarily by the wealthy to mask body odors resulting from infrequent bathing. In 1693, Italian barber Giovanni Paolo Feminis created a perfume water called Aqua Admirabilis, today best known as eau de cologne; his nephew Johann Maria Farina (Giovanni Maria Farina) took over the business in 1732.

By the 18th century the Grasse region of France, Sicily, and Calabria (in Italy) were growing aromatic plants to provide the growing perfume industry with raw materials. Even today, Italy and France remain the center of European perfume design and trade.

10.2 What Are the Different Types of Perfume?

Perfumes are classified into 5 main groups loosely based on their concentration of aromatic compounds.

(1) Parfum or extrait has a 20%-30% concentration of fragrance.

- As parfum contains a high concentration of fragrance, it is typically the most expensive.
- It is usually a heavier, oilier product than the other types of perfume, and tends to be used more sparingly.
- The smell is long–lasting–an average of 8 hours and up to 24 hours.

(2) Eau de parfum or parfum de toilet has a 15%-20% concentration of fragrance.

- Eau de parfum contains a greater concentration of alcohol and water and is generally cheaper than parfum.
- It is a lighter product and has a shorter duration of around 4 to 5 hours.
- Eau de parfum is the most common fragrance type and is the base for other fragrance types.

(3) Eau de toilette has a 5%-15% concentration of fragrance.

- Eau de toilette has a low concentration of essential oils and a high concentration of alcohol.
- It dissipates quickly and lasts 2 to 3 hours.
- It is a cheaper option and often used for daywear.

(4) Eau de cologne has a 2%-4% concentration of fragrance.

- Eau de cologne has a much lower fragrance concentration to create a very light formulation.
- It is primarily used in fragrances designed for males as an aftershave or splash-on fragrance.
- It dissipates quickly and lasts about 2 hours.
- Cologne, by definition, refers to 'eau de cologne'. However, the term 'cologne' is in common use in the English language to denote any fragrance worn by a male.

(5) Eau fraiche has a 1%-3% concentration of fragrance.

- Eau fraiche has the lowest fragrance concentration of all types of parfum, and is diluted with water rather than alcohol or oil.
- Common uses for eau fraiche include mists, splashes, and veils that are very light and dissipate within an hour.

Fragrances are also added to many cosmetics and household goods.

10.3 How Is Perfume Manufactured?

The perfume manufacturing process for natural essences involves collection, extraction, blending and ageing of the product.

(1) Collection

- In the collection stage, initial ingredients are obtained from various plant substances and the fatty extracts of animal products.

(2) Extraction

- Distillation is the most commonly used method to extract perfume. In steam or dry distillation, the material is heated to high temperatures and condensed into gas to release the desired essential oils, which are then cooled and liquefied. Water distillation is more effective for some more delicate materials, in which plant material is placed into boiling water.

- In solvent extraction, the material is added to a solvent, which forms a waxy aromatic compound that is then mixed with alcohol to release the essential oils. This method is becoming less commonly used as it is costly and time-consuming.
- Expression means the material is compressed and the oils are mechanically squeezed out. This method is primarily used for fresh fruit rinds and thermally labile components.
- In maceration, the material is soaked in carrier oils serving as solvents, which capture heavier, larger plant molecules. This method is useful for materials which require a higher yield of essential oils.
- In enfleurage, the material is drawn out into a fat or oil base and then extracted with alcohol. Enfleurage is no longer used commercially.

(3) Blending

- Oils are blended according to a particular formula using multiple different ingredients.
- The scent is mixed with alcohol; the volume of alcohol used depends on the intended type of perfume.

(4) Ageing

- It takes months to years after the scent has been blended to achieve the desired scent. This ageing period allows the ongoing blending of the selected chemicals to modify the scent.

Other perfumes are synthetically manufactured in laboratories. While the need to obtain natural ingredients is avoided, achieving the desired scent can be more challenging.

10.4 What Are the Ingredients and Chemical Structures of Perfume?

Perfumes are composed of three structural parts–the head, middle, and base notes–to provide the first impression, body, and lasting impression of the fragrance respectively after the application of a perfume. The presence of one note may alter the perception of another.

(1) The head note, or top note, provides the initial scent that forms the first impression of the perfume. It has the following features:

- It is comprised of small light molecules that have a strong fresh scent but evaporate quickly, usually 5-30 minutes after application.
- Common top notes include *Citrus* (lemon, orange), light fruits (berries), and herbs (sage, lavender).

(2) The middle note, or heart note, masks any unpleasant initial impression of the base note and provides the main body of the scent.

- It is comprised of more complex molecules than top notes and has a more mellow, rounded, and balancing scent.

- It first appears 20-60 minutes after application, and usually lasts 2-4 hours.
- Tones are made from more potent florals and spices. Common middle notes include lemongrass, rose, geranium, jasmine, nutmeg, lavender, cinnamon, and coriander.

(3) The base note adds to the middle note to boost and deepen the existing body of the scent and provide its lasting impression.

- It is comprised of large heavy molecules to provide a rich and smooth nature to the scent.
- It is typically not perceived until 30 minutes after application or during the dry-down period. Some can last over 24 hours after application.
- Common base notes include cedarwood, sandalwood, vanilla, amber, patchouli, oakmoss, and musk.

10.5 Why Is Perfume Used?

Perfume is used to give a pleasant and desirable scent to a person's body, typically with the aim of increasing self-appeal and self-confidence.

Scents are reported to enhance health and well-being by improving mood, reducing anxiety and stress, increasing cognitive function, and improving sleep.

A link has also been identified between pleasant scents and improved pain tolerance through the activation of opioid pathways.

10.6 What Are the Adverse Effects of Perfume?

Adverse effects of perfume primarily involve irritant and allergic reactions.

The use of perfume directly exposes skin to various chemicals, which are then absorbed. The chemical substances in some fragrances can cause contact allergic dermatitis in sensitised individuals. Studies have reported that one-third of the general population experienced at least one adverse health effect from fragranced products.

Common adverse effects may include:

- Headaches
- Eye, nose, and throat irritation
- Nausea
- Irritant and allergic contact dermatitis (see fragrance mix allergy)
- Asthma flares

Irritant contact dermatitis is due to repeated or excessive exposure to irritating chemicals in the fragrance, usually over a significant period of time. It presents as localised erythema at the site of exposure, dry cracked skin, and blisters and erosions in severe cases.

Allergic contact dermatitis is a delayed hypersensitivity reaction and may occur after just one, or more frequently, many exposures. Symptoms are similar to irritant contact dermatitis, but tiny quantities may be sufficient to cause allergy. It may also result in dermatitis in sites that were not directly exposed to the fragrance. Allergic contact dermatitis is confirmed by patch testing.

Cosmetic products do not legally require allergen labelling, unlike foods.

10.7 What Is the Difference Between Perfumes Marketed to Males and Females?

'Pour homme' and 'pour femme' are French terms used to denote perfumes intended for males and females respectively.

The difference is based on contemporary cultural and marketing trends. Perfumes that emit oriental, woody, and musky scents are typically marketed as masculine perfumes, whereas fruity and floral scents are typically more feminine.

10.8 What Do the Terms 'Fragrance-Free' and 'Unscented' Mean?

Fragrance-free products do not use fragrance materials for the specific purpose of imparting scent; however, they may contain fragrance ingredients intended for therapeutic uses such as to relieve muscle aches or help with sleep.

Unscented products contain chemicals that neutralise or mask the odours of other ingredients.

Reference

Mazzoni D. Fragrances and Perfumes[EB/OL].(2020-01-06)[2023-04-26]. https://dermnetnz.org/topics/.

Key Words & Phrases

fixative *n.* 定影剂；定色剂；防（香味）挥发剂；固定剂
recipe *n.* 配方；食谱；秘诀；方法；烹饪法；诀窍
fragrant oil 香油；芬芳油；香精
salve *n.* 药膏；软膏；油膏
alembic *n.* 蒸馏器（釜，罐）；净化器具
petal *n.* 花瓣
eau de Cologne 科隆香水；古龙水
parfum *n.* 香精；（法）香水
eau de parfum 淡香精；香水
eau de toilette 淡香水
dissipate *v.* 消散；驱散；（使）消失
daywear *n.* 便服；日常衣服
eau fraiche 清香水；淡香水；幽香水
expression *n.* 压出法
maceration *n.* 浸软；浸渍
Citrus *n.* 柑橘属
sage *n.* 圣人；鼠尾草（可用作调料）
spice *n.*（调味）香料
lemongrass *n.* 柠檬草；香茅
geranium *n.* 天竺葵；老鹳草
jasmine *n.* 茉莉
nutmeg *n.* 肉豆蔻
cinnamon *n.* 肉桂皮；桂皮香料
coriander *n.* 芫荽；香菜
cedarwood *n.* 杉木；雪松属木料
sandalwood *n.* 檀香木
vanilla *n.* 香草
amber *n.* 琥珀色；琥珀；黄褐色
patchouli *n.* 广藿香精油；广藿香
oakmoss *n.* 橡树苔；橡木苔
musk *n.* 麝香

anxiety *n.* 焦虑；忧虑；担心；害怕；渴望
opioid *n.* 阿片样物质；阿片类[物质]
nausea *n.* 恶心；恶心想吐
erythema *n.* 红斑
erosion *n.* 侵蚀；糜烂；腐蚀
pour homme 男士香水
pour femme 女士香水
musky *adj.* 有麝香味的；麝香的
masculine *adj.* 男子的；男性的
therapeutic *adj.* 治疗的；治病的；有助于放松精神的

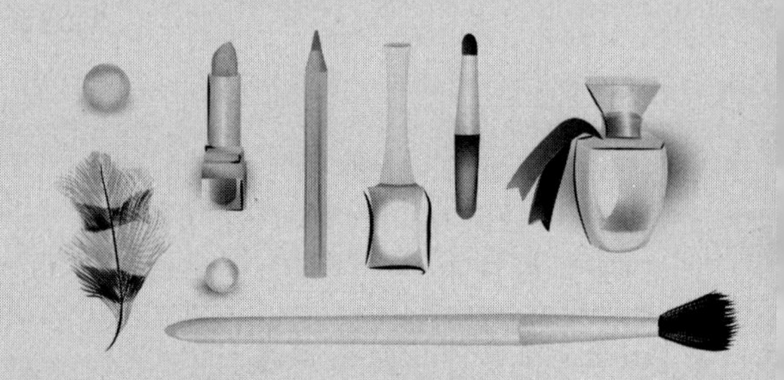

Unit 4

Ingredients in Cosmetics

Ingredients of cosmetics include solvents, oils, surfactants, thickeners, moisturizers, coloring agents, ultraviolet absorbers, inorganic and organic powders, ingredients for giving efficacies, chelating agents, preservatives, antioxidants, oxidizing and reducing agents, and aromatic essential oils *etc*.

In this unit, you will see:

① Oils;

② Surfactants;

③ Thickeners;

④ Moisturizers;

⑤ Colourants;

⑥ Ultraviolet absorbers;

⑦ Skin whitening agents;

⑧ Anti-aging agents;

⑨ Preservatives and antioxidants.

Chapter 11 Oils

Oils are wildly used in cosmetics. They can give the product form, and play the role of skin conditioning, emollient, smoothing and moisturizing *etc*. Oils are often classified by the source into natural oils，synthetic oils, semi synthetic oils, mineral oils.

11.1 Natural Oils

Animal and vegetable oils and fats are triglycerides, which are esters of three fatty acids and glycerin. Natural triglycerides contain the five most common fatty acids in various proportions: palmitic, stearic, oleic, linoleic, and linolenic acids. Many kinds are available. They are included in large amounts as a chief ingredient of cosmetics.

11.1.1 Vegetable oils

Vegetable oils, or vegetable fats, are oils extracted from seeds or from other parts of fruits. Soybean oil, grape seed oil, and cocoa butter are examples of seed oils, or fats from seeds. Olive oil, palm oil, and rice bran oil are examples of fats from other parts of fruits. In common usage, vegetable oil may refer exclusively to vegetable fats which are liquid at room temperature. Vegetable oils are usually edible.

The following are some common vegetable oils that are used in cosmetics.

(1) Coconut oil is an edible oil derived from the wick, meat, and milk of the coconut palm fruit. Coconut oil is a white solid fat, melting at warmer room temperatures of around 25 ℃ , in warmer climates during the summer months it is a clear thin liquid oil. Coconut oil contains only trace amounts of free fatty acids (about 0.03% by mass). Most of the fatty acids are present in the form of esters.

Coconut oil is an important base ingredient for the manufacture of soap. Soap made with coconut oil tends to be hard, though it retains more water than soap made with other oils and thus increases manufacturer yields. It is more soluble in hard water and salt water than other soaps allowing it to lather more easily.

(2) Olive oil is a liquid fat obtained from olives, a traditional tree crop of the Mediterranean Basin, produced by pressing whole olives and extracting the oil. It is commonly used in cooking, for frying foods or, as a salad dressing. It is also used in cosmetics, pharmaceuticals, and soaps, as a fuel for traditional oil lamps, and has additional uscs in somc rcligions.

(3) Palm oil is an edible vegetable oil derived from the mesocarp of the fruit of the oil palms. The oil is used in food manufacturing, in beauty products, and as biofuel. Palm oil accounted for about 33% of global oils produced from oil crops in 2014.

Palm oil is pervasively used in personal care and cleaning products, and it provides the foaming agent in nearly every soap, shampoo, or detergent. Around 70% of personal care products including soap, shampoo, makeup, and lotion, contain ingredients derived from palm oil. However, there are more than 200 different names for these palm oil ingredients and only 10% of them include the word “palm”.

(4) Avocado oil is an edible oil extracted from the pulp of avocados. It is used as an edible oil both raw and for cooking, where it is noted for its high smoke point. It is also used for lubrication and in cosmetics. Avocado oil is one of few edible oils not derived from seeds; it is pressed from the fleshy pulp surrounding the avocado pit. Avocado oil has a similar monounsaturated fat profile to olive oil.

(5) Tea seed oil is an edible plant oil. It is obtained from the seeds of *Camellia oleifera*. Oil analysis of cultivated varieties showed: 76%-82% oleic acid; 5%-11% linoleic acid; 7.5%-10% palmitic acid; 1.5%-3% stearic acid—the ratios are similar to that found in wild *oleifera*. The composition is similar to that of olive oil. Another analysis of several cultivars found: 82%-84%

unsaturated acids of which 68%-77% oleic acid; and 7%-14% polyunsaturated acids.

(6) Jojoba oil is the liquid produced in the seed of the *Simmondsia chinensis* (jojoba) plant, a shrub, which is native to southern Arizona, southern California, and northwestern Mexico. The oil makes up approximately 50% of the jojoba seed by mass. The terms "jojoba oil" and "jojoba wax" are often used interchangeably because the wax visually appears to be a mobile oil, but as a wax it is composed almost entirely (~97%) of mono-esters of long-chain fatty acids and alcohols (wax ester), accompanied by only a tiny fraction of triglyceride esters. This composition accounts for its extreme shelf-life stability and extraordinary resistance to high temperatures, compared with true vegetable oils. Unrefined jojoba oil appears as a clear golden liquid at room temperature with a slightly nutty odor. Jojoba oil is found as an additive in many cosmetic products, especially those marketed as being made from natural ingredients. In particular, such products commonly containing jojoba are lotions and moisturizers, hair shampoos and conditioners. The pure oil itself may also be used on skin, hair, or cuticles.

(7) Cocoa butter, also called *Theobroma* oil, is a pale-yellow, edible fat extracted from the cocoa bean. It is used to make chocolate, as well as some ointments, toiletries, and pharmaceuticals. Cocoa butter has a cocoa flavor and aroma. Its melting point is just below human body temperature. Cocoa butter contains a high proportion of saturated fats as well as monounsaturated oleic acid, which typically occurs in each triglyceride. The predominant triglycerides are POS, SOS, POP, where P=palmitic, O=oleic, and S=stearic acid residues. Cocoa butter, unlike non-fat cocoa solids, contains only traces of caffeine and theobromine. For a fat melting around body temperature, cocoa has good stability. This quality, coupled with natural antioxidants, prevents rancidity—giving it a storage life of two to five years. The velvety texture, pleasant fragrance and emollient properties of cocoa butter have made it a popular ingredient in products for the skin, such as soaps and lotions.

(8) Carnauba, also called Brazil wax and palm wax, is a wax of the leaves of the carnauba palm *Copernicia prunifera*. It is known as the "Queen of Waxes". In its pure state, it is usually available in the form of hard yellow-brown flakes. It is obtained by collecting and drying the leaves, beating them to loosen the wax, then refining and bleaching the wax.

Because of its hypoallergenic and emollient properties as well as its gloss, carnauba wax is used as a thickener in cosmetics such as lipstick, eyeliner, mascara, eye shadow, foundation, deodorant, and skincare and sun care preparations. It is also used to make cutler's resin.

(9) Candelilla wax is a wax derived from the leaves of the small Candelilla shrub. It is yellowish-brown, hard, brittle, aromatic, and opaque to translucent. The high hydrocarbon content distinguishes this wax from carnauba wax. It can be used in cosmetic industry, as a component of lip balms and lotion bars.

11.1.2 Animal oils

Animal fats and oils are lipids derived from animals: oils are liquid at room temperature, and fats are solid. Chemically, both fats and oils are composed of triglycerides. Although many animal parts and secretions may yield oil, in commercial practice, oil is extracted primarily from rendered tissue fats from livestock animals like pigs, chickens and cows. Dairy products yield animal fat and oil products such as butter.

The following are some common animal fats and oils that are used in cosmetics.

(1) Lard is a semi-solid white fat product obtained by rendering the fatty tissue of a pig. It is distinguished from tallow, a similar product derived from fat of cattle or sheep. Lard consists mainly of fats, which in the language of chemistry are known as triglycerides. These triglycerides are composed of three fatty acids and the distribution of fatty acids varies from oil to oil. In general lard is similar to tallow in its composition. Rendered lard can be used to produce soap.

(2) Tallow is a rendered form of beef or mutton fat, primarily made up of triglycerides. In the soap industry and among soap-making hobbyists, the name tallowate is used informally to refer to soaps made from tallow. Sodium tallowate, for example, is obtained by reacting tallow with sodium hydroxide or sodium carbonate. It consists chiefly of a variable mixture of sodium salts of fatty acids, such as oleic and palmitic.

(3) Lanolin also called wool yolk, wool wax, or wool grease, is a wax secreted by the sebaceous glands of wool-bearing animals. Lanolin used by humans comes from domestic sheep breeds that are raised specifically for their wool. Historically, many pharmacopoeias have referred to lanolin as wool fat; however, as lanolin lacks glycerides (glycerol esters), it is not a true fat. Lanolin primarily consists of sterol esters instead. Lanolin's waterproofing property aids sheep in shedding water from their coats. Certain breeds of sheep produce large amounts of lanolin.

Lanolin's role in nature is to protect wool and skin from climate and the environment; it also plays a role in skin hygiene. Lanolin and its derivatives are used in the protection, treatment and beautification of human skin.

A typical high-purity grade of lanolin is composed predominantly of long chain waxy esters (approximately 97% by mass) with the remainder being lanolin alcohols, lanolin acids and lanolin hydrocarbons.

(4) Beeswax is a natural wax produced by honey bees. Chemically, beeswax consists mainly of esters of fatty acids and various long-chain alcohols.

The use of beeswax in skin care and cosmetics has been increasing. A German study found beeswax to be superior to similar barrier creams, when used according to its protocol. Beeswax is used in lip balm, lip gloss, hand creams, salves, and moisturizers; and in cosmetics such as eye shadow, blush, and eye liner. Beeswax is also an important ingredient in moustache wax and hair pomades, which make hair look sleek and shiny.

(5) Mink oil is obtained by the rendering of mink fat which has been removed from pelts destined for the fur industry.

Mink oil is a source of palmitoleic acid, which possesses physical properties similar to human sebum. Because of this, mink oil is used in several medical and cosmetic products. Mink oil is also used for treating, conditioning and preserving nearly any type of leather.

Mink oil and its fatty acids are unique among animal-derived fats and oils, and can be used as a hair treatment for growing dreadlocks.

11.2 Synthetic and Semi Synthetic Oils

Synthetic oils used in cosmetics are a series of chemical compounds that are artificially made. Semi synthetic oils used in cosmetics are a series of chemical compounds that are synthesized from some natural oils. The following are some of them.

11.2.1 Fatty acid

(1) Lauric acid Lauric acid or systematically, dodecanoic acid, is a saturated fatty acid with a 12-carbon atom chain, thus having many properties of medium-chain fatty acids. It is a bright white, powdery solid with a faint odor of bay oil or soap.

Like many other fatty acids, lauric acid is inexpensive, has a long shelf-life, is nontoxic, and is safe to handle. It is used mainly for the production of soaps and cosmetics. For these purposes, lauric acid is reacted with sodium hydroxide to give sodium laurate, which is a soap. Most commonly, sodium laurate is obtained by saponification of various oils, such as coconut oil. These precursors give mixtures of sodium laurate and other soaps.

(2) Palmitic acid Palmitic acid, or hexadecanoic acid in IUPAC nomenclature, is the most common saturated fatty acid found in animals, plants and microorganisms. Its chemical formula is $CH_3(CH_2)_{14}COOH$. It is a major component of the oil from the fruit of oil palms (palm oil), making up to 44% of total fats. Meats, cheeses, butter, and other dairy products also contain palmitic acid, amounting to 50%-60% of total fats. Palmitates are the salts and esters of palmitic acid. The palmitate anion is the observed form of palmitic acid at physiologic pH (7.4).

Palmitic acid is used to produce soaps, cosmetics, and industrial mold release agents. These applications use sodium palmitate, which is commonly obtained by saponification of palm oil. To this end, palm oil, rendered from palm tree, is treated with sodium hydroxide, which causes hydrolysis of the ester groups, yielding glycerol and sodium palmitate.

(3) Stearic acid Stearic acid is a saturated fatty acid with an 18-carbon chain. The IUPAC name is octadecanoic acid. It is a waxy solid and its chemical formula is $C_{17}H_{35}CO_2H$. Stearic acid is mainly used in the production of detergents, soaps, and cosmetics such as shampoos and shaving cream products. Soaps are not made directly from stearic acid, but indirectly by saponification of triglycerides consisting of stearic acid esters. Esters of stearic acid with ethylene glycol (glycol stearate and glycol distearate) are used to produce a pearly effect in shampoos, soaps, and other cosmetic products. They are added to the product in molten form and allowed to crystallize under controlled conditions. Detergents are obtained from amides and quaternary alkylammonium derivatives of stearic acid.

(4) Oleic acid Oleic acid is a fatty acid that occurs naturally in various animal and vegetable fats and oils. It is an odorless, colorless oil, although commercial samples may be yellowish. In chemical terms, oleic acid is classified as a monounsaturated omega-9 fatty acid.

It has the formula $CH_3(CH_2)_7CH{=}CH(CH_2)_7COOH$. It is the most common fatty acid in nature. Oleic acid as its sodium salt is a major component of soap is an emulsifying agent. It is also used is an emollient.

(5) Linoleic acid Linoleic acid is an organic compound with the formula $COOH(CH_2)_7CH{=}CHCH_2CH{=}CH(CH_2)_4CH_3$. Both alkene groups are *cis*. Linoleic acid is a polyunsaturated omega-6 fatty acid. It is a colorless or white liquid that is virtually insoluble in water but soluble in many organic solvents. It typically occurs in nature as a triglyceride rather than as a free fatty acid. Linoleic acid has become increasingly popular in the beauty products industry because of its beneficial properties on the skin. Research points to linoleic acid's anti-inflammatory, acne reductive, skin-lightening and moisture retentive properties when applied topically on the skin.

11.2.2 Higher alcohols

(1) Dodecanol Dodecanol or lauryl alcohol, is an organic compound produced industrially from palm kernel oil or coconut oil. It is a fatty alcohol. Lauryl alcohol is tasteless and colorless with a floral odor. In cosmetics, dodecanol is used as an emollient.

(2) Cetyl alcohol Cetyl alcohol, also known as hexadecan-1-ol and palmityl alcohol, is a C-16 fatty alcohol with the formula $CH_3(CH_2)_{15}OH$. At room temperature, cetyl alcohol takes the form of a waxy white solid or flakes.

Cetyl alcohol is used in the cosmetic industry as an opacifier in shampoos, or as an emollient, emulsifier or thickening agent in the manufacture of skin creams and lotions. Moreover, it can also be used as a non-ionic co-surfactant in emulsion applications.

(3) Stearyl alcohol Stearyl alcohol, or 1-octadecanol, is an organic compound classified as a saturated fatty alcohol with the formula $CH_3(CH_2)_{16}CH_2OH$. It takes the form of white granules or flakes, which are insoluble in water. It has a wide range of uses as an ingredient in lubricants, resins, perfumes, and cosmetics. It is used as an emollient, emulsifier, and thickener in ointments, and is widely used as a hair coating in shampoos and hair conditioners. Stearyl heptanoate, the ester of stearyl alcohol and heptanoic acid, is found in most cosmetic eyeliners.

(4) Oleyl alcohol Oleyl alcohol or *cis*-9-octadecen-1-ol, is an unsaturated fatty alcohol with the molecular formula $C_{18}H_{36}O$ or the condensed structural formula $CH_3(CH_2)_7{-}CH{=}CH{-}(CH_2)_8OH$. It is a colorless oil, mainly used in cosmetics.

It can be produced by the hydrogenation of oleic acid esters by Bouveault-Blanc reduction, which avoids reduction of the $C{=}C$ group. The required oleate esters are obtained from beef fat, fish oil, and, in particular, olive oil. It has uses as a nonionic surfactant, emulsifier, emollient and thickener in skin creams, lotions and many other cosmetic products including shampoos and hair conditioners. It has also been investigated as a carrier for delivering medications through the skin or mucus membranes; particularly the lungs.

(5) Behenyl alcohol 1-Docosanol, also known as behenyl alcohol, is a saturated fatty alcohol containing 22 carbon atoms, used traditionally as an emollient, emulsifier, and thickener

in cosmetics, and nutritional supplement.

11.2.3 Fatty acid esters

Synthetic fatty acid esters are generally saturated and with high chemical stability. Many kinds are available. They are widely used in cosmetics as emollients, penetrants and solvents.

(1) Isopropyl myristate Isopropyl myristate (IPM) is the ester of isopropyl alcohol and myristic acid. Isopropyl myristate is a polar emollient and is used in cosmetic and topical pharmaceutical preparations where skin absorption is desired. Isopropyl myristate is also used as a solvent in perfume materials, and in the removal process of prosthetic makeup.

(2) Isopropyl palmitate Isopropyl palmitate is the ester of isopropyl alcohol and palmitic acid. It is an emollient, moisturizer, thickening agent, and anti-static agent. The chemical formula is $CH_3(CH_2)_{14}COOCH(CH_3)_2$.

(3) Ethylhexyl palmitate Ethylhexyl palmitate, or octyl palmitate, is the fatty acid ester derived from 2-ethylhexanol and palmitic acid. Ethylhexyl palmitate is commonly used in cosmetic formulations as a solvent, carrying agent, pigment wetting agent, fragrance fixative and emollient. Its dry-slip skin feel is similar to some silicone derivatives.

(4) Isononyl isononanoate Isononyl isononanoate is a synthetic ingredient used in cosmetics and skincare products to improve the texture and feel of a formulation. Isononyl isononanoate is used mainly as an emollient, texture enhancer, and plasticizer.

(5) Lauryl lactate Lauryl lactate is an ester of lauryl alcohol and lactic acid which is used as an emollient, skin conditioning agent, and exfoliant.

(6) Caprylic/Capric triglyceride Caprylic triglyceride is an ingredient used in soaps and cosmetics. It's usually made from combining coconut oil with glycerin. This ingredient is sometimes called capric triglyceride.

Caprylic triglyceride has been widely used for more than 50 years. It helps smooth skin and works as an antioxidant. It also binds other ingredients together, and can work as a preservative of sorts to make the active ingredients in cosmetics last longer. Caprylic triglyceride is valued as a more natural alternative to other synthetic chemicals found in topical skin products. Companies that claim that their products are "all natural" or "organic" often use caprylic triglyceride.

(7) Triethylhexanoin A synthetic mixture of glycerin and a triglyceride fatty acid is known as 2-ethylhexanoic acid. Triethylhexanoin functions primarily as a solvent, occlusive agent, skin-conditioning emollient, and hair conditioner.

11.2.4 Derivatives of lanolin

An estimated 8,000 to 20,000 different types of lanolin esters are present in lanolin, resulting from combinations between the 200 or so different lanolin acids and the 100 or so different lanolin alcohols identified so far.

Lanolin's complex composition of long-chain esters, hydroxyesters, diesters, lanolin alcohols, and lanolin acids means in addition to it being a valuable product in its own right, it is also the starting point for the production of a whole spectrum of lanolin derivatives, which possess wide-ranging chemical and physical properties. The main derivatisation routes include hydrolysis, fractional solvent crystallisation, esterification, hydrogenation, alkoxylation and quaternisation. Lanolin derivatives obtained from these processes are used widely in both high-value cosmetics and skin treatment products.

Hydrolysis of lanolin yields lanolin alcohols and lanolin acids. Lanolin alcohols are a rich source of cholesterol (an important skin lipid) and are powerful water-in-oil emulsifiers; they have been used extensively in skincare products for over 100 years. Notably, approximately 40% of the acids derived from lanolin are alpha-hydroxy acids (AHAs). The use of AHAs in skin care products has attracted a great deal of attention in recent years.

Lanolin and its many derivatives are used extensively in the personal care (*e.g.*, high value cosmetics, facial cosmetics, lip products). Lanolin is also used in lip balm products such as Carmex. For some people, it can irritate the lips.

(1) Acetylated lanolin alcohol Acetylated lanolin alcohol is a non-drying organic compound produced from lanolin, the fat of wool shearings, which has been reacted with acetic acid and a small amount of lye. There are synthetic variants available; however, the animal-derived product has more anti-allergenic tendencies. Acetylated lanolin alcohol is used as an emollient, to soften skin, but is mildly comedogenic, with a rating of 0-2 out of 5. For this reason, those who are prone to whiteheads and blackheads should patch test before using on a large scale.

(2) Isopropyl lanolate Isopropyl lanolate is the solid high molecular weight fraction of the isopropyl lanolate, which is produced by esterification of natural lanolin fatty acids with isopropyl alcohol. The solid content is separated from the rest of the ester mixture by fractional distillation under high vacuum.

Isopropyl lanolate has excellent characteristics in terms of colour and odour. It has a high homogeneity because most of the liquid components have been removed. It is non-greasy and gives the skin a soft and smooth feeling. Although it is not a strong W/O emulsifier, it has good compatibility with emulsions of this type, especially when a small amount of cetyl alcohol is added.

When colour and odour is of importance consideration for hand-, sport- and cleaning creams formulation, isopropyl lanolate is the right choice of lanolin derivative to achieve an emollient effect or to reduce the occlusive characteristics and stickiness of hydrocarbon bases. It is an excellent lubricant in lipsticks to reduce the frictional resistance during application and helps to achieve a glossy film. Furthermore, it can be used as binder in compact powders and as super-fattening agent for luxury soaps.

(3) PEG-30 lanolin PEG-30 lanolin is a polyoxyethylene condensate with an average chain length of 30 ethylene oxide units. The complex chemical composition of PEG-30 lanolin includes the natural lanolin esters and free hydroxyl groups, and the ether polymer chain. It is a soft, golden yellow wax with weak fruity odour.

PEG-30 lanolin is a special product because it is completely soluble in water. The addition of ethylene oxide is kept to a minimum level, so as to keep the content of lanolin as high as possible. Although it is completely water-soluble, concentrated solutions might show slight opalescence at room temperature. This insolubility is reversible upon cooling.

PEG-30 lanolin is a non-ionic detergent with manifold applications for emulsifying, solubilising, moisturising and cleaning without negative effects on the skin. It is used for hair setting and as plasticiser in shampoos, cleansing lotions, hand creams, detergent formulations and creams. It is also used for superfatting of soaps. The addition of cetyl alcohol or cetostearyl alcohol can increase both foam and stability of emulsion of products prepared with PEG-30 lanolin.

(4) PEG-75 lanolin PEG-75 lanolin is a polyoxyethylene condensate with the best pharmaceutical lanolin. It has a mean chain length of 75 ethylene oxide units. The lanolin content in PEG-75 lanolin is approximately 17%. It is a hard, pale-yellow wax with weak fruity odour.

PEG-75 lanolin is ethoxylated, to obtain not only complete water solubility, but also solutions that are crystal clear in all concentrations, both in water and in aqueous ethanol concentrations of up to 40%. The solutions are nonionic and compatible with most other solubilisers including up to 10% electrolytes solutions. The solution is only slightly affected by oxidative and reducing agents. It is stable in a pH range of 2-10. A particularly unique feature of PEG-75 lanolin is its carefully controlled manufacturing that ensures minimum viscosity variations of the aqueous solutions.

PEG-75 lanolin is particularly recommended for use in aqueous or aqueous-alcoholic lotions and solutions with high clarity. Moreover, the product has emulsifying, solubilising and emollient properties and a mild cleaning effect. Main applications include skin cleansing and after-shave lotions, as well as in shampoos and detergent formulations, where viscosity is of importance.

11.2.5 Alkanes

Hydrocarbons do not possess oxygen atoms, are saturated, and are thus nonpolar. The straight or branched chains consisting of carbon and hydrogen atoms can be combined in all cosmetics because they are scarcely oxidized and are hardly affected by pH changes and oxidant and reductant. They are useful as nonpolar oil for making emulsions. The chemical structure, melting point, and molecular weight of the hydrocarbon to be included should be examined for consistency with the purpose, properties, and feel of use of the product.

(1) Isododecane Isododecane is unsaturated 2,2,4,6,6-pentamethylheptane (mw:170.3), which is free of aromatics, is highly hydrorefined and is easy to dry. It is highly compatible with dimethicone and is a useful solvent for highly polymerized silicones. It is flammable and should be kept away from flames. It gives a very light touch, and almost no feel is felt when applied on the skin.

(2) Squalane Commercially available squalane is a hydrocarbon derived by hydrogenation of squalene, which was traditionally sourced from the livers of sharks. In contrast to squalene, due to the complete saturation of squalane, it is not subject to auto-oxidation. This fact, coupled with lower costs associated with squalane, make it desirable in cosmetics manufacturing, where it is used as an emollient and moisturizer. The hydrogenation of squalene to produce squalane was first reported in 1916.

Squalane was introduced as an emollient in the 1950s. The unsaturated form squalene is produced in human sebum and the livers of sharks. Squalane has low acute toxicity and is not a significant human skin irritant or sensitizer.

11.2.6 Silicones

Silicones consist of an inorganic silicon-oxygen backbone chain (—Si—O—Si—O—Si—O—) with two organic groups attached to each silicon center. Commonly, the organic groups are methyl. The materials can be cyclic or polymeric. By varying the—Si—O—chain lengths, side groups, and crosslinking, silicones can be synthesized with a wide variety of properties and compositions.

Silicones are ingredients widely used in skincare, color cosmetic and hair care applications. Some silicones, notably the amine functionalized amodimethicones, are excellent hair conditioners, providing improved compatibility, feel, and softness, and lessening frizz. The phenyl dimethicones, in another silicone family, are used in reflection-enhancing and color-correcting hair products, where they increase shine and glossiness. Phenyltrimethicones, unlike the conditioning amodimethicones, have refractive indices close to that of a human hair. However, if included in the same formulation, amodimethicone and phenyltrimethicone interact and dilute each other, making it difficult to achieve both high shine and excellent conditioning in the same product.

(1) Polydimethysiloxanes Polydimethylsiloxanes (PDMS) belong to a group of polymeric organosilicon compounds that are commonly referred to as silicones. PDMS are the most widely used silicon-based organic polymers, as their versatility and properties lead to many applications.

PDMS are used variously in the cosmetic and consumer product industry as well. For example, PDMS can be used in the treatment of head lice on the scalp and dimethicone is used widely in skin-moisturizing lotions where it is listed as an active ingredient whose purpose is "skin protection". Some cosmetic formulations use dimethicone and related siloxane polymers in concentrations of use up to 15%. The Cosmetic Ingredient Review's (CIR) Expert Panel, has concluded that dimethicone and related polymers are "safe as used in cosmetic formulations".

PDMS compounds such as amodimethicone, are effective conditioners when formulated to consist of small particles and be soluble in water or alcohol/act as surfactants, and are even more conditioning to the hair than common dimethicone and/or dimethicone copolyols.

(2) Cyclomethicones Cyclomethicones are a group of methyl siloxanes, a class of

liquid silicones (cyclic polydimethylsiloxane polymers) that possess the characteristics of low viscosity and high volatility as well as being skin emollients and in certain circumstances useful cleaning solvents. Unlike dimethicones, which are linear siloxanes that do not evaporate, cyclomethicones are cyclic: both groups consist of a polymer featuring a monomer backbone of one silicon and two oxygen atoms bonded together, but instead of having a very long "linear" backbone surrounded by a series of methyl groups (which produces a clear, non-reactive, non-volatile liquid ranging from low to high viscosity), cyclomethicones have short backbones that make closed or nearly-closed rings or "cycles" with their methyl groups, giving them many of the same properties of dimethicones but making them much more volatile. They are used in many cosmetic products where eventual complete evaporation of the siloxane carrier fluid is desired. In this way they are useful for products like deodorants and antiperspirants which need to coat the skin but not remain tacky afterward. Most cyclomethicone is manufactured by Dow Corning. Cyclomethicones have been shown to involve the occurrence of silanols during biodegradation in mammals. The resulting silanols are capable of inhibiting hydrolytic enzymes such as thermolysin, acetylcholinesterase, however the doses required for inhibition are by orders of magnitude higher than the ones resulting from the accumulated exposure to consumer products containing cyclomethicones.

(3) Amino-modified silicones Amino-modified silicones have a structure in which an alkylamino group or its derivative is added to a methyl group of dimethicone. They are widely used for hair care products to give a good texture. The oily type is suitable for creamy conditioners. There is also a hydrophilic type, which is modified with polyoxyethylene and polyoxypropylene and is used for liquid cosmetics such as shampoos.

11.3 Mineral Oils

Mineral oils are a series of various colorless, odorless, light mixtures of higher alkanes from a mineral source, particularly a distillate of petroleum, as distinct from usually edible vegetable oils. Mineral oil is a common ingredient in baby lotions, cold creams, ointments, and cosmetics. It is a lightweight inexpensive oil that is odorless and tasteless. It can be used on eyelashes to prevent brittleness and breaking and, in cold cream, is also used to remove creme makeup and temporary tattoos.

The following are some common mineral oils that are used in cosmetics.

(1) Vaseline is a semi-solid mixture of hydrocarbons (with carbon numbers mainly higher than 25), originally promoted as a topical ointment for its healing properties.

Most petroleum jelly today is used as an ingredient in skin lotions and cosmetics, providing various types of skin care and protection by minimizing friction or reducing moisture loss, or by functioning as a grooming aid (*e.g.*, pomade). It's also used for treating dry scalp and dandruff.

(2) liquid paraffin is a very highly refined mineral oil used in cosmetics and medicine. Cosmetic or medicinal liquid paraffin should not be confused with the paraffin used as a fuel. It is a transparent, colorless, nearly odorless, and oily liquid that is composed of saturated

hydrocarbons derived from petroleum.

Liquid paraffin is a hydrating and cleansing agent. Hence, it is used in several cosmetics both for skin and hair products. It is also used as one of the ingredients of after wax wipes.

(3) Paraffin wax (or petroleum wax) is a soft colorless solid derived from petroleum, coal or oil shale that consists of a mixture of hydrocarbon molecules containing between twenty and forty carbon atoms. It is solid at room temperature and begins to melt above approximately 37℃ . It can be used as moisturiser in toiletries and cosmetics such as vaseline.

(4) Microcrystalline waxes are a type of wax produced by de-oiling petrolatum, as part of the petroleum refining process. In contrast to the more familiar paraffin wax which contains mostly unbranched alkanes, microcrystalline wax contains a higher percentage of isoparaffinic (branched) hydrocarbons and naphthenic hydrocarbons. It is characterized by the fineness of its crystals in contrast to the larger crystal of paraffin wax. It consists of high molecular weight saturated aliphatic hydrocarbons. It is generally darker, more viscous, denser, tackier and more elastic than paraffin waxes, and has a higher molecular weight and melting point. The elastic and adhesive characteristics of microcrystalline waxes are related to the non-straight chain components which they contain. Typical microcrystalline wax crystal structure is small and thin, making them more flexible than paraffin wax. It is commonly used in cosmetic formulations.

Waxes are used for giving the form to stick-type cosmetics such as lipsticks and are also widely used in hair waxes. Those of high melting points produce waxes that can fix the hair firm. They are also added into rinse-off hair conditioning products for enhancing the springiness.

References

[1] Wikipedia org. Oil[EB/OL]. 2014[2023-05-17]. https://encyclopedia.thefreedictionary.com/Oil.

[2] Wikipedia org. Mineral oil[EB/OL]. 2014[2023-05-17]. https://encyclopedia.thefreedictionary.com/Mineral+oil.

Key Words & Phrases

palmitic acid 棕榈酸
oleic acid 油酸
linoleic acid 亚油酸
rice bran oil 米糠油
coconut oil 椰子油
Mediterranean basin 地中海盆地
mesocarp *n.* 中果皮
avocado oil 鳄梨油
Camellia oleifera 油茶
Simmondsia chinensis 希蒙德木；油蜡树
shrub *n.* 灌木
Theobroma oil 可可油
theobromine *n.* 可可碱；咖啡碱
rancidity *n.* 酸败；油脂酸败
palm *n.* 棕榈；棕榈树
Copernicia prunifera 巴西棕榈树
hypoallergenic *adj.* 低敏感性的；低过敏的
deodorant *n.* 除臭剂；解臭剂
candelilla wax 小烛树蜡；堪地里拉蜡
candelilla shrub 小烛树灌木
lard *n.* 猪油
tallow *n.* 牛脂；动物油脂
mink oil 水貂油
lauric acid 月桂酸
anti-inflammatory *adj./n.* 抗炎的（药）
granule *n.* 颗粒状物；微粒；细粒
flake *n.* 小薄片；（尤指）碎片
behenyl alcohol 山嵛醇；二十二醇
capric triglyceride 癸酸甘油三酯
triethylhexanoin *n.* 三异辛酸甘油酯
occlusive *adj.* 闭塞的；咬合的
hydroxyester *n.* 羟基酯
diester *n.* 双酯

hydrolysis *n.* 水解；水解作用
esterification *n.* 酯化反应
hydrogenation *n.* 氢化
alkoxylation *n.* 烷氧基化
quaternisation *n.* 季铵盐化
cholesterol *n.* 胆固醇
water-in-oil 油包水
acetylated lanolin alcohol 乙酰化羊毛脂醇
comedogenic *adj.* 产生粉刺的
isopropyl lanolate 羊毛酸异丙酯
homogeneity *n.* 同质性；同质；同种
W/O emulsifier W/O型乳化剂
manifold *adj.* 多的；多种多样的；许多种类的
solubiliser *n.* 增溶剂
nonpolar *adj.* 非（无）极性的
isododecane *n.* 异十二烷
squalane *n.* 鲨烷；角鲨烷
squalene *n.*（角）鲨烯；三十碳六烯
phenyl dimethicone 苯基聚二甲基硅氧烷
phenyl trimethicone 苯基聚三甲基硅氧烷
hydrolytic *adj.* 水解的
thermolysin *n.* 嗜热菌蛋白酶
acetylcholinesterase *n.* 乙酰胆碱酯酶
amino-modified silicone oil 氨基改性硅油
tattoo *n.* 文身；（在皮肤上刺的）花纹
isoparaffinic *adj.* 异构化烷烃的
naphthenic *adj.* 环烷（烃）的

Chapter 12 Surfactants

Surfactants are compounds that lower the surface tension (or interfacial tension) between two liquids, between a gas and a liquid, or between a liquid and a solid. Surfactants may act as detergents, wetting agents, emulsifiers, foaming agents, dispersants, stabilizers and solubilizers *etc.* in cosmetics.

12.1 Composition and Structure

Surfactants are usually organic compounds that are amphiphilic, meaning they contain both hydrophobic groups (their tails) and hydrophilic groups (their heads). Therefore, a surfactant contains both a water-insoluble component and a water-soluble component. Surfactants will diffuse in water and adsorb at interfaces between air and water or at the interface between oil and water, in the case where water is mixed with oil. The water-insoluble hydrophobic group may extend out of the bulk water phase, into the air or into the oil phase, while the water-soluble head group remains in the water phase.

In the bulk aqueous phase, surfactants form aggregates, such as micelles, where the hydrophobic tails form the core of the aggregate and the hydrophilic heads are in contact with the surrounding liquid. Other types of aggregates can also be formed, such as spherical or cylindrical micelles or lipid bilayers. The shape of the aggregates depends on the chemical structure of the surfactants, namely the balance in size between the hydrophilic head and hydrophobic tail. A measure of this is the hydrophilic-lipophilic balance (HLB). Surfactants reduce the surface tension of water by adsorbing at the liquid-air interface.

12.2 Classification

The "tails" of most surfactants are fairly similar, consisting of a hydrocarbon chain, which can be branched, linear, or aromatic. Fluorosurfactants have fluorocarbon chains. Siloxane surfactants have siloxane chains. Many important surfactants include a polyether chain terminating in a highly polar anionic group. The polyether groups often comprise ethoxylated (polyethylene oxide-like) sequences inserted to increase the hydrophilic character of a surfactant. Polypropylene oxides conversely, may be inserted to increase the lipophilic character of a surfactant. Surfactant molecules have either one tail or two; those with two tails are said to be double-chained.

Surfactants are usually classified according to polar head group. A non-ionic surfactant has no charged groups in its head. The head of an ionic surfactant carries a net positive, or negative charge. If the charge is negative, the surfactant is more specifically called anionic; if the charge is positive, it is called cationic. If a surfactant contains a head with two oppositely charged groups, it is termed amphoteric. Commonly encountered surfactants of each type include: anionic, cationic, zwitterionic (or amphoteric), and non-ionic.

12.2.1 Anionic surfactants

Anionic surfactants contain anionic functional groups at their head, such as sulfate, sulfonate, and carboxylates *etc*.

Anionic surfactants are one of the most frequently employed surfactants and make up about 41% of all consumed surfactants. Anionic surfactants are not only the main active components of daily chemical detergents and cosmetics, but also have a wide range of uses in many other industrial fields.

According to compositions and structures, anionic surfactants are classified to the following categories: alkylsulfate salts, polyoxyethylene alkylsulfate salts, PEG fatty acid amide MEA sulfate, alkyl methyltaurate sodium salts, alkyl methylalanine sodium salts, alkyl sarcosinate, alkyl succinate, alkyl phosphates, fatty acid salts, and acylamino acid salts.

Following are some examples for the anionic surfactants used in cosmetics.

(1) Sodium dodecyl sulfate (SDS) Sodium dodecyl sulfate (SDS) or sodium lauryl sulfate (SLS), sometimes written as sodium laurilsulfate, is an organic compound with the formula $CH_3(CH_2)_{11}OSO_3Na$. It is an anionic surfactant used in many cleaning and hygiene products. This compound is the sodium salt of the 12-carbon organosulfate. Its hydrocarbon tails combined with a polar "headgroup" give the compound amphiphilic properties and so make it useful as a detergent. SDS is also component of mixtures produced from inexpensive coconut and palm oils. SDS is a common component of many domestic cleaning, personal hygiene and cosmetic.

SDS is a component in hand soaps, toothpastes, shampoos, shaving creams, and bubble bath formulations, for its ability to create a foam, for its surfactant properties, and in part for its thickening effect.

(2) Ammonium lauryl sulfate (ALS) Ammonium lauryl sulfate (ALS) is the common name for ammonium dodecyl sulfate [$CH_3(CH_2)_{10}CH_2OSO_3NH_4$]. The anion consists of a nonpolar hydrocarbon chain and a polar sulfate end group. The combination of nonpolar and polar groups confers surfactant properties to the anion: it facilitates dissolution of both polar and non-polar materials. It is found primarily in shampoos and body-wash as a foaming agent. Lauryl sulfates are very high-foam surfactants that disrupt the surface tension of water in part by forming micelles at the surface-air interface.

(3) Sodium cetearyl sulfate Sodium cetearyl sulfate is the most frequently used in cosmetics and personal care products. It is the sodium salt of a mixture of cetyl and stearyl sulfate. It is a white to faintly yellow powder. In cosmetics and personal care products, it is used in the formulation of hair care products including shampoos, skin cleansers, and other skin care products.

(4) Sodium myreth sulfate Sodium myreth sulfate is a mixture of organic compounds with both detergent and surfactant properties. It is found in many personal care products such as soap, shampoo, and toothpaste. It is an inexpensive and effective foaming agent. Typical of many detergents, sodium myreth sulfate consists of several closely related compounds. Sometimes the number of ethylene glycol ether units (n) is specified in the name as myreth-n sulfate, for example myreth-2 sulfate.

(5) Sodium laureth sulfate Sodium laureth sulfate (SLES), also called sodium alkylethersulfate, is an anionic detergent and surfactant found in many personal care products (soaps, shampoos, *etc.*). It is derived from palm kernel oil or coconut oil. Its chemical formula is $CH_3(CH_2)_{11}(OCH_2CH_2)_nOSO_3Na$. Sometimes the number represented by n is specified in the name, for example laureth-2 sulfate. The product is heterogeneous in the number of ethoxyl groups, where n is the mean. Laureth-3 sulfate is the most common one in commercial products.

(6) Disodium lauryl sulfosuccinate Disodium lauryl sulfosuccinate (DLS) is a gentle surfactant that foams with water and cleanses the skin and hair. It is often used in cosmetics as a milder alternative to sulfates, although one doesn't systematically replace the other as they are not the same material and will not behave in the same way with other ingredients. It is derived from palm oil, which is why Lush try not to overuse it and find other options when possible.

(7) Potassium cetyl phosphate A white to beige powder that is described as the golden standard emulsifier for emulsions (oil+water mixtures) that are difficult to stabilize. It is especially popular in sunscreens as it can boost SPF protection and increase the water-resistance of the formula.

12.2.2 Cationic surfactants

Cationic surfactant is a positively charged surfactant and refers to a hydrophilic base whose molecules are dissolved in water and attached to a lipophilic base. The lipophilic base is usually a long carbon chain hydrocarbon base. Most hydrophilic bases are cations containing nitrogen atoms, and few are cations containing sulfur or phosphorus atoms. According to the chemical structure of cationic surfactants, they can be divided into amine salt type, quaternary ammonium salt type, and heterocyclic type.

Following are some examples for the cationic surfactants used in cosmetics.

(1) Laurtrimonium chloride Laurtrimonium chloride is mainly used as antistatic ingredient and emulsifier. An antistatic cosmetic ingredient reduces static electricity by neutralizing electrical charge on a surface of the hair. This helps to prevent frizz that is caused by static. As surfactant it helps to reduce surface tension of substances to be mixed or emulsified. It is often found in hair care, lotion and makeup products.

(2) Cetrimonium chloride Cetrimonium chloride, or cetyltrimethylammonium chloride (CTAC), is a topical antiseptic and surfactant. Long-chain quaternary ammonium surfactants, such as cetyltrimethylammonium chloride (CTAC), are generally combined with long-chain fatty alcohols, such as stearyl alcohols, in formulations of hair conditioners and shampoos. The

cationic surfactant concentration in conditioners is generally of the order of 1%-2% and the alcohol concentrations are usually equal to or greater than those of the cationic surfactants. The ternary system, surfactant/fatty alcohol/water, leads to a lamellar structure forming a percolated network giving rise to a gel.

Cetrimonium chloride is mainly used in hair conditioning products. It is used to increase the softness or smoothness of hair, reduce tangles and hair surface roughness. It also functions as an antistatic cosmetic ingredient and reduces static electricity by neutralizing electrical charge on the hair surface. This helps to prevent frizz that is caused by static.

(3) Steartrimonium chloride Steartrimonium chloride is mainly used in styling and perm/straightener products as it prevents or inhibits the build-up of a static. It also helps to form emulsions by reducing the surface tension of the substances to be emulsified and helps to distribute or to suspend an insoluble solid in a liquid.

(4) Behentrimonium chloride Behentrimonium chloride, also known as docosyltrimethylammonium chloride or BTAC-228, is a yellow waxlike organic compound with chemical formula $CH_3(CH_2)_{21}N(Cl)(CH_3)_3$, used as an antistatic agent and, sometimes, a disinfectant. It is commonly found in cosmetics such as conditioners, hair dye, and mousse, and also in detergents. Laboratory tests have indicated that it does not readily biodegrade.

Behentrimonium chloride is mainly used in hair conditioning products. It is used to increase the softness or smoothness of hair, reduce tangles and hair surface roughness. It also functions as an antistatic and reduces static by neutralizing electrical charge on the hair surface. This helps to prevent frizz that is caused by static.

(5) Cetrimonium bromide Cetrimonium bromide {$[(C_{16}H_{33})N(CH_3)_3]Br$; CTAB} is a quaternary ammonium surfactant. It is one of the components of the topical antiseptic cetrimide. The cetrimonium (hexadecyltrimethylammonium) cation is an effective antiseptic agent against bacteria and fungi. The closely related compounds cetrimonium chloride and cetrimonium stearate are also used as topical antiseptics and may be found in many household products such as shampoos and cosmetics. CTAB, due to its relatively high cost, is typically only used in select cosmetics.

12.2.3 Zwitterionic (or amphoteric) surfactants

Zwitterionic (amphoteric) surfactants have both cationic and anionic centers attached to the same molecule. They can change between anionic properties, the isoelectric neutral stage and the cationic properties depending on the pH value. Amphoteric surfactants have characteristics of stability against electrolytes, acids, alkalis, and hard water. Anionic, cationic and non-ionic surfactants are compatible with amphoteric surfactants. The major amphoteric surfactants are alkylamidopropylamine N-oxide (APAO), alkyldimethylamine N-oxide (AO), alkylbetaine (Bt) and alkylamidopropylbetaine (APB). Cocamidopropyl betaine, cocoamphoacetate and cocoamphodiacetate are also some commonly-used amphoteric surfactants. The amphoterics are dermatologically mild surfactants owing to their behaviour and protein-like structure. They can form complexes with anionic surfactants, show good surface-active functions over a wide range of pH and are able to reduce their irritative properties, and as a result they are mainly used as mild surfactants in cosmetics, toiletries and hand dishwashing liquids. Amphoterics surfactants have many effects such as cleansing, foaming, emulsifying, solubilizing, low toxicity, easy-biodegradation and so on. In addition, the application of amphoteric surfactants is related closely to the synergistic effects of amphoteric surfactant with other surfactants. Amphoteric surfactants can cooperate with other surfactants such as non-ionic surfactants, anionic surfactants. In a word, amphoteric surfactants form part of special surfactants available for formulators to improve or design new formulations in response to environmental, toxicity, safety and performance demands. Following are some examples for the zwitterionic (or amphoteric) surfactants used in cosmetics.

(1) Cocamidopropyl betaine (CAPB) Cocamidopropyl betaine (CAPB) is a mixture of closely related organic compounds derived from coconut oil and dimethylaminopropylamine. CAPB is available as a viscous pale-yellow solution and it is used as a surfactant in personal care products. The name reflects that the major part of the molecule, the lauric acid group, is derived from coconut oil. Cocamidopropyl betaine to a significant degree has replaced cocamide DEA.

Cocamidopropyl betaine is used as a foam booster in shampoos. It is a medium-strength surfactant also used in bath products like hand soaps. It is also used in cosmetics as an emulsifying agent and thickener, and to reduce irritation purely ionic surfactants would cause. It also serves as an antistatic agent in hair conditioners, which most often does not irritate skin or mucous membranes. However, some studies indicate it is an allergen.

(2) Lauryldimethylamine oxide (LDAO) Lauryldimethylamine oxide (LDAO), also known as dodecyldimethylamine oxide (DDAO), is an amine oxide based zwitterionic surfactant, with a C12 (dodecyl) alkyl tail. It is one of the most frequently-used surfactants of this type. Like other amine oxide based surfactants it is antimicrobial, being effective against common bacteria such as *S. aureus* and *E. coli*.

(3) Myristamine oxide Myristamine oxide is an amine oxide based zwitterionic surfactant with a C14 (tetradecyl) alkyl tail. It is used as a foam stabilizer and hair conditioning agent in some shampoos and conditioners. Like other amine oxide-based surfactants it is antimicrobial, being slightly more effective than lauryldimethylamine oxide against the common bacteria *S. aureus* and *E. coli*.

12.2.4 Non-ionic surfactants

Non-ionic surfactants have covalently bonded oxygen-containing hydrophilic groups, which are bonded to hydrophobic parent structures. The water-solubility of the oxygen groups is the result of hydrogen bonding. Hydrogen bonding decreases with increasing temperature, and the water solubility of non-ionic surfactants therefore decreases with increasing temperature.

Non-ionic surfactants are less sensitive to water hardness than anionic surfactants, and they foam less strongly. The differences between the individual types of non-ionic surfactants are slight, and the choice is primarily governed having regard to the costs of special properties (*e.g.*, effectiveness and efficiency, toxicity, dermatological compatibility, biodegradability) or permission for use in food.

Common non-ionic surfactants include the following categories: ethoxylates, fatty alcohol ethoxylates, alkylphenol ethoxylates (APEs or APEOs), fatty acid ethoxylates, ethoxylated amines and/or fatty acid amides, terminally blocked ethoxylates, fatty acid esters of polyhydroxy compounds, fatty acid esters of glycerol, fatty acid esters of sorbitol (Spans and Tweens), fatty acid esters of sucrose and alkyl polyglucosides. There are some examples for the non-ionic surfactants used in cosmetics.

(1) Nonoxynols Nonoxynols also known as nonaethylene glycol or polyethylene glycol, nonyl phenyl ether, are mixtures of nonionic surfactants used as detergents, emulsifiers, wetting agents or defoaming agents. The most commonly discussed compound nonoxynol-9 is a spermicide, formulated primarily as a component of vaginal foams and creams. Nonoxynol was found to metabolize into free nonylphenol when administered to lab animals. Arkopal-N60, with on average 6 ethylene glycol units is a related used surfactant. Nonoxynols have been used as detergents, emulsifiers and wetting agents in cosmetics, including hair products, and defoaming agents.

(2) Cocamide MEA Cocamide MEA, or cocamide monoethanolamine, is a solid, off-white to tan compound, often sold in flaked form. The solid melts to yield a pale yellow viscous clear liquid. It is a mixture of fatty acid amides which is produced from the fatty acids in coconut oil when reacted with ethanolamine. Cocamide MEA and other cocamide ethanolamines such as cocamide DEA are used as foaming agents and nonionic surfactants in shampoos and bath products, and as emulsifying agents in cosmetics.

(3) Cocamide DEA Cocamide DEA, or cocamide diethanolamine, is a diethanolamide made by reacting the mixture of fatty acids from coconut oils with diethanolamine. It is a viscous liquid and is used as a foaming agent in bath products like shampoos and hand soaps, and in cosmetics as an emulsifying agent.

(4) Glycerol monostearate Glycerol monostearate, commonly known as GMS, is a monoglyceride commonly used as an emulsifier in foods. It takes the form of a white, odorless, and sweet-tasting flaky powder that is hygroscopic. Chemically it is the glycerol ester of stearic acid.

GMS is a food additive used as a thickening, emulsifying, anticaking, and preservative agent; an emulsifying agent for oils, waxes, and solvents; a protective coating for hygroscopic powders; a solidifier and control release agent in pharmaceuticals; and a resin lubricant. It is also used in cosmetics and hair care products.

(5) Monolaurin Monolaurin, also known as glycerol monolaurate (GML), glyceryl laurate or 1-lauroyl-glycerol, is a monoglyceride. It is the mono-ester formed from glycerol and lauric acid. Its chemical formula is $C_{15}H_{30}O_4$. Monolaurin is most commonly used as a surfactant in cosmetics, such as deodorants.

(6) Sorbitan monostearate (Span 60) Span 60 is an ester of sorbitan and stearic acid and is sometimes referred to as a synthetic wax. Sorbitan monostearate is used in the manufacture of food and healthcare products as a non-ionic surfactant with emulsifying, dispersing, and wetting properties. It is also employed to create synthetic fibers, metal machining fluid, and as a brightener in the leather industry.

(7) Polysorbates Polysorbates are oily liquids derived from ethoxylated sorbitan (a derivative of sorbitol) esterified with fatty acids. Common brand names for polysorbates include Kolliphor, Scattics, Alkest, Canarcel, and Tween. For example, polysorbate 20 (Tween 20) is a polysorbate-type nonionic surfactant formed by the ethoxylation of sorbitan before the addition of lauric acid. As the name implies the ethoxylation process leaves the molecule with 20 repeat units of polyethylene glycol; in practice these are distributed across 4 different chains, leading to a commercial product containing a range of chemical species. Tween 20, Tween 40 and Tween 80 are usually used as surfactants in soaps and cosmetics, or a solubilizer such as in a mouthwash.

(8) Decyl glucoside Decyl glucoside is a mild non-ionic surfactant used in cosmetic formularies, including baby shampoo and in products for individuals with a sensitive skin. Many natural personal care companies use this cleanser because it is plant-derived, biodegradable, and gentle for all hair types.

However, in 2017, the American Contact Dermatitis society named alkyl glucosides, a class of compounds including decyl, lauryl, cetearyl, and coco glucosides, the Allergen of the Year, with decyl glucoside named as "the most common one in the class of alkyl glucosides to cause allergic contact dermatitis".

Reference

Wikipedia org. Surfactant[EB/OL]. 2014[2023-05-17]. https://encyclopedia.thefreedictionary.com/Surfactant.

Key Words & Phrases

amphiphilic *adj.* 两亲的；两性分子的
hydrophobic *adj.* 疏水的
hydrophilic-lipophilic balance 亲水亲油平衡
amphoteric *adj.* 两性的
zwitterionic *adj.* 两性离子的
non-ionic *adj.* 非离子（式）的
sodium dodecyl sulfate 十二烷基硫酸钠
organosulfate *n.* 有机硫酸盐
ammonium lauryl sulfate 十二烷基硫酸铵；月桂基硫酸铵
sodium cetyl sulfate 十六烷基硫酸钠
sodium myreth sulfate 肉豆蔻醇聚醚硫酸钠
disodium lauryl sulfosuccinate 月桂醇磺基琥珀酸酯二钠
potassium cetyl phosphate 十六烷基磷酸钾
laurtrimonium chloride 月桂基三甲基氯化铵
cetrimonium chloride 西曲氯铵；十六烷基三甲基氯化铵
lamellar *adj.* 薄片状的；层式的；成薄层的；多层（片）的
steartrimonium chloride 硬脂基三甲基氯化铵
behentrimonium chloride 山嵛基三甲基氯化铵
disinfectant *n.* 消毒剂；杀菌剂
cetrimonium bromide 十六烷基三甲基溴化铵；西曲溴铵
fungi *n.* 真菌
alkylamidopropylamine N-oxide 烷基酰胺丙基胺N-氧化物
alkyldimethylamine N-oxide 烷基二甲胺N-氧化物
alkylbetaine *n.* 烷基甜菜碱

alkylamidopropylbetaine *n.* 烷基酰胺丙基甜菜碱
cocoamphoacetate *n.* 椰油乙酸盐
cocoamphodiacetate *n.* 椰油二乙酸盐
dermatologically *adv.* 皮肤病学地
S. aureus 金黄色葡萄球菌
E. coli 大肠杆菌
myristamine oxide 肉豆蔻胺氧化物
terminally blocked ethoxylate 末端封闭乙氧基化物
polyhydroxy compound 多羟基化合物
nonoxynol *n.* 壬苯醇醚
spermicide *n.* 杀精子剂
metabolize *v.* 新陈代谢
cocamide MEA 椰油酸单乙醇酰胺
cocamide DEA 椰油酸二乙醇酰胺
glycerol monostearate 单硬脂酸甘油酯
monolaurin *n.* 甘油一月桂酸酯；单月桂酸甘油酯
sorbitan monostearate 山梨醇酐单硬脂酸酯
polysorbate *n.* 聚山梨醇酯；聚山梨酸酯；吐温类乳化剂

Chapter 13 Thickeners

A thickening agent or thickener is a substance which can increase the viscosity of a liquid without substantially changing its other properties. Thickeners may also improve the suspension of other ingredients or emulsions which increases the stability of the product. Thickeners are also used in foods, paints, inks, explosives, and cosmetics.

According to the relative molecular weight, thickeners can be divided into low molecular weight and polymeric ones.

13.1 Low Molecular Weight Thickeners

(1) Inorganic salts In some aqueous solution systems, inorganic salts such as sodium chloride, potassium chloride, ammonium chloride, sodium sulfate, sodium phosphate, disodium phosphate and pentasodium triphosphate can be used as thickeners. The most commonly used inorganic thickener is sodium chloride.

(2) Fatty alcohols and fatty acids Fatty alcohols such as lauryl alcohol, myristyl alcohol, decanol, hexanol, octanol, cetyl alcohol, stearyl alcohol and docosyl alcohol can be used as thickeners. Similarly, fatty acids such as lauric acid, linoleic acid, linolenic acid, myristic acid and stearic acid can be used as thickeners too.

(3) Other low molecular weight thickeners Some other small molecules such as ethers, esters, alkanolamides and amine oxides can be used as thickeners.

13.2 Polymeric Thickeners

According to the origins, polymeric thickeners can be divided into natural polymers, semi synthetic polymers and synthetic polymer compounds.

13.2.1 Natural polymeric thickeners

In this kind, com starch, xanthan gum, guar gum, arabic gum, carrageenan, tragacanth gum, starch, agar, shellac, sodium alginate, gelatin are often used as thickeners in cosmetics. The following are some brief introductions on some of them.

(1) Xanthan gum Xanthan gum is a polysaccharide with many industrial uses. In cosmetics, xanthan gum is used to prepare water gels. It is also used in oil-in-water emulsions to enhance droplet coalescence. Xanthan gum is under preliminary research for its potential uses in tissue engineering to construct hydrogels and scaffolds supporting three-dimensional tissue formation.

(2) Guar gum Guar gum, also called guaran, is a galactomannan polysaccharide extracted

from guar beans that has thickening and stabilizing properties useful in food, feed, and cosmetics. It is often used as thickener in toothpastes, conditioner in shampoos.

(3) Gum arabic Gum arabic, also known as "gum sudani", *Acacia gum*, and by other names, is a natural gum consisting of the hardened sap of two species of the *Acacia tree*. The term "gum arabic" does not indicate a particular botanical source. In a few cases, the so called "gum arabic" may not even have been collected from *Acacia* species, but may originate from *Combretum*, *Albizia*, or some other genus. The gum is harvested commercially from wild trees, mostly in Sudan (80%) and throughout the Sahel, from Senegal to Somalia. The name "gum arabic" was used in the Middle East at least as early as the 9th century. Gum arabic first found its way to Europe *via* Arabic ports, so retained its name. Gum arabic is a complex mixture of glycoproteins and polysaccharides predominantly consisting of arabinose and galactose. It is soluble in water, edible, and used in cosmetics as an emulsifying agent, and a suspending or viscosity increasing agent.

(4) Carrageenans or carrageenins Carrageenans or carrageenins are a family of natural linear sulfated polysaccharides that are extracted from red edible seaweeds. The most well known and still most important red seaweed used for manufacturing the hydrophilic colloids to produce carrageenan is Chondrus crispus (Irish moss) which is a dark red parsley-like plant that grows attached to the rocks. Carrageenans are widely used in shampoos and cosmetic creams as thickener.

(5) Gum tragacanth Gum tragacanth is a viscous, odorless, tasteless, water-soluble mixture of polysaccharides obtained from sap that is drained from the root of the plant and dried. The gum seeps from the plant in twisted ribbons or flakes that can be powdered. It absorbs water to become a gel, which can be stirred into a paste. The major fractions are known as tragacanthin, highly water-soluble as a mucilaginous colloid, and the chemically related bassorin, which is far less soluble but swells in water to form a gel. The gum is used in vegetable-tanned leatherworking as an edge slicking and burnishing compound, and is occasionally used as a stiffener in textiles. The gum has been used historically as a herbal remedy for such conditions as cough and diarrhea. As a mucilage or paste, it has been used as a topical treatment for burns. It can be used in cosmetics as an emulsifier, thickener, stabilizer.

(6) Agar Agar or agar-agar, is a jelly-like substance consisting of polysaccharides obtained from the cell walls of some species of red algae. Agar consists of a mixture of two polysaccharides: agarose and agaropectin, with agarose making up about 70% of the mixture. Agarose is a linear polymer, made up of repeating units of agarobiose, a disaccharide made up of D-galactose and 3,6-anhydro-L-galactopyranose. Agaropectin is a heterogeneous mixture of smaller molecules that occur in lesser amounts, and is made up of alternating units of D-galactose and L-galactose heavily modified with acidic side-groups, such as sulfate and pyruvate.

Agar exhibits hysteresis, solidifying at about 32-40℃ but melting at 85℃. This property lends a suitable balance between easy melting and good gel stability at relatively high temperatures. Since many scientific applications require incubation at temperatures close to human body temperature (37℃), agar is more appropriate than other solidifying agents that melt at this temperature, such as gelatin.

(7) Shellac Shellac is a resin secreted by the female lac bug on trees in the forests of India and Thailand. It is processed and sold as dry flakes and dissolved in alcohol to make liquid shellac, which is used as a brush-on colorant, food glaze and wood finish. Shellac functions as a tough natural primer, sanding sealant, tannin-blocker, odour-blocker, stain, and high-gloss varnish.

(8) Gelatin Gelatin or gelatine is a translucent, colorless, flavorless food ingredient, commonly derived from collagen taken from animal body parts. It is brittle when dry and rubbery when moist. It may also be referred to as hydrolyzed collagen, collagen hydrolysate, gelatine hydrolysate, hydrolyzed gelatine, and collagen peptides after it has undergone hydrolysis. It is commonly used as a gelling agent in food, beverages, medications, drug and vitamin capsules, photographic films and papers, and cosmetics.

In cosmetics, hydrolyzed collagen may be found in topical creams, acting as a product texture conditioner, and moisturizer. Collagen implants or dermal fillers are also used to address the appearance of wrinkles, contour deficiencies, and acne scars, among others. The U.S. Food and Drug Administration has approved its use, and identifies cow (bovine) and human cells as the sources of these fillers. According to the FDA, the desired effects can last for 3-4 months, which is relatively the most short-lived compared to other materials used for the same purpose.

13.2.2 Semi synthetic polymeric thickeners

Derivatives of cellulose such as methyl cellulose, ethyl cellulose, hydroxyethyl cellulose, sodium carboxymethyl cellulose and hydroxypropyl cellulose are often used as thickeners in cosmetics. The following are some brief introductions on some of them.

(1) Methyl cellulose Methyl cellulose (or methylcellulose) is a chemical compound derived from cellulose. It is sold under a variety of trade names and is used as a thickener and emulsifier in various food and cosmetic products, and also as a bulk-forming laxative. Like cellulose, it is not digestible, not toxic, and not an allergen.

Methyl cellulose is occasionally added to hair shampoos, tooth pastes and liquid soaps, to generate their characteristic thick consistency. This is also done for foods, for example ice cream or croquette. Methyl cellulose is also an important emulsifier, preventing the separation of two mixed liquids because it is an emulsion stabilizer.

(2) Ethyl cellulose Ethyl cellulose (or ethylcellulose) is a derivative of cellulose in which some of the hydroxyl groups on the repeating glucose units are converted into ethyl ether groups. The number of ethyl groups can vary depending on the manufacturer.

It is mainly used as a thin-film coating material for coating paper, vitamin and medical pills, and for thickeners in cosmetics and in industrial processes.

(3) Carboxymethyl cellulose Carboxymethyl cellulose (CMC) or cellulose gum is a cellulose derivative with carboxymethyl groups ($—CH_2—COOH$) bound to some of the hydroxyl groups of the glucopyranose monomers that make up the cellulose backbone. It is often used as its sodium salt, sodium carboxymethyl cellulose. It used to be marketed under the

name Tylose, a registered trademark of SE Tylose.

13.2.3 Synthetic polymeric thickeners

Some synthetic polymers such as polyvinyl alcohol, polyvinyl pyrrolidone, sodium polyacrylate, polyethylene oxide and carbopol resin are usually used as thickeners in cosmetics. The following are some brief introductions on some of them.

(1) Poly(vinyl alcohol)   Poly(vinyl alcohol) (PVOH, or PVA) is a water-soluble synthetic polymer. It has the idealized formula $[CH_2CH(OH)]_n$. It is used in papermaking, textile warp sizing, as a thickener and emulsion stabilizer in PVAc adhesive formulations, in a variety of coatings, and 3D printing.

Unlike most vinyl polymers, PVA is not prepared by polymerization of the corresponding monomer, since the monomer, vinyl alcohol, is thermodynamically unstable with respect to its tautomerization to acetaldehyde. Instead, PVA is prepared by hydrolysis of polyvinyl acetate, or sometimes other vinyl ester-derived polymers with formate or chloroacetate groups instead of acetate. The conversion of the polyvinyl esters is usually conducted by base-catalysed transesterification with ethanol. The properties of the polymer are affected by the degree of transesterification.

(2) Polyvinylpyrrolidone   Polyvinylpyrrolidone (PVP), also commonly called polyvidone or povidone, is a water-soluble polymer made from the monomer *N*-vinylpyrrolidone. PVP is also used in personal care products, such as shampoos and toothpastes, in paints, and adhesives that must be moistened, such as old-style postage stamps and envelopes. It has also been used in contact lens solutions and in steel-quenching solutions. PVP is the basis of the early formulas for hair sprays and hair gels, and still continues to be a component of some.

(3) Sodium polyacrylate   Sodium polyacrylate is a sodium salt of polyacrylic acid with the chemical formula $[—CH_2—CH(CO_2Na)—]_n$ and has broad applications in consumer products.

Since sodium polyacrylate can absorb and retain water molecules, it is used often in diapers, hair gels, and soaps. Sodium polyacrylate is considered a thickening agent because it increases the viscosity of water-based compounds. In diapers, sodium polyacrylate absorbs water found in urine in order to increase the capacity to store liquid and to reduce rashes.

In addition, some inorganic polymeric compounds such as magnesium aluminum silicate, bentonite can be used as thickeners in cosmetics too.

Reference

Wikipedia org. Thickening agent[EB/OL]. 2014[2023-05-17]. https://encyclopedia.thefreedictionary.com/Thickening_agent.

Key Words & Phrases

com starch 玉米淀粉

xanthan gum 黄原胶

arabic gum 阿拉伯树胶

carrageenan *n.* 卡拉胶

tragacanth gum 黄芪胶

agar *n.* 琼脂

shellac *n.* 虫胶；紫胶；紫胶片
sodium alginate 褐藻酸钠；海藻多糖；海藻酸钠
gelatin *n.* 明胶
polysaccharide *n.* 多糖；聚糖
galactomannan *n.* 半乳甘露聚糖
Acacia tree 刺槐；金合欢树
Combretum *n.* 风车子属；风车藤属；风车子
Albizia *n.* 合欢；合欢属
sahel *n.* 萨赫勒；萨赫勒地区；萨赫尔
glycoprotein *n.* 糖蛋白
arabinose *n.* 阿拉伯糖；果胶糖
galactose *n.* 半乳糖
carrageenan *n.* 角叉聚糖；角叉菜胶；卡拉胶
carrageenin *n.* 角叉胶；角叉菜胶；卡拉胶
chondrus crispus 角叉菜；鹿角菜；爱尔兰海藻
irish moss 爱尔兰苔藓；角叉菜；爱尔兰藓
tragacanthin *n.* 西黄蓍胶素；黄耆质；黄耆糖
mucilaginous *adj.* 黏的；黏液质的
bassorin *n.* 西黄蓍胶；黄蓍胶糖
mucilage *n.* 黏液；黏质；胶浆剂
agarose *n.* 琼脂糖
agaropectin *n.* 硫琼胶；琼脂胶
agarobiose *n.* 琼脂二糖；琼胶二糖
disaccharide *n.* 双糖；二糖
D-galactose *n.* D-半乳糖
3,6-anhydro-L-galactopyranose 3,6-脱水-L-半乳吡喃糖
pyruvate *n.* 丙酮酸；丙酮酸盐
resin *n.* 树脂；合成树脂
collagen *n.* 胶原蛋白
contour *n.* 外形
methyl cellulose 甲基纤维素
ethyl cellulose 乙基纤维素
carboxymethyl cellulose 羧甲基纤维素
tautomerization *n.* 互变异构化
acetaldehyde *n.* 乙醛
formate *n.* 甲酸盐
chloroacetate *n.* 氯乙酸盐
transesterification *n.* 酯交换；转酯基作用
sodium polyacrylate 聚丙烯酸钠

Chapter 14 Ingredients of Moisturizers

In the human body, water constantly evaporates from the deeper layers of the skin through an effect known as transepidermal water loss (TEWL). By regulating its water content, human skin naturally maintains a dry, easily shed surface as a barrier against pathogens, dirt, or damage, while protecting itself from drying out and becoming brittle and rigid. The ability to retain moisture depends on the lipid bilayer between the corneocytes.

Corneocytes contain small molecules called natural moisturizing factors (NMF), which absorb small amounts of water into the corneocytes thereby hydrating the skin. The natural moisturizing factor is a collection of water-soluble compounds produced from the degradation of histidine-rich proteins called filaggrin, which are responsible for aggregating keratin filaments to form keratin bundles that maintain the rigid structure of the cells in stratum corneum. When filaggrin is degraded, urea, phospholipids, pyrrolidone carboxylic acids, amino acids and some other acids are produced. These are collectively referred to as the 'natural moisturizing factor' of the skin. The components of the natural moisturizing factor absorb water from the atmosphere to ensure that the superficial layers of the stratum corneum stay hydrated. As they are water-soluble themselves, excessive water contact may leach them and inhibit their normal functions which is why prolonged contact with water makes the skin drier. The intercellular lipid layer helps prevent the loss of natural moisturizing factor by sealing the outside of each corneocyte.

Moisturizers can modify the rate of water loss and prevent damage caused by dryness. The active ingredients of moisturizers fall into two categories: occlusives and humectants.

14.1 Occlusives

Occlusives form a coating on the surface of the skin, keeping moisture from escaping. The more occlusive the formulation, the greater the effect. Ointments are more occlusive than aqueous creams, which are more occlusive than lotion. Water loss through the skin is normally about 4-8 $g/(m^2 \cdot h)$. Substances such as mineral-derived fatty compounds, animal fats and vegetable oils which are introduced in Chapter 11 can be used as ingredients of occlusive products. For example, a layer of petrolatum applied to normal skin can reduce that loss by 50%-75% for several hours.

Occlusive products are usually sticky and oily, so consumers will generally refrain from using products that appear in an oilier form (such as an ointment). Therefore, these products are generally combined as creams or lotions (which have a greater water content). These are easier to apply and are preferred by most consumers. After water evaporation, the occlusive components that remain will protect the skin and fulfill their function.

14.2 Humectants

Humectants absorb water. They can absorb the water from the air and moisturize the skin when the humidity is greater than 70%, but more commonly they draw water from the dermis into the epidermis, making skin dryer. A study published in *Skin Research and Technology* in 2001 found no link between humectants and moisturizing effect. When used in practical applications, they are almost always combined with occlusives.

Substances such as fatty alcohols, fatty acids and their salts, allantoin, ceramide and silk peptide, including components of the natural moisturizing factors, can be used as ingredients of occlusive products. The following are some brief introductions on some of them.

14.2.1 Fatty alcohols

Fatty alcohols are molecules containing hydroxyl groups. Hydroxyl groups can form hydrogen bonds with water molecules. Therefore, Fatty alcohols can keep moisture from escaping.

(1) Glycerol is a simple polyol compound. It is a colorless, odorless, viscous liquid that is sweet-tasting and non-toxic. The glycerol backbone is found in lipids known as glycerides. Due to having antimicrobial and antiviral properties it is widely used in FDA approved wound and burn treatments. Conversely, it is also used as a bacterial culture medium.

Glycerol is used in medical, pharmaceutical and personal care preparations, often as a means of improving smoothness, providing lubrication, and as a humectant. Glycerol is a component of glycerin soap. Essential oils are added for fragrance. This kind of soap is used by people with sensitive, easily irritated skin because it prevents skin dryness with its moisturizing properties. It draws moisture up through skin layers and slows or prevents excessive drying and evaporation.

(2) Sorbitol is a sugar alcohol with a sweet taste which the human body metabolizes slowly. It can be obtained by reduction of glucose, which changes the converted aldehyde group (—CHO) to a primary alcohol group (—CH_2OH). Most sorbitol is made from potato starch, but it is also found in nature, for example in apples, pears, peaches, and prunes. It is converted to fructose by sorbitol-6-phosphate 2-dehydrogenase. Sorbitol is an isomer of mannitol, another sugar alcohol; the two differ only in the orientation of the hydroxyl group on carbon 2. While similar, the two sugar alcohols have very different sources in nature, melting points, and uses.

Sorbitol is often used in cosmetics and health care like face creams, lotions, *etc.* as a humectant and thickener. Sorbitol is added to soaps, especially transparent glycerin bar soaps. It is widely used in oral hygiene formulation industries *i.e.* mouthwash, toothpaste and transparent gels as it resists fermentation by dental plaque bacteria.

(3) Propylene glycol is a viscous, colorless liquid, which is nearly odorless but possesses a faintly sweet taste. Its chemical formula is $CH_3CH(OH)CH_2OH$. Containing two alcohol groups, it is classed as a diol. It is miscible with a broad range of solvents, including water, acetone, and chloroform. In general, glycols are non-irritating and have very low volatility. In alcohol-based

hand sanitizers, it is used as a humectant to prevent the skin from drying.

(4) Panthenol (also called pantothenol) is the alcohol analog of pantothenic acid (vitamin B_5), and is thus a provitamin of B_5. In organisms, it is quickly oxidized to pantothenic acid. It is a viscous transparent liquid at room temperature. Panthenol is used in pharmaceutical and cosmetic products as a moisturizer and to improve wound healing. In pharmaceuticals, cosmetics, and personal-care products, panthenol is a moisturizer and humectant, used in ointments, lotions, shampoos, nasal sprays, eye drops, lozenges, and cleaning solutions for contact lenses.

Dipropylene glycol is a mixture of three isomeric chemical compounds, 4-oxa-2,6-heptandiol, 2-(2-hydroxy-propoxy)-propan-1-ol, and 2-(2-hydroxy-1-methyl-ethoxy)-propan-1-ol. It is a colorless, nearly odorless liquid with a high boiling point and low toxicity. Its low toxicity and solvent properties make it an ideal additive for perfumes and skin and hair care products.

14.2.2 Fatty acids and their salts

(1) Sodium lactate is the sodium salt of lactic acid, and has a mild saline taste. It is produced by fermentation of a sugar source, such as corn or beets, and then, by neutralizing the resulting lactic acid to create a compound having the formula $NaC_3H_5O_3$. Sodium lactate is sometimes used in shampoo products and other similar items such as liquid soaps, as it is an effective humectant and moisturizer. It is also used to regulate the pH value of the cosmetic formula.

(2) Sodium PCA is a natural hydrating agent derived from oils, plants and fruits. It is the sodium salt version of pyrrolidone carboxylic acid (PCA), also known as pyroglutamic acid. It is also popularly known as amino acid proline, which forms the base of all proteins. The salt form of the molecules of sodium PCA make it water-soluble.

Sodium PCA is a humectant that naturally occurs in the skin. It is also used in a variety of skincare and haircare products. Humectants maintain and secure moisture into your skin. They attract water droplets from the surrounding air and lock them into the upper layer of your skin, giving it a moisturized appearance.

While Sodium PCA is a naturally derived ingredient, it can also be created synthetically. Also, the usage of sodium PCA may vary, depending upon the source of extraction.

Its hygroscopic nature allows it to attract moisture from surrounding air and retain the moisture of your hair and skin. Sodium PCA also adds up to your natural moisturizing factor (NMF) that consists of components like lactic acid, sugar, amino acids and protein. These naturally found lipids, maintain the health and hydration of your skin. Your skin appears supple and hydrated, due to the healthy epidermis.

Skincare products contain sodium PCA at a really low concentration of 1% to 2.5%. It can effectively enter the topmost layer of your skin and pass the benefits to the inner layers. Studies demonstrated that 2% of sodium PCA in your skincare product can notably reduce the dryness of your skin.

The natural level of sodium PCA in your skin decreases with age. Skincare products containing sodium PCA can replenish the lost amounts. Being very water-absorbent, it absorbs water from the environment and while showering. It then locks it inside your skin, making it look hydrated, plump and youthful.

(3) Hyaluronic acid (HA) is a non sulfated glycosaminoglycan composed of repeated disaccharide units of D-glucuronic acid and *N*-acetyl-d-glucosamine. This natural biopolymer is particularly concentrated in the extracellular matrix of smooth connective tissue, skin dermis, eye vitreous fluid, hyaline cartilage, synovial joint fluid, intervertebral disc, and umbilical cord.

The HA molecule has interesting properties, such as versatility, biocompatibility, biodegradability, and mucoadhesiveness, which allow its use in different medical, pharmaceutical, and cosmetic applications. When HA networks are strengthened, owing to increased molecular weight and concentration, HA solutions increase viscosity and viscoelasticity. These properties allow the HA molecules to be used in cosmetics to restore hydration and elasticity, while improving the skin's appearance.

HA is found in the extracellular matrix and interfaces of collagen and elastin fibers. In aged skin, these connections are particularly absent, which may contribute to the disorganization of collagen and elastin fibers, and thus lead to skin aging. Even though the mechanism of skin aging is not fully understood, it is evident that, during this process, the dermis loses HA, thus resulting in dehydration of the skin and the appearance of wrinkles. Bukhari *et al.* concluded that HA had a high cosmetic efficiency, for example, in reducing wrinkles and aging. Besides, these authors proved that nano delivery systems containing HA enhanced its penetrability across the biological membranes.

HA is a moisturizing active ingredient widely used in cosmetic formulations (gels, emulsions, or serums) to restore the appearance of the skin.

Nowadays, commercial HA is mainly obtained through microbial fermentation. The most frequently used bacteria in the industrial production of this compound are streptococcus. However, this genus is known to have several human pathogens and, therefore, the costs of HA purification using these bacteria are high.

Therefore, genetically modified microorganisms were considered for HA production. One of such microorganisms is *Bacillus subtilis*, which is one of the most widely used models in genetic engineering, which guarantees products free from any endotoxin.

14.2.3 Other humectants

(1) Urea, also known as carbamide, is an organic compound with chemical formula $CO(NH_2)_2$. This amide has two —NH_2 groups joined by a carbonyl (C=O) functional group. It's an ingredient in some skin cream, moisturizers, hair conditioners, and shampoos.

(2) Allantoin is a chemical compound with formula $C_4H_6N_4O_3$. It is also called 5-ureidohydantoin or glyoxyldiureide. It is a diureide of glyoxylic acid. Allantoin is a major

metabolic intermediate in most organisms including animals, plants and bacteria. It is produced from uric acid, which itself is a degradation product of nucleic acids, by action of urate oxidase (uricase).

Manufacturers cite several beneficial effects for allantoin as an active ingredient in over-the-counter cosmetics, including: a moisturizing and keratolytic effect, increasing the water content of the extracellular matrix and enhancing the desquamation of upper layers of dead skin cells, increasing the smoothness of the skin; promoting cell proliferation and wound healing; and a soothing, anti-irritant, and skin protectant effect by forming complexes with irritant and sensitizing agents.

An animal study in 2010 found that based on the results from histological analyses, a soft lotion with 5% allantoin ameliorates the wound healing process, by modulating the inflammatory response. The study also suggests that quantitative analysis lends support to the idea that allantoin also promotes fibroblast proliferation and synthesis of the extracellular matrix.

A study published in 2009 reported the treatment of pruritus in mild-to-moderate atopic dermatitis with a topical nonsteroidal agent containing allantoin.

(3) Ceramides are a family of waxy lipid molecules. A ceramide is composed of sphingosine and a fatty acid. Ceramides are found in high concentrations within the cell membrane of eukaryotic cells, since they are component lipids that make up sphingomyelin, one of the major lipids in the lipid bilayer. Contrary to previous assumptions that ceramides and other sphingolipids found in cell membrane were purely supporting structural elements, ceramide can participate in a variety of cellular signaling: examples include regulating differentiation, proliferation, and programmed cell death (PCD) of cells.

Ceramides may be found as ingredients of some topical skin medications used to complement treatment for skin conditions such as eczema. They are also used in cosmetic products such as some soaps, shampoos, skin creams, and sunscreens.

Reference

Wikipedia org. Moisturizer[EB/OL]. 2014[2023-05-17]. https://encyclopedia.thefreedictionary.com/Moisturizer.

Key Words & Phrases

transepidermal *adj.* 经皮的
pathogen *n.* 病原体；病原物
brittle *adj.* 易碎的；脆性的
rigid *adj.* 固执的；僵化的
corneocyte *n.* 角化细胞；角质层细胞
natural moisturizing factor 天然保湿因子
histidine *n.* 组氨酸
filaggrin *n.* 聚丝蛋白
keratin filament 角蛋白丝；角蛋白纤维
stratum corneum 角质层；角化层
phospholipid *n.* 磷脂
pyrrolidone carboxylic acid 吡咯烷酮羧酸
dermis *n.* 真皮
allantoin *n.* 尿囊素
ceramide *n.* 神经酰胺
sorbitol *n.* 山梨醇
prune *n.* 干梅子；西梅干；李子干
fructose *n.* 果糖
sorbitol-6-phosphate 2-dehydrogenase 山梨醇-6-磷酸2-脱氢酶

mannitol *n.* 甘露醇；甘露糖醇
plaque *n.* 匾牌；牙菌斑
propylene glycol 丙二醇
panthenol *n.* 泛醇
analog *n.* 模拟；[结构]类似物
provitamin *n.* 维生素原
dipropylene glycol 一缩二丙二醇；二丙二醇
sodium lactate 乳酸钠
glycosaminoglycan *n.* 糖胺聚糖
vitreous fluid 玻璃体液
hyaline cartilage 透明软骨
synovial joint fluid 滑膜关节液
intervertebral disc 椎间盘
umbilical cord 脐带
muco adhesiveness 黏液黏附性
extracellular matrix 细胞外基质
streptococcus *n.* 链球菌
pathogen *n.* 病原体
Bacillus subtilis 枯草杆菌
endotoxin *n.* 内毒素
5-ureidohydantoin *n.* 尿囊素
glyoxyldiureide *n.* 尿囊素
diureide *n.* 二酰脲
uric acid 尿酸
nucleic acid 核酸
urate oxidase 尿酸氧化酶
uricase *n.* 尿酸氧化酶
keratolytic *adj.* 角质层分离的；角蛋白溶解的
desquamation *n.* 脱屑；剥离
soothing *adj.* 安慰性的；减轻（痛苦）的；起镇定作用的
histological *adj.* 组织（学）的
pruritus *n.* 瘙痒[症]
atopic *adj.* 特应性的；异位的
nonsteroidal *adj.* 非甾体化合物的；非类固醇的
sphingosine *n.* 鞘氨醇
sphingomyelin *n.* 鞘磷脂；神经鞘磷脂
proliferation *n.* 增生；增殖

Chapter 15 Colourants

A colourant/colour additive (British spelling) or colorant/color additive (American spelling) is a substance that is added or applied in order to change the colour of a material or surface. Colourants can be used for many purposes including printing, painting, and for colouring many types of materials such as foods and plastics. Colourants work by absorbing varying amounts of light at different wavelengths (or frequencies) of its spectrum, transmitting (if translucent) or reflecting the remaining light in straight lines or scattered.

Most colourants can be classified as dyes or pigments, or containing some combination of these. Typical dyes are formulated as solutions, while pigments are made up of solid particles generally suspended in a vehicle (*e.g.*, linseed oil). The color a colorant imparts to a substance is mediated by other ingredients it is mixed with such as binders and fillers are added, for example in paints and inks. In addition, some colourants impart colour through reactions with other substances.

Although there are a great number of colourants, only few of them are permitted to be used in cosmetics for the reasons of safety. In China, according to Safety and Technical Standards for Cosmetics in 2015, only 157 colourants are allowed to be used in cosmetics. In the United States, some noncertified naturally occurring plant and animal colorants, such as alkanet, annatto [CAS: 1393-63-1], carotene [CAS: 36-88-4], chlorophyll [CAS: 1406-65-1], cochineal [CAS: 1260-17-9], saffron [CAS: 138-55-6], and henna [CAS: 83-72-7], can be used in cosmetics. However, natural food colors, such as beet extract or powder, and turmeric, are not allowed as cosmetic colorants.

Colourants, or their constituent compounds, may be classified chemically as inorganic (often from a mineral source) and organic (often from a biological source). Chemically speaking, for now, pigments can be organic or inorganic, while dyes are only organic.

15.1 Inorganic Colourants

Inorganic colourants are usually used as pigments. There are a great number of pigments. However, some of them contain heavy metals such as lead, mercury, and cadmium that are highly toxic. The use of these pigments is now highly restricted in cosmetics. Titanium dioxide (white), zinc oxide (white), talc steatite (whitish), barium sulfate (white), mica (glossy, colorless), ferric oxide (glossy, nacreous, multicolored), coated mica (multicolored), bismuth oxychloride (white, nacreous), iron oxides 85% Fe_2O_3 (yellow to orange), umber Fe_2O_3/Fe_3O_4 (brown), sienna Fe_2O_3 (ignited) (red), Fe_3O_4 (black), chrome hydroxide green (bluish green), chrome oxide greens (green), ferric ammonium ferrocyanide (blue), ferric ferrocyanide (blue), manganese violet (violet), and ultramarine blue (violet, red, pink, green) are the most commonly used inorganic pigments in cosmetics such as skin-makeups and eye-area products. Here is an example for prussian blue.

Prussian blue (also known as Berlin blue or, in painting, Parisian or Paris blue) is a dark blue pigment produced by oxidation of ferrous ferrocyanide salts. It has the chemical formula

$Fe_4^{III}[Fe^{II}(CN)_6]_3$. It was the first modern synthetic pigment. It is prepared as a very fine colloidal dispersion, because the compound is not soluble in water. It contains variable amounts of other ions and its appearance depends sensitively on the size of the colloidal particles. Prussian blue can be used as pigments in eyebrow pencils.

15.2 Organic Colourants

Organic colorants can be classified chemically as azo dyes, anthraquinone dyes, indigoid dyes, triarylmethane dyes, heterocyclic dyes, cyanine dyes, sulfur dyes and phthalocyanine dyes *etc*. In some cases, the lakes of organic dyes, made by extending the colorant on a substrate such as aluminum hydroxide or barium sulfate, can be used in some cosmetics.

The following are some brief introductions on some of organic colourants that can be used in cosmetics.

(1) Acid red 87 [C.I. (color index) 45380; CAS Registry Number:17372-87-1]

It is also called eosine Y. It is orange powder which is soluble in water and ethanol with fluorescent blue light. Eosine Y is the form of eosin most commonly used in histology, most notably in the H&E (haematoxylin and eosin) stain. Eosin Y is also widely used in the Papanicolaou stain (or Pap stain used in the Pap test) and the Romanowsky type cytologic stains. It is also used as a colourant in cosmetics such as lipstick. But it is not recommended for eye cosmetics, facial cosmetics and nail polish.

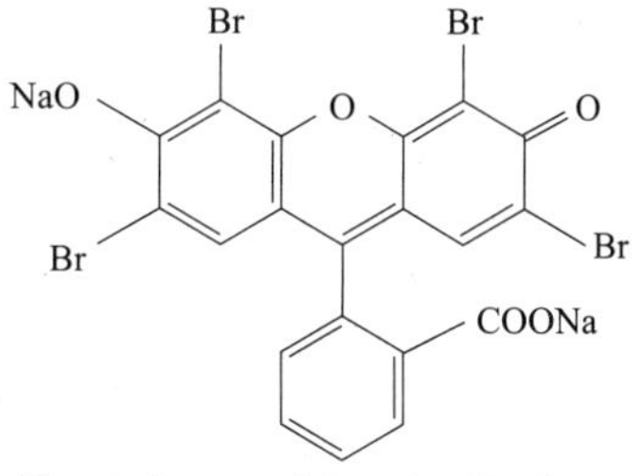

Chemical structural formula of Eosine Y

(2) Acid green 25 (C.I. 61570; CAS Registry Number: 4403-90-1)

It is orange green powder which is soluble in water and slightly soluble in acetone, ethanol and pyridine. And it can be used in cosmetics. For example, it was usually used in semi-permanent hair dye formulations as a direct dye at a maximum concentration of 0.3% in the finished cosmetic product. However, it should not be used for eye cosmetics.

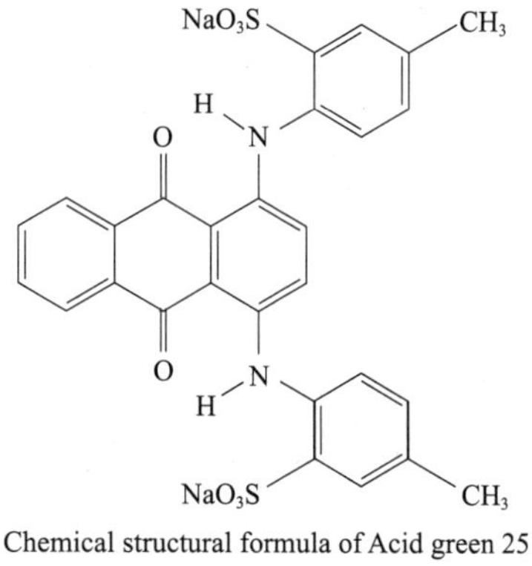

Chemical structural formula of Acid green 25

(3) Acid yellow 73 (C.I. 45350; CAS Registry Number: 518-47-8)

It is brilliant yellow and soluble in water and ethanol (with strong green fluorescent). And it can be used in cosmetics such as bath liquid, shampoo. However, it should not be used for eye, oral and lip cosmetics.

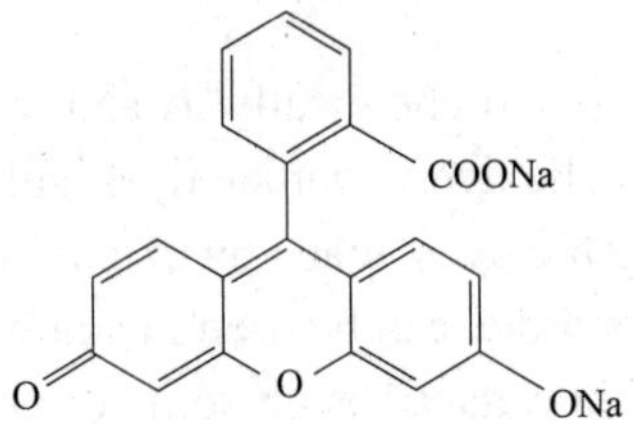

Chemical structural formula of Acid yellow 73

(4) Acid orange 7 (C.I. 15510; CAS Registry Number: 633-96-5)

It is golden yellow powder which is soluble in water and ethanol. And it can be used in cosmetics such as bath liquid and shampoo. However, it should not be used for eye, oral and lip cosmetics.

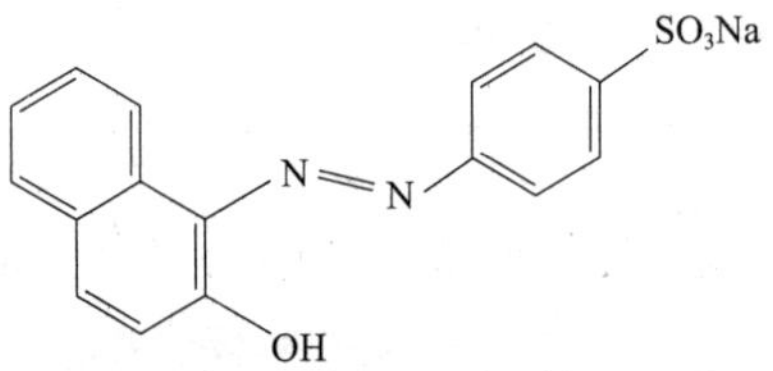

Chemical structural formula of Acid orange 7

(5) Food blue 2 (C.I.42090; CAS Registry Number: 2650-18-2)

It is colourful light blue which is soluble in water and ethanol. And it can be used in cosmetics such as bath liquid, shampoo, oral products.

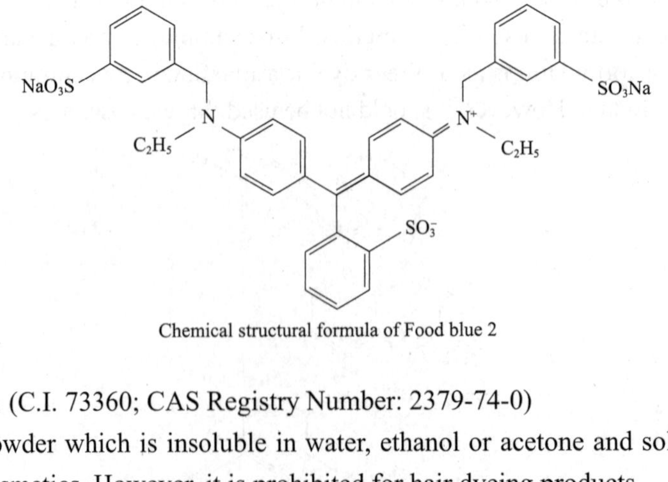

Chemical structural formula of Food blue 2

(6) Vat red 1 (C.I. 73360; CAS Registry Number: 2379-74-0)

It is pink powder which is insoluble in water, ethanol or acetone and soluble in xylene. It can be used in cosmetics. However, it is prohibited for hair dyeing products.

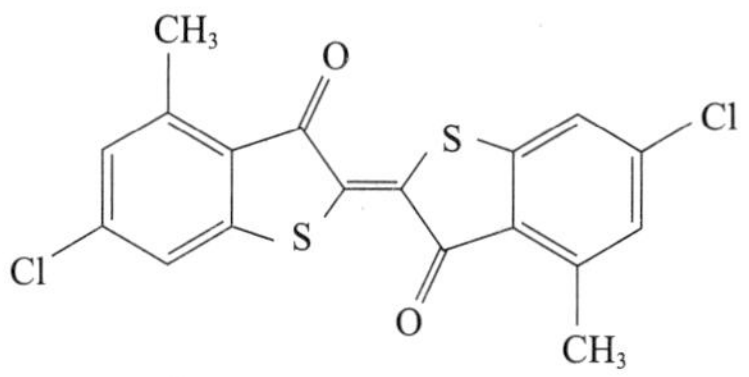

Chemical structural formula of Vat red 1

(7) Food red 1 (C.I.14700; CAS Registry Number: 4548-53-2)

It is bright light red which is soluble in water and slightly soluble in alcohol. It can be used in cosmetics. However, it should not be used for eye, oral and lip cosmetics.

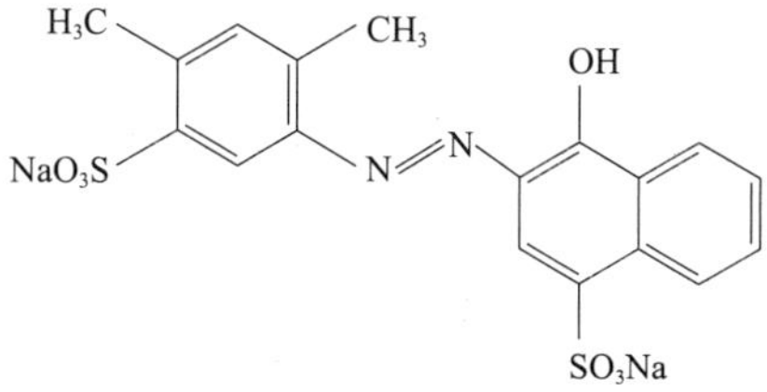

Chemical structural formula of Food red 1

(8) Food red 7 (C.I.16255; CAS Registry Number: 2611-82-7)

It is deep red powder which is soluble in water, slightly soluble in ethanol and insoluble in other organic solvents. It can be used in cosmetics such as bath liquid, shampoo, toilet water and toothpaste.

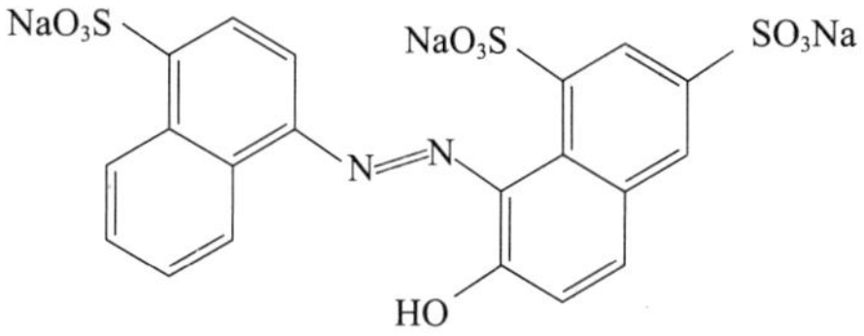

Chemical structural formula of Food red 7

(9) Food red 14 (C.I.45430; CAS Registry Number: 16423-68-0)

It is light pink which is soluble in water for cherry red. It can be used in cosmetics such as essence, toilet water, toothpaste, bath liquid, shampoo and hair oil.

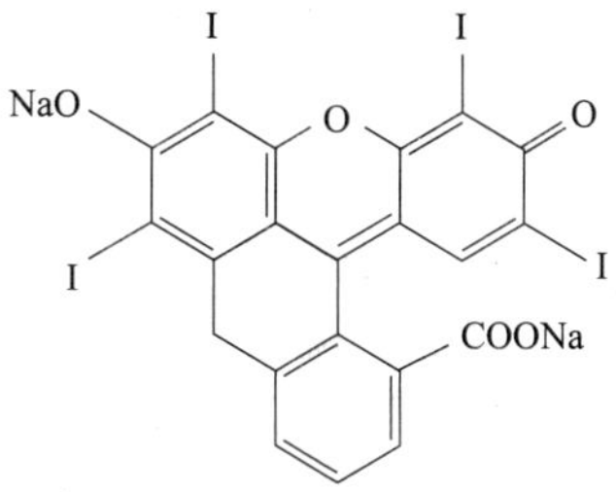

Chemical structural formula of Food red 14

(10) Food red 17 (C.I. 16035; CAS Registry Number: 25956-17-6)

It is also called fancy red. It is odourless deep red powder which is soluble in water, glycerin or propylene glycol, and slightly soluble in ethanol. It can be used in cosmetics. For example, its aluminum lake was used as a red pigment in cosmetics for lipstick and facial cosmetics. However, it should not be used for eye cosmetics.

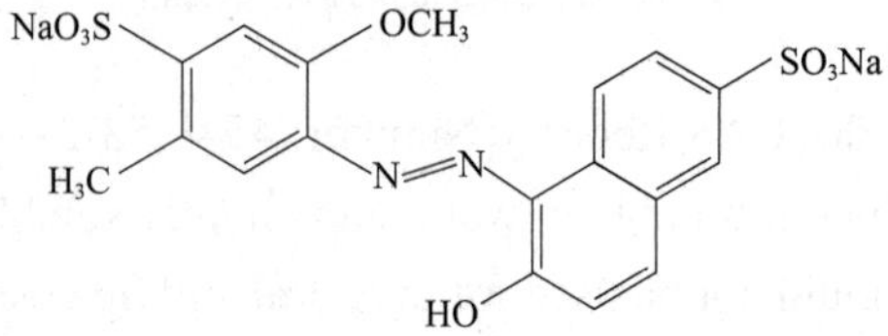

Chemical structural formula of Food red 17

(11) Food green 3 (C.I. 42053; CAS Registry Number: 2353-45-9)

It is also called fast green. It is light green which is soluble in water and ethanol for light green. It can be used in cosmetics such as bath liquid and shampoo. However, it should not be used for eye cosmetics and hair dyeing products.

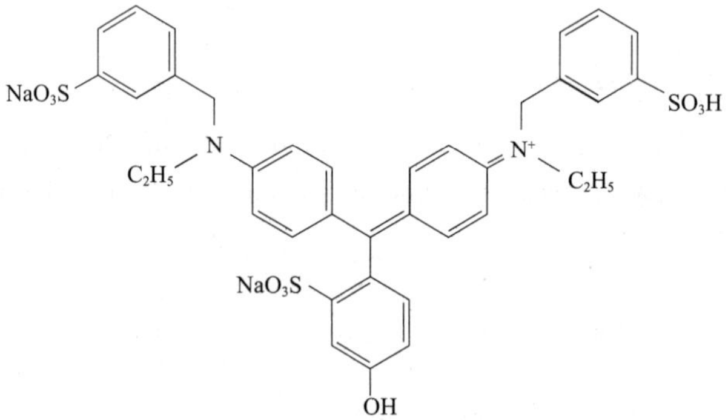

Chemical structural formula of Food green 3

(12) Food yellow 3 (C.I. 15985; CAS Registry Number: 2783-94-0)

It is also called sunset yellow. It is light yellow to colourful yellow orange. And it is soluble in water for light red clarified solution, slightly soluble in ethanol. It can be used in cosmetics. However, it should not be used for eye cosmetics.

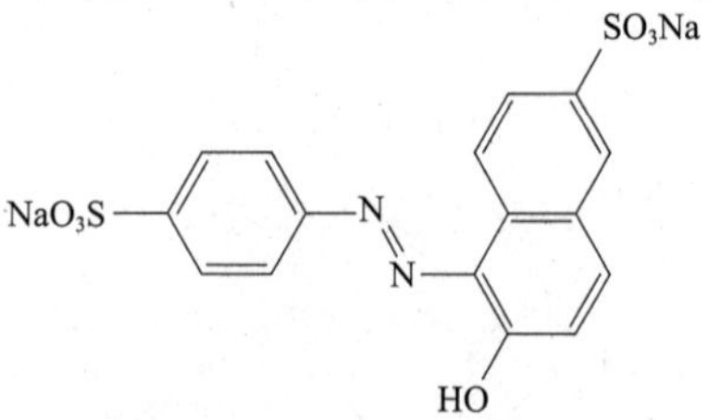

Chemical structural formula of Food yellow 3

(13) Food yellow 4 (C.I. 19140; CAS Registry Number: 1934-21-0)

It is orange yellow powder which is soluble in water for yellow, and slightly soluble in ethanol, but is insoluble in other organic solvents. It can be used in cosmetics such as essence, toilet water, bath liquid, shampoo and hair oil.

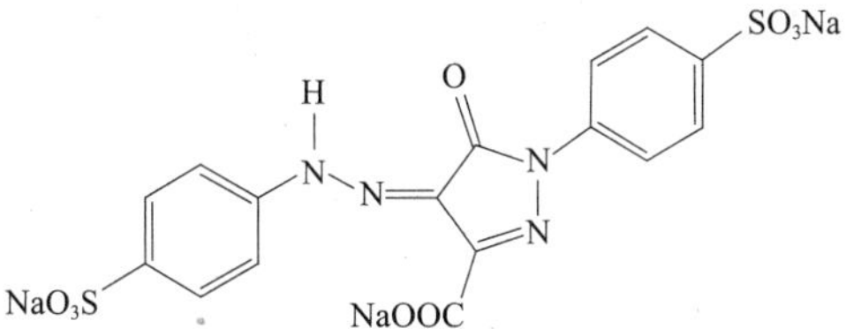

Chemical structural formula of Food yellow 4

(14) Food blue 1 (C.I. 73015; CAS Registry Number: 860-22-0)

It is blue powder which is soluble in water solution for blue, slightly soluble in alcohol. It can be used in cosmetics such as essence and toilet water. However, it should not be used for eye cosmetics.

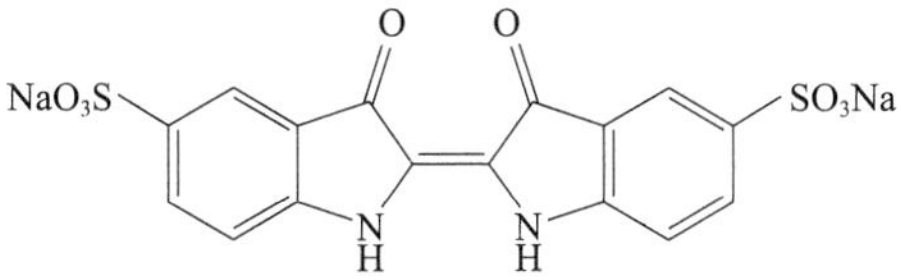

Chemical structural formula of Food blue 1

Reference

Wikipedia org. Colourant[EB/OL]. 2014[2023-05-17]. https://encyclopedia.thefreedictionary.com/Colourant.

Key Words & Phrases

dye *n.* 染料
binder *n.* 结合剂；黏合剂
alkanet *n.* 朱草；紫草根；由朱草提制的红色染料
annatto *n.* 胭脂树红；胭脂树
carotene *n.* 胡萝卜素
chlorophyll *n.* 叶绿素
cochineal *n.* 胭脂虫红颜料
saffron *n.* 藏红花粉（用作食用色素）；橘黄色
turmeric *n.* 姜黄；姜黄根粉
titanium dioxide 二氧化钛
talc steatite 滑石
barium sulfate 硫酸钡；重晶石
nacreous *adj.* 珍珠母的；珍珠（质）的；珍珠似的；有光泽的
bismuth oxychloride 氯氧化铋
umber *n.*（油漆中用的）棕土；赭土
sienna *n.* 黄土；褐土
chrome hydroxide green 氢氧化铬绿
chrome oxide green 氧化铬绿；铬绿
ferric ammonium ferrocyanide 亚铁氰化铁铵
ferric ferrocyanide 亚铁氰化铁
manganese violet 锰紫
ultramarine blue 群青蓝
prussian blue 普鲁士蓝
azo dye 偶氮染料
anthraquinone dye 蒽醌染料
indigoid dye 靛蓝染料
triarylmethane dye 三芳基甲烷染料
heterocyclic dye 杂环染料
cyanine dye 花青染料；菁染料
sulfur dye 硫化染料；硫化染料染色
phthalocyanine dye 酞菁染料
lake *n.* 色淀
eosine Y 曙红 Y
eosin *n.* 曙红（一种红色荧光染料）
histology *n.* 组织学
cytologic *adj.* 细胞学的
xylene *n.* 二甲苯

Chapter 16 Sunscreen Agents

Sunscreen agents are basically categorized into inorganic and organic UV filters which have specific mechanisms of action upon exposure to sunlight. Inorganic agents reflect and scatter light, while organic blockers absorb high-energy UV radiation. Recently, hybrid materials combining properties of organic and inorganic compounds have attracted the attention of scientists as a promising sunscreen agent. Remarkably, botanical agents, which contain large amounts of antioxidant compounds, can be used as inactive ingredients to protect the skin against adverse effects (*e.g.*, photoaging, wrinkles, and pigment).

16.1 Organic UV Filters

Organic blockers are classified into either UVA (anthranilates, dibenzoylmethanes, and benzophenones) or UVB filters [salicylates, cinnamates, para-aminobenzoic acid (PABA) derivatives, and camphor derivative], which play an important role in absorption activity of sunscreen. These agents show outstanding safety and aesthetic properties, including stability, nonirritant, nonvolatile, non-photosensitizing, and non-staining to human skin, compared to inorganic UV filters. Besides, they are mostly used in combination at levels currently allowed by the FDA to provide broad-spectrum absorption, as well as increased SPF values. Nevertheless, the combination is limited in selecting the appropriate UVA/UVB filters to avoid possible negative interactions between the combining agents. Particularly, some organic filters (*e.g.*, PABA, PABA derivatives, and benzophenones) show considerable negative effects, including eczematous dermatitis, burning sensation, and increased risk of skin cancer. Therefore, sunscreens have recently minimized or avoided the use of these compounds to protect consumers from undesirable effects. For example, the use of the two most popular organic filters, octinoxate (ethylhexul methoxycinnamate) and oxybenzone, has recently been restricted in Hawaii because of their negative effect on the coral reefs. Besides, some photo unstable filters (*e.g.*, avobenzone and dibenzoylmethanes) show a number of photoreactive results in the formation of photoproducts that can absorb in different UV regions, therefore reducing their photoprotective efficacy. Particularly, these photodegradation products can come in direct contact with the skin, thus promoting phototoxic, photosensitizing, and photoallergic contact dermatitis on the skin.

16.2 Inorganic UV Filters

Inorganic blockers have been approved to protect human skin from direct contact with sunlight by reflecting or scattering UV radiation over a broad spectra. The current agents are

ZnO, TiO_2, Fe_xO_y, calamine, ichthammol, talc, and red veterinary petrolatum. Although they are generally less toxic, more stable, and safer for human than those of organic ingredients, they are visible due to white pigment residues left on the skin and can stain clothes. Since the early 1990s, these metal oxides have been synthesized in the form of micro and nanoscale particles (10-50 nm), which can reduce the reflection of visible light and make them appear transparent throughout the skin, resulting in enhance aesthetics over the larger size. For instance, micro-size TiO_2 and ZnO have been replaced by nano-size TiO_2 and ZnO in sunscreen, eliminating undesired opaqueness and improving SPF value.

Moreover, the main disadvantage of utilization of nanoparticles (NPs) is that sunscreens tend to block shorter wavelength from UVAII to UVB rather than long radiation (visible and UVA range). In particular, most NPs can produce ROS (reactive oxygen species) radicals and are small enough to penetrate into the stratum corneum, thus causing severe skin effects with prolonged exposure, such as photoallergic contact dermatitis and skin aging. Therefore, in order to improve natural appearance as well as reduce side effects on the skin, these cosmetics using nanoparticles need to be controlled by numerous factors, including particles size and distribution, agglomeration and aggregation, and morphology and structure of the NPs. For examples, the utilization of TiO_2 and ZnO NP-coated silicon or doped elements (Al_2O_3 and Zr) can minimize ROS production and prevent negative effects as mentioned above.

16.3 Hybrid UV Filters (Organic/Inorganic Agents)

According to the literature, hybrid materials are two half-blended materials intended to create desirable functionalities and properties. They are constituted of organic components (molecule or organic polymer) mixed with inorganic components (meal oxides, carbonates, phosphates, chalcogenides, and allied derivatives) at the molecular or nanoscale. The combination creates ideal materials with a large spectrum and high chemical, electrochemical, optical transparency, magnetic, and electronic properties. Furthermore, some less toxic and biocompatible hybrid materials have been utilized as active ingredients in cosmetics due to their ability to absorb or deliver organic substances into the hair cuticle and skin layers, thereby improving skin care effect. For instance, L'Oréal and Kerastase have introduced the Intra-Cylane™ shampoo, which contains amino functionalized organosilanes hybrid substances that not only protect against hair damage, but also create hair volume expansion, better mechanical properties, and better texture. Merch KgaA and EMD Chemicals Inc. have utilized a number of hybrid compounds, such as silica microcapsules, to control the release of active ingredients that can reduce skin aging and provide high SPF.

16.4 Botanical Agents

Botanical agents are secondary metabolites produced by living organisms which play a crucial role in the growth and continuity of these organisms. It has been indicated that metabolites possess antioxidant and UV ray absorption abilities. Their featured properties are related to π-electron systems, which are mainly found in conjugated bond structures expressed in linear chain molecules and in most of aromatic compounds containing electron resonances. Certainly, there is no denying that UV radiation can generate huge amounts of ROS radicals, which leads to inflammation, neutrophil infiltrate, activates nicotinamide adenine dinucleotide phosphate (NADPH) oxidase and sebaceous gland dysfunction, and accelerates skin pigmentation and dermal matrix. In the presence of antioxidants, the ROS radicals are directly scavenged and prevented from their biological targets. As a consequence, the propagation of oxidants is limited, resulting in preventing aging.

These antioxidant compounds are obtained from vitamin C, vitamin E, and plant extracts (phenolic, carotenoids, and flavonoid compounds) (Table 16.1). In fact, a large number of botanicals have been approved as inactive agents for preserving, emulsifying, moisturizing, and smoothing sunscreen to further protect the skin. These typical varieties are aloe vera, tomatoes, pomegranate, green tea, cucumber, *Pongamia pinnata* (L.)−Indian beech tree, *Spathodea campanulata* (L.)−African tulip tree, *Dendropanax morbifera*, and *Opuntia humifusa*.

Table 16.1 Photoprotection mechanism of antioxidant compounds

Compounds	Protection mechanism
Vitamin C	−Neutralizing ROS radicals in aqueous compartments of the skin based on the oxidation capacity of ascorbate −Reducing sunburn cell formation, erythema, and immunosuppression −Inhibiting tyrosinase synthesis and maintaining hydration to protect the skin epidermis barrier −Challenging: poor skin penetration and instability
Vitamin E	−Protecting the cell membrane from oxidative stress −Inhibiting UV-induced cellular damage: photoaging, lipid peroxidation, immunosuppression, and photocarcinogenesis
Phenolic compounds	−Scavenging free radicals −Conserving proper skin structure through the regulation of matrix metalloproteinases (MMPs) −Inhibiting collagenase and elastase thus facilitating the maintenance of proper skin structure
Flavonoid compounds	−Their double bonds in flavonoid molecules provide a high ability to absorb UV −The presence of hydroxyl groups attached to aromatic rings also contributes to their ROS scavenging capacity
Carotenoids	−Physical quenching function: efficacy antioxidants for scavenging peroxide and singlet oxygen ($^1O_2$) radicals generated in during photooxidation −Absorbing of UV, visible, and blue light

16.5 Safety and Health Hazards of Sunscreen Agents

According to the literature, sunscreen agents should be safe, nontoxic, chemically inert, non-irritating, and fully protect against broad spectrum that can prevent photocarcinogenesis and photoaging. However, they also have negative effects, including contact sensitivity, estrogenicity, photoallergic dermatitis, and risk of vitamin D deficiency.

It has been reported that an increased incidence of melanoma may result from the use of sunscreen. Gorham *et al.* (2007) pointed out that some commercial sunscreens completely absorb UVB, but transmit large amounts of UVA, which may contribute to risk of melanoma in populations at latitudes greater than 40℃ . In addition, intentional long-term topical sunscreen can increase melanoma risk, especially when using high-SPF products. Thus, the labeling of sunscreen should inform consumers about the carcinogenic hazards related to sunscreen abuse.

Moreover, some sensitive ingredients in sunscreen may also be a photoallergic factor. In particular, PABA and oxybenzone are the most common ingredients causing skin disorders. The penetration and systemic toxic effects of inorganic agents at micro- or nano-size have been reported through several *in vivo* and *in vitro* analyses. Pan *et al.* (2009) demonstrated that TiO_2 NPs [(15±3.5)nm] can pass through cell membranes and impair the function of human dermal fibroblast cultures. Filipe *et al.* (2006) suggested that TiO_2 NPs (~ 20 nm) in sunscreen appear on the skin surface and in the stratum corneum regions. Therefore, it does not penetrate deeply into the skin.

On the other hand, although UVB can cause sunburn for long-term exposure, it is responsible for more than 90% of individual vitamin D production on skin. There are controversies about vitamin D deficiency due to sunscreen application. In particular, this photoprotection can lead to a significant reduction in the amount of previtamin D_3 produced by sunlight in the skin, resulting in insufficient vitamin D levels. In contrast, Fourschou *et al.* (2012) indicated that vitamin D synthesis increases exponentially with the application of thinner layers of sunscreen (<2 mg/cm^2). On the other hand, Marks *et al.* (1995) reported adequate production of vitamin D in the Australian population during the summer in most people using sunscreen or without these skin protection substances.

Reference

Ngoc L T N, Tran V V, Moon J Y, Chae M, Park D, Lee Y C. Recent Trends of Sunscreen Cosmetic: An Update Review. Cosmetics, 2019, 6: 64.

Key Words & Phrases

hybrid material 混杂材料；杂化材料
photoaging *n.* 光老化
anthranilate *n.* 邻氨基苯甲酸盐
dibenzoylmethane *n.* 二苯甲酰甲烷
benzophenone *n.* 二苯甲酮
salicylate *n.* 水杨酸盐
cinnamate *n.* 肉桂酸酯；肉桂酸；肉桂酸盐
para-aminobenzoic acid 对氨基苯甲酸
camphor *n.* 樟脑
aesthetic *adj.* 审美的；美学的；美的；有审美观点的；艺术的
eczematous *adj.* 湿疹性的
octinoxate *n.* 桂皮酸盐
ethylhexul methoxycinnamate 甲氧基肉桂酸乙基己酯
oxybenzone *n.* 羟苯甲酯
scatter *v.* 散射；分散；散开；撒播；四散
calamine *n.* 炉甘石
ichthammol *n.* 鱼石脂
red petrolatum 红色凡士林
opaqueness *n.* 不透明性
agglomeration *n.* 集聚；聚集；（杂乱聚集的）团
chalcogenide *n.* 硫属化物
silica microcapsule 二氧化硅微胶囊
metabolite *n.* 代谢物
conjugated bond 共轭键

neutrophil *n.* 中性粒细胞；嗜中性粒细胞
nicotinamide adenine dinucleotide phosphate 烟酰胺腺嘌呤二核苷酸磷酸
oxidase *n.* 氧化酶
sebaceous gland dysfunction 皮脂腺功能障碍
propagation *n.* 传播；扩展；培养
phenolic *adj.* 酚的
carotenoid *n.* 类胡萝卜素；类叶红素；脂色素
flavonoid *n.* 黄酮类化合物
pomegranate *n.* 石榴
Pongamia pinnata (L.)-Indian beech tree 印度山毛榉
Spathodea campanulata (L.)-African tulip tree 火焰树，非洲郁金香树
Dendropanax morbifera 黄漆木
Opuntia humifusa 仙人掌
ascorbate *n.* 抗坏血酸盐
immunosuppression *n.* 免疫抑制
tyrosinase *n.* 酪氨酸酶
matrix metalloproteinase 基质金属蛋白酶
collagenase *n.* 胶原酶
elastase *n.* 弹性蛋白酶
photocarcinogenesis *n.* 光致癌
estrogenicity *n.* 雌激素活性
melanoma *n.* 黑（色）素瘤
exponentially *adv.* 以指数方式
adequate *adj.* 充足的；足够的；合格的；合乎需要的

Chapter 17 Skin Whitening Agents

Skin whitening or lightening refers to the practice, deeply embedded in many ethnic groups, of using natural or synthetic substances to lighten the skin tone or provide an even complexion by reducing the melanin concentration in the skin. The use of whitening agents can be driven by medicinal necessity in the case of persons suffering from dermatological conditions linked to an abnormal accumulation of melanin (*e.g.*, melasma, senile lentigo, *etc.*) or simply by culture-specific beauty preferences. Numerous chemical substances have already been proven as effective skin whiteners, and some even display beneficial side effects (antioxidants, antiproliferative activity, protection of macromolecules such as collagen against UV radiation, *etc.*), but others have recently raised safety concerns, leading to their ban in some countries. The search for non-cytotoxic natural whitening compounds benefits from the fact that natural ingredients have become more prevalent nowadays in cosmetic formulations due to consumers' concern about synthetic ingredients and the risks they may represent for human health. In recent years, the quest for fairness has led to the identification of a number of whiteners originating from various biological sources that work as well as the synthetic ones, with few or no side effects. However, there is still a long way to go from the discovery of an active ingredient to its incorporation into cosmetics and its commercialization. In fact, ingredients' cytotoxicity, insolubility, and instability, as well as development costs, are some of the difficulties encountered when establishing a formulation. The following are some brief introductions on several chemicals which have been frequently used in whitening cosmetics.

17.1 Hydroquinone and Its Derivatives

Hydroquinone is a ubiquitous molecule. It is the oxidation product of certain aromatic compounds. As to its mode of action, various mechanisms have been suggested. The inhibition of tyrosinase (key enzyme responsible for melanin production) synthesis, the inhibitory effect on this enzyme, the destruction of melanocytes by the production of free radicals and interference with melanin-containing organelles (melanosomes) are instrumental in making this ingredient an effective depigmentation agent.

Given that hydroquinone is one of the most effective depigmentation molecules, a codrug was obtained by esterifying hydroquinone with azelaic acid. The aim is threefold: obtain a synergistic effect, increase cutaneous permeation, and increase the stability of the active ingredients by masking the most labile functions. As for cutaneous penetration, the gamble paid off in that the tests performed *in vitro* showed that the permeation of the codrug was two times greater than that obtained with the parent compounds alone.

Hydroquinone mono methyl ether (MEHQ), also known as 4-hydroxyanisole or mequinol, is less irritating than hydroquinone and is not cytotoxic to human melanocytes. The combination of mequinol (2%) with retinoic acid (0.01%) turns out to be synergistic. A clinical trial performed on a group of 595 subjects (486 women and 109 men) with solar lentigines was able to demonstrate the value of this combination. Other tests performed *in vivo* confirmed these results. However, it should be noted that this combination, though effective, did not produce any fewer undesirable effects. Erythema, burning, tingling, desquamation, and pruritus may be observed.

17.2 Retinoic Acid or Tretinoin or Vitamin A Acid

Initially used in combination with hydroquinone as a penetration factor, its inherent efficacy soon became evident. Tretinoin has several action mechanisms. It directly intervenes in the melanogenesis process by inhibiting tyrosinase induction and dispersing the pigments in the keratinocytes on the one hand, and is capable of accelerating epidermal turnover on the other hand. The undesirable effects observed are erythema and desquamation. Vitamin A acid, a molecule extensively metabolized in vertebrates, is thought to be the active metabolite. Fatty acid esterification, oxidation, dehydrogenation, and beta-glucuronosylation reactions are induced. The teratogenic nature of retinoic acid has been clearly demonstrated. For this reason, this molecule is banned in the cosmetics industry.

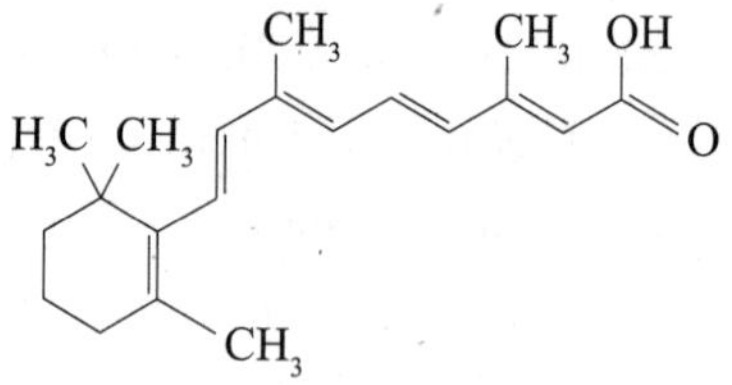

CH_3 CH_3 OH
H_3C CH_3 O
CH_3

Tretinoin

17.3 Alpha Hydroxy Acid

Also rather inappropriately called fruit acids, the alpha hydroxy acids (AHA) have been in use for a long time for alleviating the effects of photoaging. The main ones are glycolic acid, lactic acid, malic acid, citric acid, tartaric acid, and mandelic acid.

Depending on their application rates, these acids are used for different purposes. At low rates (<1%), they are used as pH adjusters. At higher concentrations (around 10%), they act as hydrating agents capable of retaining moisture in the stratum corneum; above 50%, the reduction of inter-corneocyte cohesion is reflected in a keratolytic effect of importance in the anti-aging field. The peeling effect makes it possible to smooth wrinkles. The most commonly used concentrations in cosmetics are around 5% to 10%. However, it is possible to find preparations with an alpha hydroxy acid content of 30%. Such is the case, for example, with Clairial Peel® (Le Plessis-Pâté, France) gel from SVR Laboratories, which contains 20% glycolic acid, 10% citric acid, and 5% vitamin C.

Glycolic acid is definitely the most commonly used alpha hydroxy acid of all. Studies recently conducted *in vitro* seem to indicate that the influx of H^+ ions into the basal layer keratinocytes induces an increase in the proliferation of cells of this category. It should also be noted that the production of collagen, elastin, and mucopolysaccharides is stimulated. Glycolic acid can be used for superficial peels. Based on the percentage of glycolic acid in the preparation, a distinction is made between peels that can be done at home (5% to 20% glycolic acid, pH=4-6) and peels that require a dermatologist (50% to 70% glycolic acid, pH=1-2). The highest concentrations induce an epidermolysis down to the basal layer of the epidermis.

Pyruvic acid can be used at a concentration of 40%-50% by a dermatologist to reach the granular layer of the epidermis.

When one looks at cosmetic peels, one also notes that these ingredients are not regulated and hence there is a regulatory gap. It is, therefore, necessary to rely on the wisdom of the manufacturer who, depending on the percentage that he chooses, will position his product either with a cosmetic status or as a medical device.

Because alpha hydroxy acid-based preparations can increase the sensitivity of the skin to UV radiation, we would reiterate that applying such preparations and then exposing oneself to the sun is totally contraindicated. This leads to an increased rate of erythema development and sunburn cell formation.

Salicylic acid, a beta hydroxy acid, can also be used for superficial peels. It should be noted that the percentages used in the cosmetics field are restricted by law because of their potential toxicity. When it is used for purposes other than preservation, it can be used up to a limit of 3% in the case of products that are to be rinsed off and used for facial hair, or 2% in the case of other products. It must be borne in mind that using preparations with high concentrations of salicylic acid can lead to acute poisonings (salicylism). This can occur after using 6% preparations on an area corresponding to 40% of the body surface. The literature reports 25 cases of poisonings (including 4 deaths) from 1966 to the present day. In terms of its mode of action, it is known that salicylic acid is capable of reducing keratinocyte proliferation and solubilizing the intercellular cement responsible for the cohesion of the stratum corneum cells. It is thus traditionally used in dermatology to treat diseases of the epidermis. Preparations containing salicylic levels ranging from 10% to 40% are used for this purpose.

The actual efficacy of preparations containing only low percentages of salicylic acid is questionable. It could be demonstrated *in vivo* that the keratolytic effect of salicylic acid was observable with a low percentage of use (2%-preparation with a pH of 7) for a minimum exposure time of 6 h. This active ingredient was chosen for Anti-blemish solutions® gel (Estée Lauder, New York, USA) and Depiderm® cream (Uriage, Neuilly sur Seine, France).

Dermatologists on the other hand use more concentrated (20% to 25%) preparations. They use 20% to 30% ethanol solutions or 50% salves. At these percentages, it is possible to reach the granular layer of the epidermis. Rare cases of poisoning or salicylism have been reported. Clinical tests conducted on around twenty Latin American women refute the efficacy of this type of peel. In women with melasma who were treated on half of their faces with 4% hydroquinone preparations (2 applications per day) with or without a 20% to 30% salicylic acid peel (1 peel every 2 weeks for an 8 week study period), no significant differences were observed between the two treated sides.

An element that must be taken into account is the incorporation of other ingredients in the formulation. Past studies have shown a synergistic effect when salicylic acid and propylene glycol are combined. This moisturizer, which is found in a great many commercial formulations, helps improve the results obtained.

17.4 Ascorbic Acid and Its Derivatives

Ascorbic acid or vitamin C is readily degraded by oxidation, especially in aqueous media. For this reason, it is preferable to use more stable derivatives such as ascorbyl palmitate and magnesium-l-ascorbyl-2-phosphate (MAP). Vitamin C and its derivatives act as reducers and block the chain of oxidations transforming tyrosine into melanin at different points. Furthermore, the interaction of ascorbic acid with copper, an essential cofactor in tyrosinase activity, explains the tyrosinase inhibitor effect observed *in vitro*. Although less effective than hydroquinone, ascorbic acid does not have the harmful effects of the latter. This active ingredient is not used alone. It is always used in combination with another ingredient. Its use is not regulated and vitamin C is found in cosmetics in concentrations ranging from 4% to 20%.

17.5 Retinol and Retinaldehyde

In the 1990s, retinol was widely used in anti-aging cosmetics. The concentration used generally ranged from 0.04% to 0.07% retinol equivalent. In keratinocytes, retinol is oxidized to retinaldehyde and then to retinoic acid, a molecule that is banned in cosmetics in Europe. It should also be pointed out that studies conducted using radio-labeled molecules showed that 7% of the dose applied to the skin could be detected at the systemic level. Pregnant women must therefore exercise caution when using depigmentation products containing ingredients of this kind. This point needs to be emphasized because pregnancy mask arising in pregnant women with dark complexions is a compelling reason for them to resort to depigmentation products. We feel that this measure is justified in spite of the fact that a consortium of industrialists including L'Oreal, Unilever, Johnson & Johnson, BASF, Henkel, DSM Nutritional Products, Beiersdorf and Shiseido showed in 2006 that the daily application of topical preparations (at a dose of 30,000 IU/d for 21 days) on an area equivalent to 1/6 of the body surface did not lead to a significant increase in endogenous plasma levels of vitamin A and its metabolites.

Furthermore, retinoids are known to induce the production of pro-inflammatory cytokines and are therefore irritating molecules. Lastly, the "photomutagenicity" of retinol requires that certain precautions be taken in case of exposure to the sun. A possible solution would be to combine retinol with a mixture of UV filters. Johnson & Johnson laboratories came up with this idea in 2008. It was thus possible to test a retinol-based cosmetic labeled with a Sun Protect Factor (SPF) of 30 on a group of 30 volunteers. Even though the results obtained in terms of reduction of photoaging signs may be positive, the message that this type of product conveys seems less than ideal. An optimum level of photo-protection will not be ensured for the entire day unless skin care cosmetics are applied at the rate of 2 mg/cm^2 and re-applied every 2 h. Skin care cosmetics displaying an SPF value pose the risk of clouding the public health message clearly indicating that a sunscreen product must be applied liberally and on a regular basis during the period of exposure.

17.6 Arbutin and Aloesin

Arbutin is a hydroquinone glycoside with 2 isoforms, 4-hydroxyphenyl-α-glucopyranoside and 4-hydroxyphenyl-β-glucopyranoside. It is obtained from various plants in the Ericaceae (bearberry, strawberry tree, huckleberry, heather), Saxifragaceae, Asteraceae, Rosaceae, Lamiaceae, and Apiaceae families.

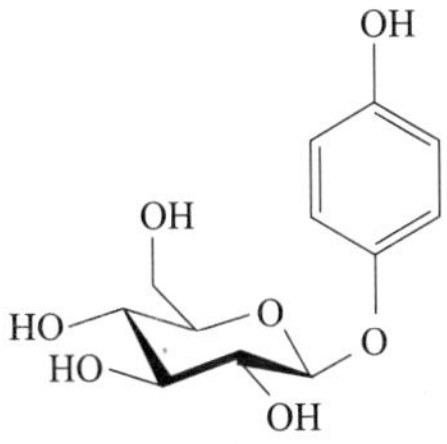

Arbutin

Bearberry (*Arctostaphyllos uva-ursi*), traditionally used in herbal therapy to treat urinary tract disorders, is without doubt the plant most often used in the cosmetics industry for formulating skin lightening products. The percentage of arbutin present in a plant varies considerably according to the species (17% in the leaves of *Arctostaphyllos uva-ursi*, 5% in the leaves of *Origanum majorana*), the time of collection, and the growing conditions. Certain authors have shown that supplying hydroquinone in the growing medium induces an increase in the biomass of the plant. The small South African shrub *Myrothamnus flabellifolia* is characterized by the richness of its leaves in arbutin (27% of the dry weight) and is therefore also of interest in this field.

Numerous studies show that arbutin is just as effective as hydroquinone, but less toxic. The alpha isomer has the greatest inhibitory activity against mammalian tyrosinases. This agent, which is photostable yet readily degraded by heat, would need to be incorporated at cold temperatures into the chosen excipient.

The Scientific Committee on Consumer Safety (SCCS) assessed the use safety of alpha arbutin (for concentrations greater than 2% in facial creams and greater than 0.5% for body lotions) and of beta arbutin (for concentrations greater than 7% in facial creams). A study conducted on 10 women with melasma revealed a significant reduction in melanin levels after a one-month treatment with a 1% preparation. However, it must be pointed out that cutaneous hydrolysis of the glycoside takes place, in which hydroquinone is released.

A synthetic derivative, deoxyarbutin {4-[(tetrahydro-2H-pyran-2-yl)oxy] phenol} would have the advantage over botanical arbutin of being more effective and less cytotoxic. This is not the opinion of the SCCS, which decided in 2016 that this ingredient is not safe to use in finished products at a use rate greater than 3%. The amount of hydroquinone released as the finished products age means that they cannot be marketed with absolute safety.

Aloesin is a low molecular weight glycoprotein extracted from various species of plants in the genus *Aloe*, in which it can be found in significant amounts. Such is the case with *Aloe ferrox*. A dry extract of *Aloe ferrox* leaves may contain up to 25% aloesin.

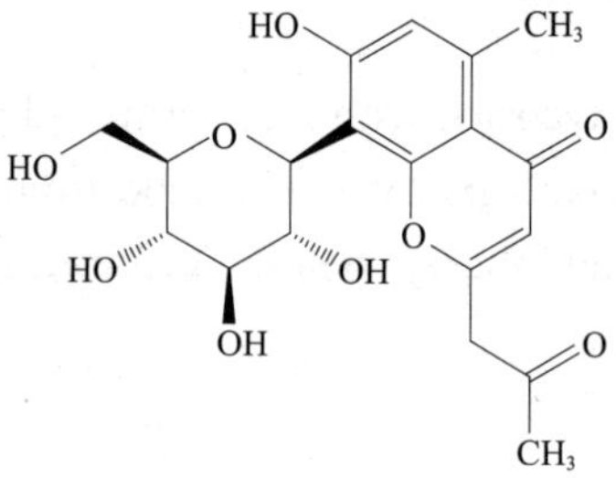

Aloesin

This molecule of human, animal, or fungal origin acts as a competitive inhibitor of tyrosinase. Efficacy is proportional to the concentration used. From the standpoint of its toxicology profile, aloesin is evidently safe to use. This molecule, which is non-genotoxic and has a high No Observable Adverse Effect Level (NOAEL) [2000mg/(kg · d) *via* the oral route], is also a good candidate for making foods or food supplements with the aim of preventing diabete.

17.7 Glabridin

Glabridin is present in the hydrophobic fraction of licorice root extract (licorice) and is capable of reducing the activity of tyrosinase on melanocytes in culture, and of inhibiting the induction of pigmentation by UVB and erythema formation in guinea pigs. Its estrogenic effect is a drawback. Even though this effect is less than that of physiological estrogens (for example, doses 10 times greater than that of 17 beta-estradiol are required to bring about an observable increase in creatine kinase activity in female rats), this aspect must be considered.

Other flavonoids present in licorice root extract such as glabrene, isoliquiritigenin, licuraside, isoliquiritin, and licochalcone A are also of interest due to their tyrosinase inhibiting nature. Lastly, we would point out that liquiritin may turn out to be of value in treating melasma. Although it does not inhibit tyrosinase, this molecule makes it possible to lighten skin tone by melanin dispersal. Tests performed on around twenty women suffering from melasma demonstrated the positive effect of liquiritin at a substantial concentration of 20%, with reduction of pigmentation intensity (result observed in 70% of the women after 4 weeks of treatment) and reduction of the size of the lesions (result observed in 60% of the women after 4 weeks of treatment).

The leaves of the white mulberry (*Morus alba*, Moraceae) have been used for many years in traditional medicine in China, Korea, Japan, and Thailand. Fever-reducing, liver-protecting, and blood pressure-lowering properties are attributed to them. The polyphenols contained in the leaves have depigmentation properties that have been demonstrated *in vitro*. In 2006, investigators in Taiwan Province, China tested a number of Chinese herbs used in folk medicine. An inhibitory effect on human tyrosinase was demonstrated for a number of extracts extracted by alcohol at 95 ℃ . This inhibitory effect was compared to the one obtained with a standard molecule, arbutin. With an inhibitory effect comparable to that of arbutin, the extract of white mulberry turned out to be the one of greatest value. It takes 100μg/mL of *Morus alba* extract *versus* 138μg/mL of arbutin to achieve a ca. 70% inhibition of tyrosinase.

17.8 Kojic Acid

Kojic acid is a molecule present in a number of fermented foods or beverages of Japanese origin. It is formed by fermentation by various species of fungi such as *Aspergillus*, *Penicillium*, and others.

Discovered in 1907 by Saito in cultures of the fungus *Aspergillus* oryizae, this acid was soon found to have many practical applications.

Used as a food additive to preserve the natural color of foods such as fresh fruits and vegetables, crustaceans, *etc*., kojic acid is a common ingredient in depigmentation preparations.

It intervenes in the depigmentation process *via* various mechanisms. It acts as a chelator of divalent ions, as a free radical trapper, and as a tyrosinase inhibitor. Its cytotoxic effect and instability over time have led certain teams to work on improving it. For example, it has been conjugated with phenylalanine. Because it acts as an iron chelator, anti-wrinkle properties have been attributed to it as well. Depending on the periods considered, Japanese and Europeans have expressed conflicting opinions. In 2008, the Scientific Committee on Consumer Products (SCCP) pleaded in favor of the use safety of preparations containing 1% kojic acid, even though risks of sensitization are conceivable. Subsequent genotoxicity tests have lent support to this opinion. A literature review conducted in 2010 also concluded that it was safe to use this agent at concentrations of less than 1%.

As far as its efficacy is concerned, it is less in comparison to the standard molecule, hydroquinone. Tests on volunteers have shown that a 4% hydroquinone-based preparation is 5 times more effective than a combined preparation of kojic acid (0.75%) and vitamin C (2.5%).

17.9 Azelaic Acid

Azelaic acid (AzA) is an organic compound with the formula $HOOC(CH_2)_7COOH$. In 1978, the tyrosinase-inhibiting activity of certain lipid fractions, mainly C_9-C_{11} dicarboxylic acids, was demonstrated for the first time *in vitro*. The interest in azelaic acid for treating pigmented lesions thus ensued. This acid is produced naturally by a yeast, *Malassezia furfur*. Its inhibitory activity against tyrosinase is reflected in the appearance of depigmented maculae on the skin of subjects suffering from a mycosis, pityriasis versicolor. This fungus produces lipoxygenases that are capable of acting on the unsaturated fatty acids present on the skin surface. In culture, this fungus is capable of oxidizing oleic acid into azelaic acid. There is unanimous agreement regarding its efficacy and absence of undesirable effects of note. A placebo-controlled clinical study conducted on 52 women with dark or pigmented skin (phototypes Ⅳ to Ⅵ) suffering from melasma demonstrated the superiority of a cream containing 20% azelaic acid. The women found that their skin was smoother and were thus satisfied overall. However, some undesirable effects (burning, tingling) were reported. It seems that the preparations containing 20% azelaic acid and 4% hydroquinone are equivalent *in vivo*. The anti-inflammatory, anti-keratinizing and bacteriostatic activity of azelaic acid justifies its use in treating diseases such as rosacea or acne. Azelaic acid apparently behaves differently, depending on the characteristics of the cells concerned. A cytotoxic effect on human melanocytes and a much higher capacity to penetrate abnormal cells than normal cells indicated possibilities for use in treating melanoma at one time. However, these hopes were dashed. From a practical standpoint, very few cosmetics use this acid. However, mention can be made of the product Melascreen® from Ducray laboratories (Boulogne Billancourt, France).

17.10 Some Lesser Known Extracts or Molecules for Skin Lightening

The search for natural melanogenesis inhibitors in recent years has demonstrated that plant extracts could be potential sources of new whitening ingredients as potent as synthetic whitening agents. The following are some examples.

Artocarpus heterophyllous (Moraceae) is a tree found in the tropical and subtropical regions of Asia. It is grown for its edible fruit in Vietnam. Its bark has numerous therapeutic properties: anti-inflammatory, anti-oxidant, and anti-aging. After a methanol extract demonstrated interesting anti-tyrosinase properties, a purification to detect the most effective molecules was deemed necessary and led to the isolation of artocarpanone. The IC50 (50% inhibiting concentration) of this molecule is indeed 22 times lower than that of the chosen standard molecule, kojic acid. Another species, *Artocarpus xanthocarpus*, is also a source of depigmenting molecules (alboctalol, steppogenin, norartocarpetin) worth exploiting. These molecules have IC50 values that are 50 times lower than that of kojic acid. A dimeric stilbene with anti-tyrosinase activity could also be isolated from *Artocarpus gomezianus*. The safety of such a molecule still needs to be demonstrated.

In 2010, an Italian team from Messina demonstrated the value of an ethanol extract of *Betula pendula* leaves. This extract shows a weak inhibitory activity, its efficacy being 50 times less than that of kojic acid. However, the value of this ingredient lies in its mechanism of action. It is not a competitive tyrosinase inhibitor, but a molecule capable of sequestering Cu^{2+} ions, which are essential cofactors for tyrosinase activity.

In 2016, Robertet laboratories (Grasse, France) started investigating the subject of depigmentation and demonstrated the valuable properties of an extract of *Populus nigra* buds in an ethanol/hexane mixture. A tyrosinase inhibitory effect two times lower than that of kojic acid was demonstrated under these conditions. However, it must be pointed out that hexane is not a solvent of choice for a cosmetic use and that in order for any further developments to take place, the extraction solvent must be looked at again.

In 2015, Indian researchers in a division of biotechnologies investigated the potential of 13 edible fruits from the northwestern Himalayans. Anti-aging and anti-tyrosinase activities were sought in particular. A dozen of the fruits tested exhibited a tyrosinase inhibitory effect of varying degrees. The acetone extract of *Ziziphus nummularia* that was retained caught our attention as well. The latter also has anti-collagenase and anti-elastase properties. Nevertheless, it must be pointed out that here as well, the extraction solvent is not compatible with a cosmetic use.

A collaboration among various research teams working in Indonesia, Japan, and Sudan, countries in which having a light complexion is of paramount importance, revealed the influence of the respective plant part on the level of efficacy attained. While the extraction solvent does indeed play a major role, the plant part used must also be taken into account. In the case of *Curcuma xanthorrhiza*, a plant widely used in traditional medicine, considerable differences will be observed depending on the part chosen. The rhizome extract in ethyl acetate is ten times less effective than the kojic acid standard, whereas the bract extract in the same solvent is 100 times less so, which clearly eliminates any possibility of skin lightening applications.

Phyla nodiflora herbal teas are used in Chinese folk medicine to treat inflammatory skin conditions. The active molecule, eupafolin, is a flavonoid. Its anti-inflammatory effect and its skin lightening effect have been demonstrated. According to tests performed on keratinocytes and melanocytes, eupafolin is a molecule that is non-toxic to skin cells. Its lightening action lies in its property of being a tyrosine inhibitor (its efficacy is ca. 65% that of the standard molecule) and in its capacity to interfere with the synthesis of this enzyme.

Finally, it must to note that some other substances found to be of whitening effects are banned in cosmetic products. For example, mercury-containing products are hazardous and have been banned in most countries (since 1976 in Europe and since 1990 in the USA): accumulation of this heavy metal may lead to chronic complications including mercurialentis, photophobia, irritability, muscle weakness, and nephrotoxicity. Mercury poisoning is also observed in newborns as the metal is easily conveyed *via* the placenta and breastmilk. Furthermore, long-term application of mercurial derivatives has adverse effects such as mercury accumulation darkening the skin and nails. However, the use of some mercury salts as preservatives (thiomersal, phenylmercury salts) is still authorized.

References

[1] Couteau C, Coiffard L. Overview of Skin Whitening Agents: Drugs and Cosmetic Products. Cosmetics, 2016, 3: 27.

[2] Burger P, Landreau A, Azoulay S, Michel T, Fernandez X. Skin Whitening Cosmetics: Feedback and Challenges in the Development of Natural Skin Lighteners. Cosmetics, 2016, 3: 36.

Key Words & Phrases

melasma *n.* 黄褐斑
senile lentigo 老年性雀斑
skin whitener 皮肤美白剂
antiproliferative activity 抗增殖活性；抗增殖
non-cytotoxic *adj.* 非细胞毒性
melanocyte *n.* 黑素细胞
organelle *n.* 细胞器
depigmentation *n.* 脱色；褪色
azelaic acid 杜鹃酸；壬二酸
cutaneous penetration 皮肤穿透
in vitro 在生物体外
4-hydroxyanisole *n.* 4-羟基茴香醚
mequinol *n.* 甲氧苯酚；对甲氧酚
retinoic acid 视黄酸；维甲酸；维生素A酸
lentigines *n.* 雀斑痣；（尤指）老年斑
tingling 刺痛感
tretinoin *n.* 维甲酸；维生素A酸
teratogenic *adj.* 致畸形的
mandelic acid 扁桃酸
inter-corneocyte cohesion 角质细胞间内聚力
keratinocyte *n.* 角蛋白形成细胞
elastin *n.* 弹性蛋白
mucopolysaccharide *n.* 黏多糖；糖胺聚糖
superficial peel 表面剥离
epidermolysis *n.* 表皮松解
pyruvic acid 丙酮酸
granular layer 颗粒层
acute poisoning 急性中毒
salicylism *n.* 水杨酸反应
keratinocyte proliferation 角质形成细胞增殖
intercellular cement 胞间黏合质；细胞间胶质
magnesium-1-ascorbyl-2-phosphate 1-抗坏血酸-2-磷酸镁
retinol *n.* 视黄醇
endogenous plasma 内源性血浆
metabolite *n.* 代谢物；代谢产物
cytokine *n.* 细胞因子；细胞激素
photomutagenicity *n.* 光致突变性
hydroquinone glycoside 对苯二酚糖苷
isoform *n.* 亚型；同工型；同型；异构体
ericaceae *n.* 杜鹃花科
bearberry *n.* 熊果；熊莓
huckleberry *n.* 美洲越橘；美洲越橘树
heather *n.* 帚石楠
saxifragaceae *n.* 虎耳草科；虎耳草属
asteraceae *n.* 菊科；菊科植物
rosaceae *n.* 蔷薇科
lamiaceae *n.* 唇形科
apiaceae *n.* 伞形科
urinary tract disorders 尿路疾病
mammalian tyrosinase 哺乳动物酪氨酸酶
excipient *n.* 赋形剂
cutaneous *adj.* 皮肤（上）的
deoxyarbutin *n.* 脱氧熊果苷
aloesin *n.* 芦荟素
Aloe ferrox 开普芦荟
fungal *adj.* 真菌的；真菌引起的
diabete *n.* 糖尿病
glabridin *n.* 光甘草定
licorice *n.* 甘草精
estrogenic *adj.* 动情（激素）的；雌激素的
glabrene *n.* 光甘草素
isoliquiritigenin *n.* 异甘草素
licuraside *n.* 异甘草素葡萄糖洋芫荽糖苷
isoliquiritin *n.* 异甘草苷
licochalcone A 甘草查尔酮A
liquiritin *n.* 甘草苷
lesion *n.*（因伤病导致皮肤或器官的）损伤；损害
white mulberry 白桑树
Morus alba 桑树；桑科植物
intervene *v.* 介入；出面；插嘴；打断；阻碍；干扰
chelator *n.* 螯合剂
free radical trapper 自由基捕捉器
Malassezia furfur 糠秕马拉色菌
maculae *n.* 缺陷；斑点；暗斑；气门斑（蜱螨）
mycosis *n.* 真菌病
pityriasis versicolor 汗斑；花斑癣；杂色糠疹；花斑糠疹；变色糠疹
lipoxygenase *n.* 脂加氧酶；脂氧化酶
placebo-controlled 安慰剂对照；安慰剂对照组；安慰剂控制
anti-keratinizing 抗角质化
bacteriostatic activity 抑菌活性
melanogenesis *n.* 黑色素生成
Artocarpus heterophyllus 木菠萝，菠萝蜜
moraceae *n.* 桑科

artocarpanone *n.* 桂木二氢黄素
Artocarpus xanthocarpus 黄果菠萝蜜
alboctalol *n.* 白桑八醇
steppogenin *n.* 草大戟素
norartocarpetin *n.* 降桂木生黄亭
dimeric stilbene 二聚二苯乙烯
Artocarpus gomezianus 长圆叶菠萝蜜
Betula pendula 欧洲白桦;垂枝桦
Populus nigra 黑杨；欧洲黑杨；钻天杨
Ziziphus nummularia 铜钱枣
anti-collagenase 抗胶原酶
anti-elastase 抗弹性蛋白酶
Curcuma xanthorrhiza 印尼莪术
rhizome *n.* 根茎；根状茎
bract *n.* 苞；苞片；托叶
Phyla nodiflora 过江藤
eupafolin *n.* 楔叶泽兰素；泽兰叶黄素
mercurialentis *n.* 汞中毒性晶状体变色
photophobia *n.* 畏光；羞明；恐光症
nephrotoxicity *n.* 肾毒性；中毒性肾损害
placenta *n.* 胎盘
thiomersal *n.* 硫柳汞
phenylmercury *n.* 苯汞；苯基汞

Chapter 18 Anti-aging Agents

The anti-aging market was growing at an approximate 8% compound annual growth rate between 2018 and 2022. As the competition increases among cosmetic brands from the anti-aging market, new products claim to contain the ultimate innovations in order to stand out, often advertising new active ingredients. The most used anti-aging agents are vitamin A, vitamin E, coenzyme Q10 and some peptides. The following are some brief introductions on them.

18.1 Vitamin A, Vitamin E and Coenzyme Q10

The topical application of fat-soluble vitamins A and E and coenzyme Q10 has various beneficial effects on the skin. Therefore, these three groups are important ingredients in the cosmetic industry. The widespread use of vitamin E over the past several decades is mostly associated with its antioxidant activity. Vitamin E is used in cosmetics as a cosmetically active ingredient (occlusive, humectant, emollient, and miscellaneous agent) or as a stabilizer of other unstable components of the cosmetic product. Because of its antioxidant activity, topically applied vitamin E is effective in the treatment of skin conditions and diseases caused by oxidative stress, including UV-induced erythema and edema, sunburns, and lipid peroxidation. It is also an effective anti-ageing agent. Vitamin E is most commonly found in cosmetics in its active form, α-tocopherol, or more stable esterified form, tocopheryl acetate, which requires hydrolysis to the active form upon skin penetration. Despite differing data on the extent of this conversion in the skin, most studies disclose the higher antioxidant activity of α-tocopherol compared to its esters. Vitamin E may be found in a wide range of concentrations, from 0.0001% to 36% in cosmetic products on the market.

Retinoids are effective in the topical treatment of acne, hyperpigmentation, psoriasis, and skin-aging, and are therefore active ingredients in a variety of cosmetic products, especially as anti-ageing agents. The most common vitamin A forms found in cosmetics include retinol and its esters, retinyl palmitate and acetate, as well as β carotene. Analogously to vitamin E esters, vitamin A esters also require hydrolytic conversion to retinol, which is further metabolized to retinal and then to the active form–retinoic acid. Therefore, retinoid activity after topical application depends on the metabolic closeness to the active form and decreases in the following order: retinoic acid > retinal > retinol > retinyl ester. Due to the possible risk of teratogenicity, retinoic acid is banned in cosmetic products in the EU. Despite their poor activity, retinyl esters, especially retinyl palmitate, are commonly used in cosmetics due to their stability. Due to safety reasons, the Scientific Committee on Consumer Safety, Secretariat at the European Commission, Directorate General for Health and Food Safety recommend a maximum retinoid concentration of 0.05% retinol equivalents (RE) in body lotions and 0.3% RE in hand and face creams, as well as other leave-on or rinse-off products for cosmetics in the EU. However, cosmetics with significantly higher retinoid contents are found on the EU market.

Coenzyme Q10 is an endogenous nonvitamin lipophilic antioxidant, which is often analytically evaluated alongside fat-soluble vitamins, due to its lipophilic structure and activities in the human body. Coenzyme Q10 is also an important antioxidant in the skin. However, its skin levels decline with age and exposure to UV irradiation. Topical coenzyme Q10 application is effective in the replenishment of its skin levels and thus provides skin protection and prevents skin inflammation, UV-induced erythema, and skin cancer. Coenzyme Q10, in its ubiquinone form, is a popular ingredient in anti-ageing cosmetics, in which it is usually found in concentrations $\leqslant$ of 0.05%.

The efficacy of cosmetic products is directly associated with their quality. As discussed above, the efficacy depends on the form of the active ingredient (*e.g.*, vitamin A or E esters), and also on their content, which is generally low (<1%). Another important challenge is the instability of these compounds, causing possible losses during manufacture and storage, leading to their even lower contents or complete loss. Therefore, a prerequisite for their quality control is appropriate analytical methodology. Several analytical methods for the determination of a single retinoid or retinoids in different forms in topical formulations may be found in the literature, including two methods for the quality control of specific vitamin A forms commonly found in cosmetics.

18.2 Peptides

18.2.1 Introduction

Peptides and proteins are amino acid polymers. Peptides are short amino acid chains. Naturally occurring human peptides are known for cellular communication. The first peptide synthesis was published in 1901 by Fischer and Fourneauin. Fischer described the first peptide as a glycyl-glycine and in his lectures explained more peptide structure like dipeptides, tripeptides and polypeptides. Years followed, scientists synthesized new peptides, identified more natural peptides, and learned more about their functions. Beside the growing knowledge about natural and synthetic peptides, different synthetic peptides were developed. Copper glycine-histidine-lysine (Cu-GHK) was developed in 1973 by Loren Pickard. In the late 80s, the first copper peptide was incorporated into skin care products. Even then, peptide development proceeded slowly until the beginning of 2000, when palmitoyl pentapeptide-4 was established. Since then, research and industry have developed many short, stable, and synthetic peptides that have a role in extracellular matrix synthesis, pigmentation, innate immunity and inflammation. These peptides are used for collagen stimulation, wound healing, wrinkle smoothing, as well as antioxidative, antimicrobial, and whitening effects.

Topical cosmeceutical peptides can be classified as signal peptides, carrier peptides, neurotransmitter inhibitor peptides, and enzyme inhibitor peptides.

Cosmeceutical peptides should have certain features in order to obtain good effects. Historically, it has always been assumed that because of the skin barrier, the molecular weight of peptides should be less than 500 Da, otherwise the peptide would not be able to pass the barrier. The melting point should be below 200 ℃ , water solubility should be >1 mg/mL and there should be no or few polar centers.

Newer studies have shown that larger molecules can traverse the skin barrier, especially in the case of dry and aged skin. Synthetic peptides consist of amino acids chains which can be now modified in various ways for different functions like increased skin penetration, and increased special receptor binding, stability, and solubility.

There are many peptides which can be used in cosmetic products, but little *in vivo* efficacy data is available. In addition, substance mixtures are on the market and tested in cosmetic formulations, so that in many cases the actual effect of individual peptides on the skin remains unclear. Claims of efficiency by cosmeceuticals are restricted to the improvement of the skin appearance. Improving cosmeceutical function might lead to re-classification from cosmetic to drug category, which is often not desirable. This often limits possibilities for development.

Research on peptides should aim to identify the peptide's mode of action, and define it for cosmetic and/or pharmaceutical use. The prerequisites for an effective active substance must be carefully examined before it is used. Study designs should be developed carefully, and maximal results should be generated. With today's methods, receptor activation, efficacy, and mechanistic information can be identified. Interesting and meaningful *in vivo* studies can be developed. The development of active peptides has opened a new field in cosmeceutical and pharmaceutical skin care in the last decade. The following is a summary of clinical studies with peptides which confirm their efficacy on human skin.

18.2.2 Peptides that trigger the signaling cascade

A number of peptides are able to trigger a signaling cascade. They are released from the extracellular matrix, and are also called collagen stimulators. With these peptides, the proliferation of collagen, elastin, proteoglycan, glycosaminoglycan and fibronectin is increased. As a consequence, pigmentation of photo-damaged skin and fine lines and wrinkles are reduced with the regeneration of the skin matrix cells. Skin elasticity increases, and skin appears smoother and firmer. Synthetic peptides modeled on repair signaling sequences like the following described in this section, have been developed to rejuvenate skin.

(1) Carnosine and *N*-acetylcarnosine　Carnosine is a dipeptide (Sequence: *β*-Ala-His) and a well-documented aqueous antioxidant with wound healing activity, and it is naturally present in high concentrations in muscle and brain tissues. Carnosine is a scavenging reactive oxygen species as well as an *α*-*β* unsaturated aldehyde formed from peroxidation of cell membrane fatty acids during oxidative stress. The low molecular weight water soluble unmodified dipeptide *β*-Ala-His has very little affinity for skin and does not penetrate beyond the first layer of the stratum corneum. The lipophilic peptide palmitoyl *β*-Ala-His, however, diffuses into the stratum

corneum, epidermal, and dermal skin layers. No systemic activity has been observed.

In two double-blind, randomized, controlled, split-face studies of four weeks each, changes in periorbital wrinkles in women (aged 30-70) were observed (Study 1, 42 volunteers; Study 2, 35 volunteers). Tested products containing niacinamide, the peptides pal-KT and pal-KTTKS, and carnosine, ameliorated periorbital skin, enhancing smoothness and diminishing larger wrinkle depth.

A double-blind irradiation study comparing a complex consisting of different active ingredients (SPF 50, photolyase, endonuclease, 8-oxoguanine glycosylase, carnosine, arazine, and ergothionine) in available products with DNA repair, antioxidant and growth factor ingredients, found the formulation to be effective in reducing pyrimidine dimers, protein carbonylation, and 8-oxo-7,8-dihydro-2′-deoxyguanosine in human skin biopsies. The formulation also appeared to enhance the genomic and proteomic integrity of skin cells after continual UV exposure. Hence, this formulation could be regarded as potentially lowering the risk of UV-induced cutaneous aging, and non-melanoma skin cancer.

During a six-month study, 20 healthy volunteers (Photo type Ⅱ or Ⅲ) were treated with carnosine and *N*-acetylcarnosine formulations. Carnosine and *N*-acetylcarnosine alone in a water solution obtained 3.6% and 7.3% reduction of erythema compared to the control. Both peptides showed antioxidant capacity, with a higher significance in conjunction with vehicles improving the substances' skin penetration capabilities. *N*-acetylcarnosine was mentioned as an interesting hydrophilic antioxidant for dermatological purposes.

(2) Trifluoroacetyl-tripeptide-2 Trifluoroacetyl-tripeptide-2 (Sequence: TFA-Val-Try-Val-OH) was evaluated in two *in vivo* split face studies. One study examined its anti-wrinkle and anti-sagging effects along the jawline of 10 volunteers (56 days) *via* fringe projection profilometry; and the other study targeted skin firmness, elasticity, and viscoelasticity *via* cytometry on 13 healthy volunteers (28 days). According to the studies, trifluoroacetyl-tripeptide-2 has progressive effects on wrinkles, firmness, elasticity and sagging.

(3) Tripeptide-10 citrulline Tripeptide-10 citrulline (Sequence: Lys-α-Asp-Ile-Citrulline), a decorin-like tetrapeptide, Decorinyl™ is used to specifically target collagen fiber organization. Puig *et al.* published results of an assessor blinded, placebo-controlled, parallel group study with 43 healthy volunteers (aged 40-58). Tripeptide-10 citrulline showed uniformity in fibril diameter, and increased skin suppleness from better collagen fiber cohesion.

(4) Palmitoyl tripeptide-1 Palmitoyl tripeptide-1, also called pal-GHK and palmitoyl oligopeptide (Sequence: Pal-Gly-His-Lys), is a messenger peptide for collagen renewal. Comparable to retinoic acid with regards to its activity, it does not trigger irritation. Collagen and glycosaminoglycan synthesis are stimulated, the epidermis is reinforced, and wrinkles are diminished. This peptide is suggested to act on TGFβ (transforming growth factor-β) to stimulate fibrillogenesis. It is used in cosmetic anti-wrinkle skincare and make-up products. In a study with 15 women, a cream containing palmitoyl tripeptide-1 was applied twice daily for four weeks, leading to statistically significant reductions in wrinkle length, depth and skin roughness. Another study applied both vehicle and palmitoyl tripeptide-1 to the skin of 23

healthy female volunteers for four weeks, documenting a small but statistically significant increase in skin thickness (~4%, compared to the vehicle alone).

A combination of pal-GHK tripeptide and pal-GQPR tetrapeptide is marketed as an anti-wrinkle compound with the trade name Matrixyl™3000. A blind, randomized clinical study with 28 volunteers twice daily applying cream including the active compound to half their face and one of their forearms and a placebo cream to the other half of the face and other forearm confirmed anti-wrinkle efficacy, reduction of wrinkle depth, volume and density, skin roughness and complexity, as well as a decrease of the area occupied by deep wrinkles, and an increase in skin tone.

(5) Palmitoyl tripeptide-3/5 Palmitoyl tripeptide-3/5 mimics the effects of an extracellular matrix protein, thrombospondin-1 (TSP-1), a naturally occurring molecule that increases TGFβ activity. In animal models and human dermal fibroblasts cell culture tests, TSP-1 acts locally to improve wound healing, and is believed to be active in the post-natal development of skin structures. The short sequence Lys-Arg-Phe-Lys of the TSP-1 protein is responsible for TGFβ stimulation. TGFβ in turn, causes a persistent increase in the amounts of Type Ⅰ and Type Ⅲ collagen that dermal fibroblasts produce. Palmitoyl tripeptide-3/5 (Sequence: Pal-Lys-Val-Lys bistrifluoracetae salt) (SYN®-COLL) stimulates collagen production in *in vitro* and in vivo studies through the growth factor TGFβ.

Animal studies indicate that palmitoyl tripeptide-3/5 may increase collagen synthesis. *In vitro* studies show that palmitoyl tripeptide-3/5 can prevent collagen breakdown by interfering with MMP1 and MMP3 collagen degradation. Palmitoyl tripeptide-3/5 seems to boost collagen synthesis, but decreases collagen breakdown. Additional data show that palmitoyl tripeptide-3/5 is roughly 3.5 times more effective at reducing the appearance of wrinkles than the placebo. *In vivo* results of a palmitoyl tripeptide-3/5 (10×10^{-6} to 25×10^{-6}) cream formulation demonstrated a dose-dependent wrinkle reduction, measured by PRIMOS (phaseshift rapid *in vivo* measurement of skin) surface topography.

In an efficacy study performed on 60 Chinese volunteers (84 days, applied twice daily), palmitoyl tripeptide-3/5 confirmed its anti-wrinkle efficacy and reduced skin roughness better than control groups, placebo and pal-KTTKS-containing creams.

(6) Palmitoyl tripeptide-38 Epithelial regeneration in skin is achieved by the constant turnover and differentiation of keratinocytes. Epidermal and dermal stem cell compartments are fundamental for the continuous renewal of the skin. Adult stem cells are the unique source for skin tissue renewal. Plants also have stem cells and plant-derived stem cell extracts are now used in topical products for their potential anti-aging and anti-wrinkle effects.

A dermocosmetic product containing apple stem cell extract, urea, creatine and palmitoyl tripeptide-38 [Sequence: Pal-Lys-Met(O_2)-Lys-OH] was applied on the face twice daily for 28 days, and assessed by clinical and instrumental evaluation in 32 women with sensitive skin bearing crow's feet wrinkles. The treatment results showed a significant increase in dermal density and elasticity, as well as anti-wrinkle effects. The anti-aging serum seems to improve aging skin signs with the first visible results achieved after one week treatment.

To determine the effectiveness of a multi-ingredient anti-aging moisturizer, an open label clinical trial was conducted with 37 female subjects of ages 35-60. The effective ingredients of the moisturizer for the facial skin included *Astragalus membranaceus* root extract, a peptide blend including palmitoyl tripeptide-38, standardized rosemary leaf extract (ursolic acid), tetrahexyldecyl ascorbate, and ubiquinone. Results were favorable in both product efficacy measurements and aesthetic self-assessment questionnaires, with subjects judging the product as being mild and well-tolerated.

(7) Palmitoyl pentapeptide-4 Palmitoyl pentapeptide-4, (Matrixyl®) (Sequence: Pal-Lys-Thr-Thr-Lys-Ser-OH or pal-KTTKS-OH) is a small, highly specific biologically active peptide which has been reported to stimulate the production of elastin, fibronectin, glucosaminoglycan and collagens (specifically Types Ⅰ, Ⅲ and Ⅳ), support of the extracellular matrix, and wound healing. KTTKS structure is related to the precursor of collagen Type Ⅰ (or procollagen Type Ⅰ). The stimulatory effect of KTTKS on collagen Types Ⅰ and Ⅲ, and fibronectin, seems to relate mainly to the biosynthetic pathway, rather than the export or degradation pathways. KTTKS has a molecular weight of 563.64 Da, the longer pal-KTTKS is 802.05 Da.

In a placebo-controlled double blind study, pal-KTTKS (0.005%) formulation was applied to the right periocular area twice daily for 28 days. As demonstrated by optical profilometry, this resulted in a quantitative decrease in fold depth, fold thickness, and skin rigidity, by 18%, 37%, and 21%, respectively. These results were confirmed in two other placebo-controlled double blind studies in women (42 and 35 subjects) with moderate to distinct periorbital wrinkles.

A double-blind, placebo-controlled, split face, left-right randomized trial involving 93 subjects was carried out to assess the clinical efficacy of pal-KTTKS, with fine line or wrinkle improvement as the parameter of interest. In another four month-long double-blind study, 49 women were directed to apply either pal-KTTKS or vehicle twice daily to their faces. The results showed that pal-KTTKS exhibited significant improvement in skin roughness, wrinkle volume, and wrinkle depth, compared with the vehicle. Data associated pal-KTTKS with an increase in elastin fiber density and thickness, as well as improved collagen Ⅳ regulation at the dermal-epidermal junction.

(8) Palmitoyl tetrapeptide-7 Palmitoyl tetrapeptide-7 (Rigin™) (Sequence: Pal-Gly-Gln-Pro-Arg or pal-GQPR) is a fragment of immunoglobulin G. Palmitoyl tetrapeptide-7 decreases IL-6 secretion in a basal setting, and serves as an anti-inflammatory after exposure to UVB-irradiation. *In vivo* reflectance confocal microscopy studies indicated that a blend of palmitoyl oligopeptide and palmitoyl tetrapeptide-7 enhanced the extracellular matrix structure compared to placebo. Sixty healthy photoaged volunteers (aged 45-80) were tested over 12 months with a formulation containing palmitoyl tetrapeptide-7 and another active ingredient. A reduction of facial wrinkles was documented by this long-term use. Better skin appearance was related to the deposition of fibrillin-rich microfibrils in the papillary dermis of treated skin.

(9) Palmitoyl hexapeptide-12 Palmitoyl hexapeptide-12 (Sequence: Pal-Val-Gly-Val-Ala-Pro-Gly), which is the peptide in biopeptide-EL, creates a response in the dermis of the skin that stimulates collagen and elastin fibroblasts, developing fibronectin and glycosaminoglycans.

It is believed to work by reducing the production of interleukin-6 (IL-6) by key skin cells, keratinocytes and fibroblasts. IL-6 is a molecule that promotes inflammation, which, in turn, leads to faster degradation of the skin matrix, and thus contributes to the development of wrinkles, and loss of skin firmness and elasticity. By reducing the levels of IL-6 and possibly other inflammation mediators, palmitoyl is thought to slow down the degradation of the skin matrix, and may also stimulate its replenishment.

A one-month double-blind study was conducted on 10 female volunteers, aged 32-56, who performed twice daily applications of a light emulsion containing 4%, or a placebo. Palmitoyl hexapeptide treatment improved elasticity, tone, skin fatigue and firmness.

(10) Acetyl tetrapeptide-9/11 Acetyl tetrapeptide-9 (Dermican™) (Sequence: N-Acetyl-Gln-Asp-Val-His) is reported to stimulate collagen Type Ⅰ and lumican synthesis, whereas acetyl tetrapeptide-11 (Sequence: N-Acetyl-Pro-Pro-Tyr-Leu) (Syniorage™) stimulates keratinocyte cell growth and syndecan-1 synthesis. Clinical studies documented that treatment with acetyl tetrapeptide-9 (17 female volunteers) or acetyl tetrapeptide-11 (19 female volunteers) led to thicker and firmer skin. Both peptides were more effective compared to placebos.

(11) Tetrapeptide-21 Tetrapeptide-21, also named GEKG (Sequence: Gly-Glu-Lys-Gly), was derived from ECM (extracellular matrix) proteins. Its amino acid sequence is glycine-glutamic acid-lysine-glycine. This peptide demonstrated in *in vitro* studies, an increase of collagen (Type Ⅰ) production on the protein level and mRNA level, hyaluronic acid synthase 1 production, and a strong increase in fibronectin (GEKG Conc. 0.001%). Increase in collagen (COL1A1), procollagen, hyaluronic acid and fibronectin, as well as skin elasticity, were measured in a double-blind, randomized, placebo-controlled study (10 women/eight weeks). A placebo-controlled study with 30 subjects was carried out to analyze the effect of GEKG on facial wrinkles. GEKG significantly decreased skin roughness.

In a comparison study with GEKG *vs.* pal-KTTKS with 60 subjects, elasticity increased after two daily treatments over eight weeks by 41.3% for GEKG, while pal-KTTKS showed an improvement of 35.6%.

(12) Tetrapeptide PKEK Tetrapeptide PKEK (Sequence: Pro-Lys-Glu-Lys) can exert skin whitening effects. For PKEK development, Lys-Glu-Lys (KEK) was modified with proline to stabilize the peptide structure. Study results showed that PKEK reduces interleukin-6, interleukin-8, and tumor necrosis factor-α, as well as cyclooxygenase gene expression in UV light-stressed keratinocytes.

The treatment of human keratinocytes with PKEK significantly reduced UVB-stimulated mRNA expression of interleukin IL-6, IL-8 and TNF-α and, most importantly, proopiomelanocorticotropin (POMC). In a randomized, double-blind, vehicle-controlled study with PKEK treatment once daily, punch biopsies of 10 healthy volunteers were taken after four weeks. PKEK treatment significantly inhibited UVB-induced upregulation of genes encoding for IL-1α, IL-6, IL-8, TNF-α as well as POMC and tyrosinase in skin areas pretreated with PKEK.

A second study was performed as a half-face *in vivo* efficacy test with 39 Caucasian women. Facial pigment spots were significantly faded after six weeks when PKEK was combined with the skin whitener sodium ascorbyl phosphate. PKEK or sodium ascorbyl phosphate alone led to less pronounced fading of the pigment spots than a combination of the two. An *in vivo* hand cream efficacy study confirmed the above result after application for eight weeks to the back of the hands of 19 Caucasians.

In the fourth study, 27 Japanese women were separated in two groups which treated faces twice daily with either sodium ascorbyl phosphate only, or with a PKEK + sodium ascorbyl phosphate formulation, for eight weeks. Application of PKEK + sodium ascorbyl phosphate significantly reduced skin pigmentation by 26%, and sodium ascorbyl phosphate by 18%, according to the SCINEXA score. It was confirmed that PKEK has the capacity to reduce UVB-induced skin pigmentation in all study models, and it may be suitable as a skin tone-modulating agent in cosmetic products.

(13) Hexapeptide-11 Hexapeptide-11 (Pentamide-6, Sequence: Phe-Val-Ala-Pro-Phe-Pro) was originally isolated from *Saccharomyces yeast* fermentation, but was later synthesized because of purity issues. Hexapeptide-11 has the ability to influence the onset of senescence in intrinsically and extrinsically aged fibroblasts, and extrinsically aged dermal papillae cells *in vitro*. Gorouhi and Maibach reviewed a placebo-controlled study with 25 healthy volunteers, treating their skin twice daily for four weeks. Initial skin elasticity and deformation response were improved.

(14) Hexapeptide-14 Hexapeptide-14 (palmitoyl hexapeptide-14) has been reported to stimulate cell migration, collagen synthesis, and fibroblast proliferation and scaffolding. A 12-week study with 29 volunteers showed results of reduced fine lines and wrinkles by palmitoyl hexapeptide-14. Results were compared with tretinoin and no irritations were noted.

18.2.3 Carrier peptides

Carrier peptides deliver or stabilize trace elements like copper and manganese, necessary for wound healing and enzymatic progress. These peptides are involved in copper or manganese transport into skin cells. Additionally, they are obtained by binding copper with a tripeptide.

(1) Copper tripeptide Copper tripeptide (Cu-GHK, lamin®) complex (Sequence: Copper Gly-l-His-l-Lys) is one of the most well-examined peptides. It plays a role in the extracellular matrix, and is released in wounds or inflammation to support healing. It acts as signal and carrier peptide, promotes regular collagen, elastin, proteoglycan, and glycosaminoglycan synthesis, and provides anti-inflammatory and antioxidant responses. In cosmetic applications, Cu-GHK is used in anti-aging, anti-wrinkle, after-sun, skin renewal, skin moisturizer, hair growth stimulating products.

Cu-GHK stimulates cellular regulatory molecules and regenerates, and heals skin and other tissues. Stem cells treated with GHK regenerated and expressed more stem cell markers. GHK and Gly-Gly-His (GGH) reduce TNF-α induced cytokines IL-6, thus ensuring better wound healing. Pickart *et al.* described that GHK significantly increased the expression of DNA repair genes, while 47 genes are stimulated and five genes are suppressed. GHK is involved with different mechanisms of action and can apparently promote regeneration, healing, and repair. It furthermore achieves good effects against the aging processes.

Cu-GHK can stimulate hair growth. The copper tripeptide complex ensures follicular enlargement and helps covering the follicle with a downy hair; the effect is comparable with that of Minoxidil. The results of a hair transplant showed significant improvement following application with a copper tripeptide product. Topical Cu-GHK products stimulate collagen synthesis on the scalp, strengthen existing hair, and encourage hair growth.

Several studies confirming Cu-GHK efficacy in various areas have been performed. These include: increasing keratinocyte proliferation, improving appearance, firmness, elasticity, skin thickness, wrinkles, spotty hyperpigmentation and light damage, skin collagen, strengthening proteins of skin protection barrier, and improvement of skin appearance.

Experiments have shown that *in vitro* Cu-GHK increases and stimulates the synthesis of collagen, glycosaminoglycans and other extracellular matrix molecules. Several placebo-controlled clinical trials have confirmed the observed effects. A topically applied cream with Cu-GHK was shown to stimulate dermal skin procollagen synthesis. Synthesis induced by the copper tripeptide was significantly superior to vitamin C, tretinoin, or melatonin in comparison.

Similarly, a study of 20 women compared the skin's production of collagen after applying creams containing Cu-GHK, vitamin C, or retinoic acid to thighs daily for one month. New collagen production was determined by skin biopsy samples using immunohistological techniques. After one month, Cu-GHK increased collagen in 70% of those treated, *versus* 50% treated with vitamin C, and 40% treated with retinoic acid.

Leyden *et al.* confirmed in two different studies (12 weeks of application by 71 or 41 women) the clinically beneficial effects of Cu-GHK formulations on both aged and sun damaged skin. The Cu-GHK face cream and an eye cream reduced the visible signs of skin aging and caused an increase in skin density and thickness. The researchers observed improved skin elasticity and skin humidity, significant smoothing of the skin by stimulating the synthesis of collagen, a significant improvement of the skin contrast, and diminished wrinkles.

In another study Cu-GHK cream was applied twice a day for 12 weeks on 67 women. The Cu-GHK cream improved the appearance of photo-damaged, aged skin. By histological analyses of biopsies, it was re-confirmed that the use of topically applied Cu-GHK products intensified skin thickness in the range of the epidermis and dermis, and that keratinocyte proliferation of the skin was greatly stimulated.

(2) Manganese tripeptide-1 Besides the well-researched copper tripeptide, there is also one clinical study focusing on manganese tripeptide-1 (Sequence: GHK-Mn^{2+}). This study evaluated the effects of a manganese peptide complex in the treatment of various signs of cutaneous facial photo-damage. During a 12-week period, subjects were instructed to apply a facial serum formulation containing the manganese peptide complex twice a day. This resulted in their skin's photo-damage ranking shifting from moderate to mild. The most significantly improved parameters were associated with hyperpigmentation, while no significant cutaneous

inflammation was reported.

18.2.4 Neurotransmitter inhibitor peptides

Muscle contraction is another strategy to reduce common ageing signs like fine lines and wrinkles. Muscles are contracted by neurotransmitter release from neurons. The neurotransmitter begins a cascade with protein-protein interactions that culminate in the fusion of neurotransmitter loaded vesicles with the neuron membrane.

Topical synthetic peptides which imitate the amino acid sequence of the synaptic protein SNAP-25 were shown to be specific inhibitors of the neurosecretion. For this reason, these peptides are also called neurotransmitter inhibitor peptides. These cosmeceutical peptides penetrate skin and relax muscles, causing the reduction and softening of wrinkles and fine lines.

(1) Acetyl hexapeptide-3 Acetyl hexapeptide-3 (Argireline®) has the sequence Acetyl-Glu-Glu-Met-Gln-Arg-Arg-NH_2, and is reported to inhibit the release of neurotransmitters, which is followed by anti-wrinkle, moisturizing effects. It seems to improve the firmness and tone of the skin. Acetyl hexapeptide-3 is a copy of the synaptosomal-associated protein 25, which competes for a position in the SNARE complex, and destabilizes its formation without breaking any of its constituent parts. Additionally, it inhibits catecholamine secretion. An increase in peptide concentration and the permeation of peptides, and a decrease in iontophoretic permeability coefficients, are affected by a number of parameters that can be optimized for effective transdermal peptide delivery.

Blanes-Mira described a placebo-controlled study in which acetyl hexapeptide-3 (10%) *vs.* placebo creams were applied twice daily (10 women, 30 days). Acetyl hexapeptide-3-treated skin areas showed a 30% improvement in wrinkles in the eye area. The anti-wrinkle efficacy in the acetyl hexapeptide-3 group was measured as 48.9%, compared with 0% in the placebo group. All parameters of roughness were decreased in the same study. The fold depth in the acetyl hexapeptide-3 group (48.9%) was significantly reduced (60 subjects).

(2) Pentapeptide-3 Pentapeptide-3 (Vialox®) (Sequence: Gly-Pro-Arg-Pro-Ala) is derived from snake venom. It is an antagonist of the acetylcholine receptor, and blocks nerves at the post-synaptic membrane, leading to muscle relaxation. Clinical studies demonstrated a reduction of wrinkles by 49%, and lesser skin roughness (47%) after a treatment of 28 days.

(3) Pentapeptide-18 Pentapeptide-18 (Leuphasyl®) (Sequence: Tyr-d-Ala-Gly-Phe-Leu) mimics the natural mechanism of enkephalins, and inhibits neuronal activity and catecholamine release. Its action can be described as having botox-like effects; and it demonstrates a proven efficacy for reducing fine lines and wrinkles, moisturizing the skin, and improving firmness and skin tone.

A cream with pentapeptide-18 (0.05%) was compared in a study (43 women) with acetyl hexapeptide-3 (0.05%) and the combination of both peptides. Wrinkle reductions were measured for pentapeptide-18, acetyl hexapeptide-3, and a combination, at 11.64%, 16.26%, and 24.62%, respectively. The pentapeptide-18 study mentioned a synergistic effect or an increase of efficiency, by its association with acetyl hexapeptide-3. The cellular pathway of both peptides is different.

Three concentrations (0.5%, 1%, and 2%) of Leuphasyl formulations were applied on 20 volunteers for two months, at the level of mimic muscles in the eyebrows zone and in the periorbital zone. To further evaluate anti-wrinkle effects, an *in vivo* study with 22 females with an average age of 51, was designed. Volunteers were injected botulinum toxin in the periorbital region combined with a topical treatment twice daily for six months. In comparison with the control, acetyl hexapeptide-3 and pentapeptide-18 induced a 20% and 11% decrease on glutamate release, respectively, while their combination caused a 40% reduction. In addition to the measured effects of each individual compound, this indicates a synergistic activity. The topical treatment demonstrated a potentiated effect for botulinum toxin treatment, and extended the anti-wrinkle benefit on the skin.

(4) Tripeptide-3 Tripeptide-3 (Sequence: β-Ala-Pro-Dab-NHBn-2-Acetate), also named dipeptide diamino-butyroyl benzylamide diacetate or SYN®-AKE, mimics the effect of waglerin-1, a peptide that is found in the venom of the viper *Tropidolaemus wagleri*. Tripeptide-3 acts at the postsynaptic membrane, and is a reversible antagonist of the acetylcholine receptor.

Tripeptide-3 has been tested in different preparations for topical application in animal models in concentrations ranging from 1% to 4%. The results have been claimed as smoothing the appearance of mimic wrinkles and expression lines, shortly after applying the preparation. Research results show that tripeptide-3 can reduce the appearance of wrinkles after 28 days by up to 52% when a 4% topical solution is used. The results from a three-month study (37 female volunteers) demonstrated that tripeptide-3 formulation treatment provided both immediate and long-term improvements in the appearance of fine and coarse wrinkles.

18.2.5 Enzyme inhibitor peptides

Enzyme inhibitor peptides directly or indirectly inhibit enzymes. Soy oligopeptides, silk fibroin peptide and rice peptides act on the skin cells. The above-mentioned peptides inhibit enzymes like tTAT-superoxide dismutase, stimulate hyaluronan synthase 2, or in the case of soy oligopeptides, inhibit proteinases.

This category of peptides shows promising results, however only very few or no *in vivo* studies have been conducted. Hence, their relevance is still unclear.

(1) Soybean peptides Soy oligopeptides are obtained from soybean proteins, consisting of 3-6 amino acids, mainly in the size range of 300-700 kDa. Various biological activities of soybean oligopeptides have been identified, such as antioxidant, blood pressure lowering, and blood lipid lowering effects. Topically applied soy oligopeptides data showed significantly increased Bcl-2 protein expression and decreased cyclobutane pyrimidine dimers-positive cells, sunburn cells, apoptotic cells, p53 protein expression, and Bax protein expressions in the epidermis of UVB-irradiated foreskin. Topically used soy oligopeptides seem to protect human

skin (nine healthy male volunteers) against UVB-induced photo damage.

A pseudo-randomized study (10 women, *in vivo/in vitro*) with soybean peptide showed a significant increase in glycosaminoglycan and collagen synthesis. This study showed the anti-aging potential of a soy peptide.

(2) Silk fibroin peptide Silk fibroin peptide is derived from the silkworm *Bombyx mori*. Scientific results show inhibitory inflammation and enhance the anti-inflammatory activity of tTAT-superoxide dismutase, which was previously reported to effectively penetrate various cells and tissues, and exert anti-oxidative activity in a mouse model of inflammation. No human *in vivo* efficacy data are currently available.

(3) Rice peptides After a specially processing rice bran protein, low molecular weight peptides (<3000 Da) were obtained. Black rice oligopeptides were measured at approximately 1300 Da. These oligopeptides were noted for inhibiting MMP (matrix metalloproteinase) activity and stimulated hyaluronan synthase 2 gene expression in human keratinocytes in a dose-dependent manner. Three new identified rice bran protein peptides had a C-terminal tyrosine residue, and exhibited significant inhibitory effects against tyrosinase-mediated monophenolase reactions. Additionally, one peptide called CT-2 (Leu-Gln-Pro-Ser-His-Tyr) potently inhibited melanogenesis in mouse melanoma cells without causing cytotoxicity, which might be of interest for melanin-related skin conditions.

Rice bran protein is a potent source of tyrosinase inhibitory peptides. Formulations containing niosomes entrapped with rice bran bioactive compounds show clinical anti-aging properties. However, there is currently no *in vivo* data for purified black rice oligopeptides available.

18.2.6 Peptides derived from structural protein digestion Keratin-based peptides

Topical formulations containing keratin-based peptides are described to have hemostatic, moisturizing, repair-promoting and potentially radio-protective properties. A placebo-controlled *in vivo* study (nine healthy females with dry skin types III to V) tested keratin-based peptide (molecular weight <1000 Da) (3%) *vs.* deionized water (3%) in a topical hand cream application which was compared to untreated areas. Biophysical results of keratin-based peptide treatment indicated skin improvement in water-holding capacity, hydration and elasticity. Results also indicated that the keratin-based peptide cream treatment can prevent some of the damaging effects associated with surfactant exposure. Additional study results (six and 16 healthy females) with keratin-based peptide (3%) tested two different formulations (aqueous solution/internal wool lipid liposome suspension). It was shown that topically applied keratin-based peptide formulations again improved the integrity and water-holding capacity of the skin barrier. Combining the keratin-based peptide with internal wool lipids led to additional beneficial effects. The participant numbers were very low and future studies should be carried out with more volunteers. No detailed information on the mode of action of these peptides was available.

References

[1] Temova Rakuša Ž, Roškar R. Quality Control of Vitamins A and E and Coenzyme Q10 in Commercial Anti-ageing Cosmetic Products. Cosmetics, 2021, 8: 61.
[2] Schagen, S.K. Topical Peptide Treatments with Effective Anti-aging Results. Cosmetics, 2017, 4: 16.
[3] Gorouhi F, Maibach H. Role of Topical Peptides in Preventing or Treating Aged Skin. International Journal of Cosmetic Science, 2009, 31: 327.

Key Words & Phrases

coenzyme Q10　辅酶Q10
lipid peroxidation　膜脂质过氧化；脂质过氧化作用
retinoid　*n.* 类视黄醇
retinyl palmitate　棕榈酸视黄酯；棕榈酰视黄酯
teratogenicity　*n.* 致畸性；致畸
endogenous　*adj.* 内源；内生的
ubiquinone　*n.* 泛醌；辅酶Q
prerequisite　*n.* 先决条件；前提；必备条件
palmitoyl pentapeptide-4　棕榈酰五肽-4
innate immunity　先天免疫
inflammation　*n.* 发炎；炎症
collagen stimulation　胶原蛋白刺激
signal peptide　信号肽
carrier peptide　载体肽
neurotransmitter inhibitor peptide　神经递质抑制肽
enzyme inhibitor peptide　酶抑制剂肽
proteoglycan　*n.* 蛋白聚糖；蛋白多糖
fibronectin　*n.* 纤连蛋白
carnosine　*n.* 肌肽
double-blind　双盲的
randomize　*v.* 使随机化；（使）作任意排列
split-face　左右面
periorbital wrinkle　鱼尾纹
niacinamide　*n.* 尼克酰胺；烟酰胺
photolyase　*n.* 光裂合酶；光修复酶
endonuclease　*n.* 核酸内切酶；内切核酸酶
8-oxoguanine glycosylase　8-氧桥鸟嘌呤糖基化酶
arazine　*n.* *N*-乙酰基-*S*-法尼基-L-半胱氨酸
ergothionine　*n.* 麦硫因
protein carbonylation　蛋白质羰基化
8-oxo-7,8-dihydro-2’-deoxyguanosine　8-氧代-7,8-二氢-2’-脱氧鸟苷
biopsies　*n.* 活组织检查
jawline　*n.* 下颌的轮廓；下巴的外形
fringe projection profilometry　条纹投影轮廓术
tripeptide-10　三肽-10
citrulline　*n.* 瓜氨酸
decorin-like　类饰胶蛋白聚糖
skin roughness　皮肤粗糙度
skin thickness　蒙皮厚度
thrombospondin-1　血小板应答蛋白-1
fibroblast　*n.* 成纤维细胞
post-natal　产后的；分娩后的
placebo　*n.* 安慰剂
epithelial　*adj.* 上皮的；皮膜的
creatine　*n.* 肌酸
dermal density　真皮密度
serum　*n.* 血清；乳清
Astragalus　*n.* 距骨；紫云英属
ursolic acid　乌苏酸；熊果酸
tetrahexyldecyl ascorbate　四己基癸醇抗坏血酸酯
profilometry　*n.* 轮廓仪；轮廓分析；轮廓测定法；轮廓形貌测量；轮廓术
reflectance confocal microscopy　反射共焦显微术
fibrillin-rich microfibril　富含原纤维蛋白的微纤维
skin firmness　皮肤紧致度
lumican　*n.* 光蛋白聚糖
syndecan-1　黏结蛋白聚糖-1
hyaluronic acid synthase 1　透明质酸合酶1
proline　*n.* 脯氨酸
interleukin-6　白细胞介素-6
tumor necrosis factor-α　肿瘤坏死因子-α
cyclooxygenase　*n.* 环加氧酶；环氧合酶
punch biopsies　穿孔活检
Caucasian　*n.* 高加索人
sodium ascorbyl phosphate　抗坏血酸磷酸酯钠；维生素C磷酸酯钠盐
Saccharomyces　*n.* 酵母属；酵母菌属；酵母菌
fermentation　*n.* 发酵
cell migration　细胞迁移
follicular　*adj.* 滤泡的；卵泡的；小囊的
minoxidil　*n.* 米诺地尔
immunohistological　*adj.* 免疫组织的
neurotransmitter　*n.* 神经递质
catecholamine secretion　儿茶酚胺分泌
permeation　*n.* 渗透；通透

iontophoretic permeability coefficient 离子导入渗透系数
transdermal *n.* 经皮给药的；透过皮肤吸收的
snake venom 蛇毒；蛇毒蛋白；蛇毒素
acetylcholine *n.* 乙酰胆碱
enkephalin *n.* 脑啡肽
leuphasyl *n.* 五肽
botulinum toxin 肉毒杆菌毒素；肉毒毒素
venom *n.* 毒液
viper tropidolaemus wagleri 瓦氏毒蛇
soy oligopeptide 大豆低聚肽
silk fibroin peptide 丝素肽
rice peptide 大米肽
hyaluronan synthase 2 透明质酸合成酶 2
cyclobutane pyrimidine 环丁烷嘧啶
p53 protein p53 蛋白
silkworm *Bombyx mori* 家蚕
rice bran protein 米糠蛋白
terminal tyrosine 末端酪氨酸
niosome *n.* 泡囊；非离子表面活性剂囊泡
keratin-based peptide 角蛋白基肽
hemostatic *adj.* 止血的

Chapter 19 Preservatives and Antioxidants

19.1 Preservatives

A preservative is a synthetic/chemical or natural ingredient with antimicrobial properties that is added to cosmetics and personal care products to maintain microbiological quality. Preservatives are used to increase the shelf life of these formulations, inhibiting the growth of microorganisms. Thus, they prevent the deterioration caused by bacteria (Gram-positive and Gram-negative), moulds and yeasts that can cause disease or simply disrupt the smooth appearance of the final product. A good preservative should provide broad-spectrum activity, *i.e.*, eliminate all types of microorganisms.

The fundamental property of any preservative is that it should be effective at low concentrations and not toxic for humans. The "ideal" preservative for use in cosmetics should be soluble in water and stable at any temperature and pH conditions that are used in the manufacturing process. The preservatives should be also colorless and odorless, or should not add color or odor to the product, and should not react with other ingredients. Combinations of preservatives (blends) could be a good solution to reduce the possible side effects associated with the chemical preservation method.

Although there are a great number of preservatives, only few of them are permitted to be used in cosmetics for the reasons of safety. In China, according to Safety and Technical Standards for Cosmetics in 2015, only 51 preservatives are allowed to be used in cosmetics. The following are some brief introductions on some of them.

(1) Bronopol Bronopol (INN; chemical name 2-bromo-2-nitro-1,3-propanediol) is an organic compound that is used as an antimicrobial. It is a white solid although commercial samples appear yellow.

The use of bronopol in personal care products (cosmetics, toiletries) has declined since the late 1980s due to the potential formation of nitrosamines. While bronopol is not in itself a nitrosating agent, under conditions where it decomposes (alkaline solution and/or elevated temperatures) it can liberate nitrite and low levels of formaldehyde and these decomposition products can react with any contaminant secondary amines or amides in a personal care formulation to produce significant levels of nitrosamines.

Manufacturers of personal care products are therefore instructed by regulatory authorities to avoid the formation of nitrosamines which might mean removing amines or amides from the formulation, removing bronopol from a formulation, or using nitrosamine inhibitors. Bronopol has been restricted for use in cosmetics in Canada.

(2) 5-Bromo-5-nitro-1,3-dioxane 5-Bromo-5-nitro-1,3-dioxane is an antimicrobial

chemical compound which was used in cosmetics since the mid-1970s as preservative for shampoos, foam bath, etc. Maximum concentration is 0.1%.

(3) Benzalkonium chloride Benzalkonium chloride, also known as alkyldimethylbenzylammonium chloride (ADBAC) and by the trade name Zephiran, is a type of cationic surfactant. It is an organic salt classified as a quaternary ammonium compound. ADBACs have three main categories of use: as a biocide, a cationic surfactant, and a phase transfer agent. ADBACs are a mixture of alkylbenzyldimethylammonium chlorides, in which the alkyl group has various even-numbered alkyl chain lengths. Especially for its antimicrobial activity, benzalkonium chloride is an active ingredient in many consumer products: personal care products such as hand sanitizers, wet wipes, shampoos, soaps, deodorants and cosmetics.

(4) Benzethonium chloride Benzethonium chloride, also known as hyamine is a synthetic quaternary ammonium salt. This compound is an odorless white solid, soluble in water. It has surfactant, antiseptic, and anti-infective properties, and it is used as a topical antimicrobial agent in first aid antiseptics. It is also found in cosmetics and toiletries such as soap.

(5) Benzoic acid, its salts and esters Benzoic acid is a white (or colorless) solid with the formula $C_6H_5CO_2H$. It is the simplest aromatic carboxylic acid. The name is derived from gum benzoin, which was for a long time its only source. Benzoic acid occurs naturally in many plants and serves as an intermediate in the biosynthesis of many secondary metabolites. Benzoic acid and its salts can be used as cosmetic preservatives.

(6) Benzylhemiformal Benzylhemiformal is a formaldehyde releaser. A formaldehyde releaser, formaldehyde donor or formaldehyde-releasing preservative is a chemical compound that slowly releases formaldehyde. Formaldehyde releasers are added to prevent microbial growth and extend shelf life. They are found in cosmetics, toiletries, cleaning agents, adhesives, paints, lacquers and metalworking fluids. Formaldehyde releasers are often used as an antimicrobial preservative in cosmetics.

(7) Chlorhexidine (INN) and its digluconate, diacetate and dihydrochloride Chlorhexidine (CHX) [commonly known by the salt forms chlorhexidine gluconate and chlorhexidine digluconate (CHG) or chlorhexidine acetate] is a disinfectant and antiseptic that is used for skin disinfection before surgery and to sterilize surgical instruments. It may be used both to disinfect the skin of the patient and the hands of the healthcare providers. Chlorhexidine is used in disinfectants (disinfection of the skin and hands), cosmetics (additive to creams, toothpaste, deodorants, and antiperspirants), and pharmaceutical products (preservative in eye drops, active substance in wound dressings and antiseptic mouthwashes). CHG is active against Gram-positive and Gram-negative organisms, facultative anaerobes, aerobes, and yeasts. It is particularly effective against Gram-positive bacteria (in concentrations $\geqslant$ 1μg/L). Significantly higher concentrations (10 to more than 73μg/mL) are required for Gram-negative bacteria and fungi. Chlorhexidine is ineffective against polioviruses and adenoviruses. The effectiveness against herpes viruses has not yet been established unequivocally.

(8) Chlorobutanol (INN) Chlorobutanol (trichloro-2-methyl-2-propanol) is a preservative, sedative, hypnotic and weak local anesthetic similar in nature to chloral hydrate. It has

antibacterial and antifungal properties. Chlorobutanol is typically used at a concentration of 0.5% where it lends long term stability to multi-ingredient formulations.

(9) 1-(4-chlorophenoxy)-1-(imidazol-1-yl)-3,3-dimethylbutan-2-one Climbazole is a topical antifungal agent commonly used in the treatment of human fungal skin infections such as dandruff and eczema. It is most commonly found as an active ingredient in OTC (over the counter) anti-dandruff and anti-fungal products, including shampoos, lotions and conditioners. It may be accompanied by other active ingredients such as zinc pyrithione or triclosan.

(10) 1,3-Bis(hydroxymethyl)-5,5-dimethylimidazolidine-2,4-dione (DMDM) DMDM hydantoin is an antimicrobial formaldehyde releaser preservative with the trade name Glydant. DMDM hydantoin is an organic compound belonging to a class of compounds known as hydantoins. It is used in the cosmetics industry and found in products like shampoos, hair conditioners, hair gels, and skin care products. DMDM hydantoin slowly releases formaldehyde and works as a preservative by making the environment less favorable to microorganisms.

(11) 4-Hydroxybenzoic acid and its salts and esters 4-Hydroxybenzoic acid, also known as *p*-hydroxybenzoic acid (PHBA), is a monohydroxybenzoic acid, a phenolic derivative of benzoic acid. It is a white crystalline solid that is slightly soluble in water and chloroform but more soluble in polar organic solvents such as alcohols and acetone. 4-Hydroxybenzoic acid is primarily known as the basis for the preparation of its esters, known as parabens. Parabens are a class of widely used preservatives in cosmetic and pharmaceutical products. Chemically, they are a series of parahydroxybenzoates or esters of parahydroxybenzoic acid. Parabens are effective preservatives in many types of formulas. These compounds, and their salts, are used primarily for their bactericidal and fungicidal properties. They are found in shampoos, commercial moisturizers, shaving gels, personal lubricants, topical/parenteral pharmaceuticals, suntan products, makeup, and toothpaste.

(12) 4-Chloro-*m*-cresol 4-Chloro-*m*-cresol is a monochlorinated *m*-cresol. It is a white or colorless solid that is only slightly soluble in water. As a solution in alcohol and in combination with other phenols, it is used as an antiseptic and preservative.

(13) 2-Phenoxyethanol Phenoxyethanol is the organic compound with the formula $C_6H_5OC_2H_4OH$. It is a colorless oily liquid. It can be classified as a glycol ether and a phenol ether. It is a common preservative in vaccine formulations. Phenoxyethanol is used as a perfume fixative; an insect repellent; an antiseptic; a solvent for cellulose acetate, dyes, inks, and resins; a preservative for pharmaceuticals, cosmetics and lubricants; an anesthetic in fish aquaculture; and in organic synthesis. Phenoxyethanol is an alternative to formaldehyde-releasing preservatives.

(14) Pyrithione zinc (INN) Zinc pyrithione (or pyrithione zinc) is a coordination complex of zinc. It has fungistatic (that is, it inhibits the division of fungal cells) and bacteriostatic (inhibits bacterial cell division) properties and is used in the treatment of seborrhoeic dermatitis. Zinc pyrithione can be used to treat dandruff and seborrhoeic dermatitis. It also has antibacterial properties and is effective against many pathogens from the Streptococcus and Staphylococcus genera.

19.2 Antioxidants

Antioxidants are molecules that can oxidize themselves before or instead of other molecules. They are compounds or systems that can interact with free radicals and stop a chain reaction before vital molecules are harmed. Antioxidants are used in food, cosmetics, beverages, pharmaceuticals, and even the feed industry. They can be used as health supplements and active ingredients as well as stabilizers. Antioxidants can be synthetic or natural, and both are used in cosmetic products. Synthetic antioxidants [*e.g.*, butylated hydroxyanisole (BHA), butylated hydroxytoluene (BHT), and propyl gallate] are widely used because they are inexpensive to produce. However, research suggests that excessive consumption of synthetic antioxidants may pose health risks. Despite the fact that synthetic antioxidants dominate the market, demand for natural antioxidants has increased in recent years and is expected to continue. This pattern can be explained by a growing consumer preference for organic and natural products that contain fewer additives and may have fewer side effects than synthetic ingredients.

Natural antioxidants used in the cosmetic industry include various substances and extracts derived from a wide range of plants, grains, and fruits, and are capable of reducing oxidative stress on the skin or protecting products from oxidative degradation. One of the major causes of oxidative stress that accelerates skin aging is reactive oxygen species (ROS). Intrinsic aging is associated with the natural process of aging, whereas extrinsic aging is associated with external factors that affect the aging process (*e.g.*, air pollution, UV radiation, and pathogenic microorganisms). Photoaging is most likely the primary cause of ROS production. Several potential skin targets have been discovered to interact with ROS (*e.g.*, lipids, DNA, and proteins). Antioxidant molecules can be enzymes or low-molecular-weight antioxidants that donate an electron to reactive species, preventing the radical chain reaction, which prevents the formation of reactive oxidants, or behave as metal chelators, oxidative enzyme inhibitors, or enzyme cofactors. Antioxidants can also be used as stabilizers, preventing lipid rancidity. Lipid oxidation occurs not only in cosmetics but also in the human body. Thus, when antioxidants are present in a product, they may serve multiple functions. The number of radicals increases during the initiation phase of lipid oxidation. Molecular oxygen and fatty acid radicals react during the propagation phase, resulting in the formation of hydroperoxide products. Hydroperoxides are unstable and can degrade to produce radicals, which can accelerate the propagation reaction. The termination phase is dominated by radical reactions. Antioxidants can inhibit lipid oxidation by reacting with lipid and peroxy radicals and converting them to more stable, non-radical products. Additionally, antioxidants can deplete molecular oxygen, inactivate singlet oxygen, eliminate peroxidative metal ions, covert hydrogen into other antioxidants, and dissipate UV light. Plants are well known for producing natural antioxidant compounds that can reduce the amount of oxidative stress caused by sunlight and oxygen. Plant extracts are used in a variety of patents and commercial cosmetic products. Green tea, rosemary, grape seed, basil grape, blueberry, tomato, acerola seed, pine bark, and milk thistle are some of the plant extracts commonly found in cosmetic formulations. Polyphenols, flavonoids, flavanols, stilbenes, and terpenes are natural antioxidants found in plant extracts (including carotenoids and essential oils). Antioxidants are classified as primary or natural antioxidants and as secondary or synthetic antioxidants according to their function. Mineral antioxidants (such as selenium, copper, iron, zinc, and manganese), vitamins (C and E), and phytoantioxidants are examples of primary antioxidants. Generally, a mineral antioxidant is a cofactor of enzymatic antioxidants. Secondary or synthetic antioxidants capture free radicals and stop the chain reaction. BHA, BHT, propyl gallate, metal chelating agents, tertiary butylhydroquinone, and nordihydroguaiaretic acid are examples of secondary antioxidants. The use of plant antioxidants is increasing and may eventually replace the use of synthetic antioxidants. A natural antioxidant can be a single pure compound/isolate, a combination of compounds, or plant extracts; these antioxidants are widely used in cosmetic products. Innate antioxidants act as oxygen free radical scavengers (singlet and triplet), ROS, peroxide decomposers, and enzyme inhibitors. Polyphenols and terpenes are the most common phytoantioxidants; this distinction is based on their molecular weight, polarity, and solubility. Polyphenols have benzene rings with —OH groups attached. The number and position of —OH groups on the benzene ring determine their antioxidant activity. Phenolic groups influence protein phosphorylation by inhibiting lipid peroxidation. The most abundant polyphenols are flavonoids and stilbenes, and the most abundant terpenes are carotenoids, which act as singlet oxygen quenchers. The following are some brief introductions on vitamins and polyphenols.

19.2.1 Vitamins

The consumption and absorption of vitamins and antioxidants, primarily through diet and, essentially, through the use of manufactured supplements, is critical to human health. The skin is our largest organ, and as our external environmental barrier, it is at the forefront of the fight against damaging free radicals from external sources. ROS are formed by ultraviolet light and environmental pollutants. Free radicals are highly reactive molecules with an unpaired electron that cause damage to the molecules and tissues around them. Free radicals cause the most significant damage to biomembranes and DNA. It is believed that using vitamins and antioxidants in cosmetics on a topical basis can help to protect from and possibly repair the damage caused by free radicals. Furthermore, some vitamins may be beneficial to the skin due to their effects, such as reduction in pigmentation and bruising, activation of collagen production, keratinization refinement, and anti-inflammatory effects. The following are some brief introductions on vitamin C and vitamin E.

(1) Vitamin C　Vitamin C is a powerful antioxidant that can neutralize oxidative stress via an electron donation/transfer process. In addition to regenerating other antioxidants in the body, such as alpha-tocopherol, vitamin C can reduce the amounts of unstable species of oxygen, nitrogen, and sulfur radicals (vitamin E). Furthermore, research with human plasma has shown that vitamin C is effective for preventing lipid peroxidation caused by peroxide radicals. Additionally, vitamin C promotes iron, calcium, and folic acid absorption, which prevents allergic reactions, and a decrease in the intracellular vitamin C content can lead to immunosuppression. Vitamin C is required for the synthesis of immunoglobulins, the production

of interferons, and the suppression of interleukin-18 (a regulating factor in malignant tumors) production. When applied topically, vitamin C can neutralize ROS caused by solar radiation and environmental factors such as smoke and pollution. Vitamin C has proven be effective for the treatment of hyperpigmentation, melasma, and sunspots. This appears to be related to its ability to obstruct the active site of tyrosinase–the enzyme that limits melanogenesis. Tyrosinase catalyzes the hydroxylation of tyrosine in 3,4-dihydroxyphenylalanine, resulting in the formation of a precursor molecule of melanin. Furthermore, vitamin C promotes keratinocyte cell differentiation and improves dermal-epidermal cohesion.

(2) Vitamins E Because of the ability to reduce lipid peroxidation, vitamin E has numerous health benefits for the eyes and cardiovascular system. Numerous cutaneous benefits have been demonstrated when vitamin E is applied topically. The most important property of vitamin E is its strong antioxidant capacity. The term "protector" has been used to describe the actions of vitamin E and its derivatives because of their ability to quench free radicals, particularly lipid peroxyl radicals. Several studies have indicated that they can reduce UV-induced erythema and edema. Clinical improvement in the visible signs of skin aging has been linked to reductions in both skin wrinkling and skin tumor formation. Tocopherol and its acetyl ester derivative, tocopherol acetate, have been studied extensively. While tocopherol is the most active form of vitamin E, topically applied vitamin E esters have also been shown to penetrate the epidermis.

19.2.2 Polyphenols

Polyphenols have been extensively studied and are reported to have antioxidant and anti-inflammatory properties. Polyphenolic compounds are found in various plants, including tea leaves, grape seeds, blueberries, almond seeds, and pomegranate extract. The beneficial properties of polyphenols have been supported by several studies on skin cells and on human skin; thus, these compounds are increasingly being incorporated into cosmetic and medicinal products. The main polyphenols in green tea are catechins gallocatechin, epigallocatechin, and epigallocatechin-3-gallate (EGCG). Research indicates that EGCG inhibits UVB-induced hydrogen peroxide release from cultured normal epidermal keratinocytes and suppresses MAPK (mitogen-activated protein kinase) phosphorylation. Furthermore, EGCG reduces inflammation by activating NFkB (nuclear factor kappa-B). Tea polyphenols can also prevent UVB-induced phosphatidyl-inositol 3-kinase activation (IP3K). On a molecular level, oral green tea administration to SKH-1 mice increased the number of UV-induced p53- and p21-positive cells, as well as the number of apoptotic sunburn cells. In addition to reducing the amount of ROS in the skin, tea polyphenols provide photoprotection by counteracting UVB-induced local and systemic immunosuppression.

References

[1] Maurício E, Rosado C, Duarte M P, Verissimo J, Bom S, Vasconcelos L. Efficiency of Nisin as Preservative in Cosmetics and Topical Products. Cosmetics, 2017, 4: 41.

[2] Hoang H T, Moon J Y, Lee Y C. Natural Antioxidants from Plant Extracts in Skincare Cosmetics: Recent Applications, Challenges and Perspectives. Cosmetics, 2021, 8: 106.

Key Words & Phrases

Gram-positive　革兰氏阳性
Gram-negative　革兰氏阴性
mould　*n.* 霉；霉菌
yeast　*n.* 酵母；酵母菌
bronopol　*n.* 溴硝丙二醇；溴硝醇
nitrosamine　*n.* 亚硝胺
nitrosating agent　亚硝化剂
contaminant　*n.* 致污物；污染物
benzalkonium chloride　苯扎氯铵
biocide　*n.* 抗微生物剂；杀虫（菌）剂
hand sanitizer　手部消毒剂；洗手液
wet wipe　湿巾；湿纸巾
benzethonium chloride　氯化苄乙氧铵；苄索氯铵
hyamine　*n.* 季铵盐
gum benzoin　安息香树胶
benzylhemiformal　*n.* 苄基半缩甲醛
chlorhexidine　*n.* 洗必泰；双氯苯双胍己烷
chlorhexidine gluconate　葡萄糖酸洗必泰
chlorhexidine digluconate　氯己定二葡糖酸盐
sterilize　*v.* 灭菌；消毒
facultative anaerobes　兼性厌氧菌；兼性厌氧；兼性厌氧微生物
aerobe　*n.* 需氧菌；好氧微生物
polioviruses　*n.* 脊髓灰质炎病毒；小儿麻痹病毒
adenovirus　*n.* 腺病毒；腺病毒科
herpes virus　疱疹病毒
chlorobutanol　*n.* 三氯叔丁醇；氯丁醇；氯代丁醇
sedative　*n.* 镇静剂
hypnotic　*n.* 催眠药；安眠药
anesthetic　*n.* 麻醉药；麻醉剂
chloral hydrate　水合氯醛；水合三氯乙醛
antifungal　*n.* 抗真菌剂；抗霉剂
climbazole　*n.* 甘宝素；氯咪巴唑；咪菌酮；苯咪丁酮
triclosan　*n.* 三氯羟基二苯醚；三氯苯氧氯酚；三氯生
4-chloro-*m*-cresol　4-氯间甲酚
seborrhoeic dermatitis　溢脂性皮炎；脂溢性皮炎；脂溢性皮肤炎
free radical　自由基
beverage　*n.* 饮料
butylated hydroxyanisole　丁基羟基茴香醚
butylated hydroxytoluene　丁基羟基甲苯
propyl gallate　没食子酸丙酯
pathogenic　*adj.* 致病性的；致病的；病原的
lipid rancidity　油脂酸败
milk thistle　水飞蓟；乳蓟
flavanol　*n.* 黄烷醇
terpene　*n.* 萜烯；萜
tertiary butylhydroquinone　叔丁基对苯二酚
nordihydroguaiaretic acid　去甲二氢愈创木酸
innate　*adj.* 与生俱来的；天生的；先天的
free radical scavenger　自由基清除剂
protein phosphorylation　蛋白质磷酸化
singlet oxygen quencher　单线态氧猝灭剂
tocopherol acetate　生育酚乙酸酯
catechin gallocatechin　儿茶素-棓儿茶素
epigallocatechin　*n.* 表没食子儿茶素
epigallocatechin-3-gallate　表没食子儿茶素-3-没食子酸酯
phosphorylation　*n.* 磷酸化
phosphatidyl-inositol 3-kinase activation　磷脂酰肌醇3-激酶激活

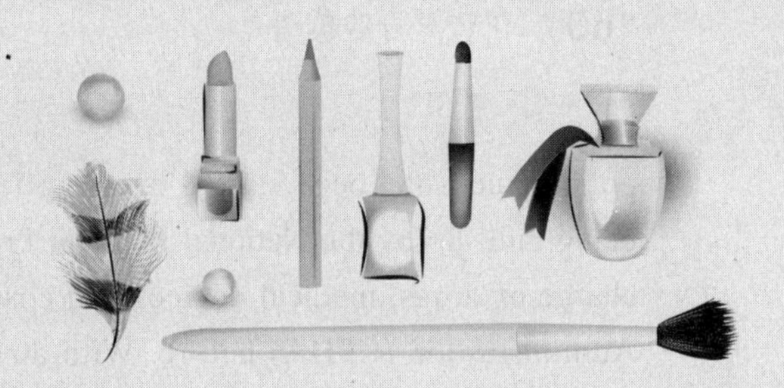

Unit 5

Regulations on Cosmetics

Cosmetics are daily chemical industrial products that are applied to any part of the human body surface. Improper use of cosmetics, or addition of prohibited chemicals, or excessive use of restricted substances, may irritate the skin, cause irreversible damage to the skin, and even cause cancer. Therefore, to ensure safety and efficacy, cosmetic products must be regulated and controlled.

Different countries have different regulations on cosmetics. In this unit, you will see:

① Cosmetics regulation of China;

② Cosmetics regulation of European Union;

③ Cosmetics regulation of United States.

Chapter 20 Regulations on the Supervision and Administration of Cosmetics in China

20.1 Introduction

In June 2020, the long-waited "Regulations on the Supervision and Administration of Cosmetics" (CSAR) was finally issued by the State Council of China, and this regulation was implemented from 1 January 2021. CSAR is the first-time revision and replacement of the "Regulations on Hygiene Supervision of Cosmetics" (CHSR), which was published in 1989. During the past 30 years, substantial changes have happened both in the industry and in consumer needs, and the market has increased significantly. CSAR, the very basic and fundamental legislation for cosmetics in China, must adapt to the rapid growth and the new trends of cosmetics nowadays.

The project of CSAR was formally initiated as early as 2013. In order to get technically prepared and establish a solid foundation for this legislation, a series of research work and policy study had been started even earlier. To support CSAR, a whole regulatory system is being built up by the National Medical Product Administration (NMPA), the administration in charge of drugs, medical devices and cosmetics, which was known as China Food and Drug Administration (CFDA) before. With an overview and discussion of the regulatory system, we can get a better understanding of the new policies in CSAR, as well as the latest research progress and new prospects in cosmetic science in China.

20.2 CSAR and Its Regulatory System

There are 6 chapters and 80 articles in CSAR, while the previous CHSR has only 35 articles. The 6 chapters include general provisions, ingredients and products, production and distribution, supervision and administration, legal liability, and supplementary provisions. Key points concerning cosmetics are specified in CSAR, for example, the definition, product classification, ingredient management, registration and notification, requirements for production, post-market supervision and inspection, and the roles and corresponding responsibilities in cosmetics-related activities.

There is a shift of regulatory focus in CSAR when compared to CHSR, which can be reflected in the change of names. CHSR emphasizes more on the hygiene qualification of cosmetic products, while in CSAR, advanced management philosophy and measures are introduced or further promoted, such as key responsibilities of enterprises (by introducing the concepts of registration person and notification person), classified risk management, encouragement of innovation, safety evaluation, and efficacy substantiation. If a keyword in CHSR is "hygiene", the new will be "safety" and "quality". Similar changes once happened when the technical regulation for cosmetics, the "Hygienic Standards for Cosmetics", was revised into "Safety and Technical Standards for Cosmetics" (STSC) in 2015.

As the fundamental regulation, CSAR provides a basis for detailed regulations and technical documents. NMPA has made a legislation plan to build up the CSAR related regulatory system, covering both the procedure rules and technical guidance. In brief, these subordinated regulations and documents can be functionally divided into four units. The first group is about registration and notification, with the procedural and technical requirements for both ingredients and final products. The second is about production and distribution, taking producers and enterprises as the main objects of administration. The third part is about toothpaste. Moreover, the last one is for standardized management of labeling. In addition, more regulations and documents are also being drafted or revised.

20.3 Definition, Scope and Classification of Cosmetics

20.3.1 Definition and scope

In the regulations concerning cosmetics, it is important to set a definition first to help judge product classification and limit the applicable range. The definition in China does not cover teeth, mucous membranes or any other parts of the oral cavity. However, according to CSAR, toothpaste shall be regulated in a similar way to general cosmetics. Thus, toothpaste is also a product category under the regulation of NMPA. Soaps, except for those claiming the efficacy of special cosmetics, are free of supervision by CSAR.

The definition of cosmetics in CSAR remains unchanged when compared to CHSR, and cosmetics are divided into special cosmetics and general cosmetics in both CHSR and CSAR. Special cosmetics are regulated with registration and must get approval before production or importation, with passing technical evaluation from the National Institutes for Food and Drug Control (NIFDC), a subordinated institution of NMPA. General cosmetics can be directly put into the market with the completion of a notification.

Although the definition of cosmetics remains the same, the interpretation of this definition and the scope of cosmetic products have actually been changed. In CHSR, there are nine special cosmetics, including products for/as hair dye, hair perm, sunscreen, depilatory, deodorant, spot corrector (including skin whitener), hair growth, breast care and body beauty (help to keep body shape). In CSAR, only five out of the nine still remain as special cosmetics, including products for/as hair dye, hair perm, spot corrector/skin whitener, sunscreen and preventing hair loss (instead of products for hair growth). Moreover, there is a new category in special cosmetics called "cosmetics with new efficacy claim".

As mentioned above, four categories will no longer belong to special cosmetics in the future. Depilatories are likely to be classified as general cosmetics since the mechanism is generally clear, and the product risk is relatively low. Existing deodorants are mostly designed for the armpit with several mechanisms: some function by covering the odor, some by absorbing sweat, and some by repressing perspiration. These products will also be classified as general cosmetics in the future. Products for breast care and body beauty, which were expected to help stimulate breast tissue growth or keep body shape, actually have a useful purpose beyond the definition of cosmetics and will no longer be cosmetics. As for the products for hair growth, three levels of claims were usually used in the past: (1) preventing hair breakage, with which the products will be classified as general cosmetics in the future; (2) preventing hair loss, will be as special cosmetics; (3) promoting hair growth, will be no longer cosmetics for suggesting a medical effect. Since 2014, NIFDC has been working on the study and comparison of cosmetic regulations worldwide. From an investigation and comparison of product classification in this study, it is easy to conclude that products with strong physiological effects and high risks (such as products for breast care, body beauty and hair growth) are mostly classified as drugs or quasi-drugs.

20.3.2 Borderline between cosmetics and drugs

In reality, some drugs can also be applied to the skin or other external parts of the human body and have functions similar to cosmetics. It is a common practice to set a borderline between cosmetics and drugs. For example, a guidance document and specific manual were published in the EU to clarify the applicable scope of cosmetic regulations and help distinguish cosmetics from pharmaceutical products and medical devices, as well as from toys, biocides and other articles. In recent years, the concept of "cosmeceutical" is getting increasingly popular, which is often interpreted by consumers as cosmetics with medical effects. In fact, the status of "cosmeceutical" is still controversial. For example, as the US Food and Drug Administration (FDA) declares, the Federal Food, Drug, and Cosmetic Act (FD&C Act) do not recognize any such category as "cosmeceutical".

Compared to drugs, cosmetics are designed for some limited purposes. They differ in the use of ingredients, risk characteristics and management, tolerance of adverse reactions, regulatory requirements, supervision systems and other important aspects. In addition, cosmetics are designed for normal people, while drugs are for those with health issues, and some should only be used under doctors' instructions. When mistaking drugs for cosmetics, one can underestimate the possible adverse effects and thus, an extra risk to users can be generated; when taking cosmetics for drugs, the precious chance to seek professional medical care can be delayed.

In January 2019, NMPA clarified it illegal to claim "medicated cosmetics" or "medical skincare products" in China. This principle will be maintained in the implementation of CSAR. Foreign quasi-drugs and drugs can only be imported as cosmetics under the premise of meeting the definition and requirements in China, including the legal compliance of the product itself and a probable adaption of labeling information.

20.3.3 Categorization and catalog

With a vast diversity of cosmetic products, it will be helpful to use a categorization system to reflect the characteristics of each product. For example, The US FDA employs a set of category codes to help describe a cosmetic product in the Voluntary Cosmetic Registration Program (VCRP). In Japan, there is a list of about 56 admitted efficacy claims for cosmetics, which could also be a description of product features.

On 29 July 2020, NMPA published a draft of "Rules and Catalog for Categorization of Cosmetics" online for public comments. With this regulation, a cosmetic product can be described in five dimensions and get a unique category code. The five dimensions are efficacy claim, applied area, product form, users and exposure way (rinse-off or leave-on). Detailed items and descriptions are listed in these dimensions term-by-term, with a numerical code for each one. In this catalog, industrial development and newly emerging techniques are also taken into account. For example, the efficacy of "restoring and protection" is collected considering the increasing consumption demands, and freeze-dried powder, which is becoming increasingly popular in production these years, is also collected as a product form. Once none of the numerical codes can cover a practical case, a capital letter code shall be picked up, which means "other". Any appearance of a capital letter in efficacy claim, applied area or users can be an indication of "cosmetics with new efficacy claim", the newly added special cosmetics in CSAR.

To give an example, according to the draft, a facial lotion with functions of sunscreen and moisture can get a code of 0409-05-02-01-02, which means "moisture and sunscreen-face-lotion-general population-leave on"; a hair shampoo made especially for infants can get a code of 01-01-03-02-01, which means "clean-hair-liquid-babies and infants (under 3 years old)-rinse off".

The category code is a new invention in the supervision and administration of cosmetics in China. A brief sketch of the product information can be immediately delineated with the code, and this system will help in accurate statistics and analysis. Furthermore, with reading the category code and some other submitted information, for example, the formula information, a high-throughput automated judgment of regulatory compliance can be possible, which is hoped to replace some labored work in the technical evaluation of cosmetics in the future.

20.4 Management of Cosmetic Ingredients

20.4.1 Ingredient lists in STSC

The safety of cosmetic products highly depends on the use of ingredients. In the EU, prohibited substances, restricted substances (including hair dyes), allowed colorants, allowed preservatives and allowed ultra-violet (UV) filters are, respectively, collected in the annexes of the "Regulation (EC) No 1223/2009 on Cosmetic Products". In the US, substances prohibited or restricted in cosmetic products are listed in the Code of Federal Regulations Title 21 (21CFR). FDA also pays attention to the management of colorants. Allowed colorants and related requirements are also specified in 21CFR. Sunscreen is recognized as a drug in America, and a list of sunscreen active ingredients is collected in the sunscreen OTC monograph. In some East Asian countries where consumption demands for skin whitening are strong, such as Japan and Korea, there is also a positive list of whitening agents.

In China, there are also technical lists of cosmetic ingredients. In STSC, high-risk ingredients are collected in the lists of prohibited substances, restricted ingredients, as well as allowed preservatives, UV filters, colorants and hair dyes. The use of these ingredients must strictly meet the requirements and technical standards specified in STSC. The current STSC was published in 2015 and is open to revision all the time.

Moreover, spot corrector/skin whitener product is the biggest share in all the special cosmetics in China. As a result, it is important to consider the necessity to build a list of whitening agents in STSC in the future, where Japan and Korea have already provided some experience.

20.4.2 New cosmetic ingredients

Another key point in the management of cosmetic ingredients in China is to distinguish between "existing ingredient" and "new ingredient". New cosmetic ingredients refer to the natural or artificial ingredients used in cosmetics for the first time within China. To better help identify new ingredients, an Inventory of Existing Cosmetic Ingredients in China (IECIC) was published in 2014 and revised in 2015, generating a collection of 8783 items of ingredients (some are collected as a group of ingredients, and besides the 8783 items, ingredients collected in the restricted or positive lists in STSC are also part of IECIC). IECIC is only an objective collection of cosmetic ingredients already used in China, different from a "positive list of cosmetic ingredients". One, who intends to use the ingredients in it, should take a safety evaluation before using.

Ingredients excluded by IECIC are regarded as new ingredients. According to CHSR, new ingredients can only be used after approval. In CASR, the policy is optimized by dividing new ingredients into different risk levels: ingredients that function as a preservative, UV filter, colorant, hair dye or spot corrector/skin whitener are considered to be relatively high-risk and will be regulated with a registration-based system by NMPA continuously; others can be immediately used after notification to NMPA. On the basis of scientific development, NMPA can submit an application to adjust the range of high-risk ingredients.

Innovative management for the use of new ingredients in CSAR is the three-year period of monitoring after registration or notification. Within the three years, the registration person or notification person shall submit a feedback report to NMPA about the use and safety situations every year, and any emergency shall be reported immediately. When a certain safety issue occurs, if any, this registration or notification can be withdrawn by NMPA. In order to protect the interest of enterprises and encourage the development of new ingredients, within the monitoring period, the ingredient will still be regarded and managed as a new ingredient: any other person who intends to use the ingredient shall complete the registration or notification of all independently, or obtain use permission from any previous registration person or notification person. Ingredients successfully passing the three-year monitoring period will have the chance to be incorporated into IECIC.

20.5 Technical Requirements about Safety and Efficacy

20.5.1 Safety evaluation

The safety of cosmetics is strictly required since they are daily used products by a large population. In the EU, the Scientific Committee on Consumer Safety (SCCS) published an "SCCS Notes of Guidance for the Testing of Cosmetic Ingredients and Their Safety Evaluation", with the 10th revision published in 2018, to give technical instruction for safety evaluation of cosmetics. With the trend of the 3R principle, animal tests for cosmetic ingredients and products have been banned in the EU, as well as in some other countries or regions. Study on alternative methods has become a common topic of concern these years, and SCCS has also emphasized replacement methodology in the recent versions of guidance. Moreover, in recent years, a new technical system called next-generation risk assessment (NGRA) has emerged. NGRA aims to incorporate new approach methodologies into an integrated strategy for risk assessment of cosmetic ingredients and has newly become a topic of interest in the work of International Cooperation of Cosmetics Regulation (ICCR).

Safety evaluation for cosmetics has also developed rapidly in China. With an announcement published by the former CFDA in 2013, toxicological tests of final products are no longer compulsive requirements for general domestic cosmetics if a safe conclusion can be derived from risk assessment. To better help improve the understanding and practical use of risk assessment and safety evaluation in China, NIFDC has held a series of workshops and seminars and has built broad cooperation with foreign governments, international organizations and research institutions. In addition, NIFDC has been increasing the input in research and verification work of alternative methods. Since 2016, six new alternative methods have been adopted into STSC as standard test methods, and more methods are under development at full speed. In 2018, NIFDC initiated a workgroup on research and validation of cosmetic alternative methods in China, together with several advanced institutions in this field, which will further accelerate the development, verification/validation and usage of alternative methods.

According to Article 21 of CASR, before registration or notification (of new ingredients and cosmetic products), the registration person or notification person shall perform a safety evaluation by themselves or by entrusting a professional agency. In the future, safety evaluation will play a more important role, and it is necessary to establish a scientific and practical technical system in China. On 29 July 2020, NMPA published a draft of notes of guidance for cosmetic safety evaluation online for public comments. Learning from the SCCS guidance and other advanced technical documents, this guidance covers the general principles and requirements about evaluation work, evaluator, risk assessment procedure and toxicological research, and it provides pragmatic guidelines for the safety evaluation work of both ingredients and final products.

Unlike the EU, there is no compulsive animal test ban in China. Both animal tests and alternative methods are collected in STSC, and enterprises can choose any when applicable. In addition, in the future, imported general products meeting certain prerequisites and with a satisfying conclusion from safety evaluation can also be exempted from toxicological tests of final products, the same as the domestic ones.

20.5.2 Efficacy and claim substantiation

Besides the requirement of safety, efficacy realization, and proper claims are also focus on cosmetics. As a basic principle to protect the right to know of consumers and fair competition, claims and labeling contents should be based on the actual functions and necessary substantiation. For example, according to Commission Regulation No. 655/2013 of the EU, claims for cosmetic products, whether explicit or implicit, shall be supported by adequate and verifiable evidence. In the US, cosmetics must comply with the labeling requirements of the FD&C Act and the Fair Packaging and Labeling Act (FP&L Act).

Human trials, consumer use tests, experimental tests and research literature are the main types of evidence for efficacy and claim. In some authorities, official or semi-official methods are provided to give better guidance for efficacy claim substantiation. For example, in Japan, the Japanese Cosmetic Science Society (JCSS) has published guidelines for the evaluation of quasi-drug whitening products and anti-wrinkle products, both of which are human test methods. In Korea, the National Institute of Food and Drug Safety Evaluation (NIFDS) has published a series of documents for the efficacy substantiation of functional cosmetics, including products for preventing hair loss (human test), whitening products (human test and *in vitro* methods), sunscreen products (human test), and anti-wrinkle products (human test and *in vitro* methods). The method for sun protection factor (SPF) of sunscreen products is relatively mature, and there is also a standard collected by the International Organization of Standards (ISO), which was recently renewed and published as ISO 24444: 2019.

In the past, requirements about cosmetic efficacy were relatively limited in China. Among all the special and general cosmetics, sunscreen product was the only one with a requirement for clinical tests (human trial). Test methods for SPF value, water-resistant performance, and protection factor of ultra-violet A (PFA) value are collected in STSC. For products for hair growth, breast care and body beauty, it was required to specify the functional ingredients and submit efficacy related evidence. However, it was hard to evaluate the real efficacy without detailed requirements or standards for the data and information submitted.

According to Article 22 of CASR, there shall be sufficient scientific evidence for efficacy and claim of cosmetics; the registration person or notification person shall publish the summary of evidence on a website designated by NMPA and accept public supervision. In addition, according to the “Management Measures for Cosmetic Labeling” drafted by NMPA, claims on the label shall be based on efficacy substantiation.

On 5 November 2020, NMPA published a draft of “Efficacy and Claims Substantiation Standards for Cosmetics” online for public comments. According to this guidance, functions or claims that can be directly detected by sensory perception, such as optesthesia and olfaction, can be exempted from efficacy substantiation. Some examples can be a cleanser, perfume, hair dye, hair perm, depilatory and deodorant. Moreover, those that function *via* physical mechanisms, such as covering, adhering or rubbing, can also be exempted, as long as it is clearly clarified in the labeling. Examples include skin whitening products by physical covering, exfoliators by physical rubbing and pore cleansers by physical adhering and pulling.

This guidance also gives a basic principle of evidence required for each certain efficacy or claim. A human trial is compulsory in some cases, for example, products for preventing hair loss, spot corrector/skin whitener, sunscreen, acne product and products claiming skin-restoring or claiming tear-free. For the “cosmetics with new efficacy claim”, the new special cosmetics in CASR, methods for efficacy substantiation should be based on the actual functions to be claimed. On 12 November 2020, methods for evaluating hair loss prevention and for spot corrector/skin whitener were published online for public comments by NIFDC, which are expected to be supplemented into STSC soon.

Reference

Su Z, Luo F Y, Pei X R, Zhang F L, Xing S X, Wang G L. Final Publication of the "Regulations on the Supervision and Administration of Cosmetics" and New Prospectives of Cosmetic Science in China. Cosmetics, 2020, 7: 98.

Key Words & Phrases

National Medical Products Administration　国家药品监督管理局
general provision　一般规定
legal liability　法律责任
registration person　注册人
notification person　通知人
substantiation　*n.* 实证；证实；证明；具体化；实任制
subordinated regulations　附属法规
special cosmetic　特殊用途化妆品
depilatory　*n.* 脱毛剂
armpit　*n.* 腋窝
freeze-dried powder　冻干粉
sketch　*n.* 素描；速写；草图；简报；概述
delineate　*v.* 勾画；描述；描画；解释
prohibited substance　违禁物质；禁用物质
restricted substance　限用物质
allowed colorant　允许用的着色剂
allowed preservative　允许用的防腐剂
ultra-violet (UV) filter　紫外线（UV）过滤器
monograph　*n.* 专论；专题文章；专著
emergency　*n.* 突发事件；紧急情况
safety evaluation　安全性评价；安全评价
entrust　*v.* 委托；交托；托付
pragmatic　*adj.* 实用的；讲求实效的；务实的
explicit　*adj.* 明确的；清楚明白的；易于理解的；(说话)清晰的；直言的；坦率的；不隐晦的；不含糊的
implicit　*adj.* 含蓄的；不直接言明的；成为一部分的；内含的；完全的；无疑问的
human trial　人体试验
mature　*adj.* 成熟的
clinical test　临床试验
water-resistant performance　防水性能
optesthesia　*n.* 视觉
olfaction　*n.* 嗅觉
exempt　*v.* 豁免；免除
compulsory　*adj.* 强制性的；（因法律或规则而）必须做的；强制的；强迫的
skin-restoring　皮肤修复
tear-free　无泪

Chapter 21 Legislative Aspects of Cosmetic Safety in the European Union

21.1 Introduction

In the last few years, Europe has updated its approach to cosmetic safety by recasting the now 40 year old EU Cosmetics Directive (Directive 76/768/EEC) into the simpler, EU Regulation No. 1223/2009 which was fully applicable from July 2013. This forms part of the smart regulations initiative, including general "maintenance" activities. One benefit is that the introduction of Regulation EC No. 1223/2009 on cosmetic products, instead of the Directive 76/768/EEC, means that the same (translated into national languages) legal text became binding in all member states, and thus at least one major simplification is achieved for the EU market: the 28 national legal frameworks for the directive are substituted with the regulation. With respect to consumer safety, the regulation states "A cosmetic product made available on the market shall be safe for human health when used under normal or reasonably foreseeable conditions of use…". In this respect, the intention does not differ from that of the preceding Directive. It is fair to recognize that with this regulation, Europe offers the most highly developed legislation concerning cosmetics, consumer safety in general and contact dermatitis in particular. Of course, the regulation is just one part of the entity; it requires also that those placing products onto the market abide by the legislation and do their utmost to develop safe cosmetics, an activity that can only be assured if there is adequate regulatory oversight and monitoring.

A second piece of European legislation also has an impact on the safety of cosmetic ingredients, and that is the Regulation, Evaluation and Authorisation of Chemicals, usually referred to simply as REACH. This will not be reviewed in detail here, almost all substances used (in almost any quantity) in the EU are subject to this regulation, whose aim is to ensure human (including occupational) and environmental safety. There are a number of exceptions, including a few common substances (*e.g.*, water) and some natural materials; guidance on this can be found *via* the European Chemicals Agency.

21.2 The Requirements of the EU Regulation

The EU Cosmetics Regulation No. 1223/2009 and much related material can be found from the official website of the European Union. The European trade associations impacted by this legislation include the European Federation for Cosmetic Ingredients (EFfCI) and Cosmetics Europe; their websites have specific pages detailing the import of the legislation, as well as practical guidance to manufacturers on how to deal with it.

The EU Cosmetics Regulation has several annexes: Annex I details the required content of the Cosmetic Product Safety Report. This is set out verbatim below since it offers a very clear outline of what the safety assessment must encompass and properly document.

The cosmetic product safety report shall, as a minimum, contain the following:

PART A—Cosmetic product safety information

(1) Quantitative and qualitative composition of the cosmetic product The qualitative and quantitative composition of the cosmetic product, including chemical identity of the substances (incl. chemical name, INCI, CAS, EINECS/ELINCS, where possible) and their intended function. In the case of perfume and aromatic compositions, description of the name and code number of the composition and the identity of the supplier.

(2) Physical/chemical characteristics and stability of the cosmetic product The physical and chemical characteristics of the substances or mixtures, as well as the cosmetic product. The stability of the cosmetics product under reasonably foreseeable storage conditions.

(3) Microbiological quality The microbiological specifications of the substance or mixture and the cosmetic product. Particular attention shall be paid to cosmetics used around the eyes, on mucous membranes in general, on damaged skin, on children under three years of age, on elderly people and persons showing compromised immune responses. Results of preservation challenge test.

(4) Impurities, traces, information about the packaging material The purity of the substances and mixtures. In the case of traces of prohibited substances, evidence for their technical unavoidability. The relevant characteristics of packaging material, in particular purity and stability.

(5) Normal and reasonably foreseeable use The normal and reasonably foreseeable use of the product. The reasoning shall be justified in particular in the light of warnings and other explanations in the product labelling.

(6) Exposure to the cosmetic product Data on the exposure to cosmetic product taking into consideration the findings under Section 5 in relation to

① The site(s) of application;

② The surface area(s) of application;

③ The amount of product applied;

④ The duration and frequency of use;

⑤ The normal and reasonably foreseeable exposure route(s);

⑥ The targeted (or exposed) population(s). Potential exposure of a specific population shall also be taken into account.

The calculation of the exposure shall also take into consideration the toxicological effects to be considered (*e.g.*, exposure might need to be calculated per unit area of skin or per unit of body weight). The possibility of secondary exposure by routes other than those resulting from direct application should also be considered (*e.g.*, non-intended inhalation of sprays, non-intended ingestion of lip products, *etc.*).

Particular consideration shall be given to any possible impacts on exposure due to particle sizes.

(7) Exposure to the substances Data on the exposure to the substances contained in the cosmetic product for the relevant toxicological endpoints taking into account the information under Section 6.

(8) Toxicological profile of the substances Without prejudice to Article 18, the toxicological profile of substance contained in the cosmetic product for all relevant toxicological endpoints. A particular focus on local toxicity evaluation (skin and eye irritation), skin sensitisation, and in the case of UV absorption photo-induced toxicity shall be made.

All significant toxicological routes of absorption shall be considered as well as the systemic effects and margin of safety (MoS) based on a no observed adverse effects level (NOAEL) shall be calculated. The absence of these considerations shall be duly justified.

Particular consideration shall be given to any possible impacts on the toxicological profile due to

- particle sizes, including nanomaterials,
- impurities of the substances and raw material used, and
- interaction of substances.

Any read-across shall be duly substantiated and justified. The source of information shall be clearly identified.

(9) Undesirable effects and serious undesirable effects All available data on the undesirable effects and serious undesirable effects to the cosmetic product or, where relevant, other cosmetic products. This includes statistical data.

(10) Information on the cosmetic product Other relevant information, *e.g.*, existing studies from human volunteers or the duly confirmed and substantiated findings of risk assessments carried out in other relevant areas.

PART B—Cosmetic product safety assessment

(1) Assessment conclusion Statement on the safety of the cosmetic product in relation to Article 3.

(2) Labelled warnings and instructions of use Statement on the need to label any particular warnings and instructions of use in accordance with Article 19(1)(d).

(3) Reasoning Explanation of the scientific reasoning leading to the assessment conclusion set out under Section 1 and the statement set out under Section 2. This explanation shall be based on the descriptions set out under Part A. Where relevant, margins of safety shall be assessed and discussed.

There shall be inter alia a specific assessment for cosmetic products intended for use on children under the age of three and for cosmetic products intended exclusively for use in external intimate hygiene.

Possible interactions of the substances contained in the cosmetic product shall be assessed. The consideration and non-consideration of the different toxicological profiles shall be duly justified. Impacts of the stability on the safety of the cosmetic product shall be duly considered.

(4) Assessor's credentials and approval of Part B

Name and address of the safety assessor;

Proof of qualification of safety assessor;

Date and signature of safety assessor.

Annex Ⅱ is a list of substances that are prohibited from use in cosmetic products available in Europe. Annex Ⅲ tabulates substances that may be used subject to certain restrictions (such as for application to hair only, not for oral hygiene). Annex Ⅳ lists colouring agents, Annex Ⅴ is the positive list of preservatives and Annex Ⅵ is the positive list of UV filters. Note that following detailed reviews of hair dye safety, hair dyes listed in Annex Ⅲ may be used, but these are not exclusive, as others may still be used. The ultimate aim is to have a positive list of hair dyes which will be housed in Annex Ⅳ.

The cosmetic industry uses thousands of substances for which there is no specific restriction in the Regulation other than the requirement to be safe as used, with only a minority of ingredients being mentioned in these annexes. The European inventory of cosmetic ingredients (CosIng) is a non-exhaustive listing of them. The requirement for each cosmetic product to have a complete list of ingredients was introduced in 1995 by the sixth amendment to the Cosmetic Directive, although fragrances were excluded, except for the need to use the word "parfum". However, specific content labelling of 26 fragrance allergens was introduced in March 2005 by the seventh amendment; labelling is required if any of the 26 fragrances is present at levels $>10\times10^{-6}$ for leave-on products or $>100\times10^{-6}$ for rinse-off products. The aim is that individuals who are already fragrance allergic have the opportunity to avoid exposure. It should be noted that the trigger levels for labelling were suggested by the European Parliament as a pragmatic solution, as truly safe levels for most fragrance allergens are largely unknown. The list of fragrance allergens that must be labelled has been reviewed. Implementation of the scientific advice is thought to be imminent, despite being subject to a prolonged period of consultation.

Before a cosmetic ingredient is introduced to an annex in the Cosmetics Regulation, scientific evaluation must be provided by an independent European Commission advisory committee. Initially, this was known as the Scientific Committee for Cosmetology (SCC), but from 1997-2004 it became the Scientific Committee for Cosmetics and Non-Food Products (SCCNFP), then from 2004-2009 the Scientific Committee for Consumer Products (SCCP) and it is now designated the Scientific Committee for Consumer Safety (SCCS). The opinions of this independent committee of experts have been made available on the European Commission website. The SCCS also regularly updates its "Guidelines for the Safety Evaluation of Cosmetic Ingredients" to accommodate scientific and technological progress. The latest version of these guidelines can be found on the European Commission website here.

One of the most notable aspects of cosmetic ingredient and product evaluation in the EU is that, since 2013, it must not be completed using new *in vivo* experimentation on animals. However, for many ingredients, historic *in vivo* data exists. In addition, where ingredients have substantial use beyond the cosmetic industry, the requirements of REACH dictate that *in vivo* data is generated. Accordingly, the SCCS notes of guidance also cover this type of toxicological data in addition to *in vitro* and in silico methods. Thus, whether an ingredient is being evaluated by the SCCS or by a company toxicologist/safety assessor, a broad spectrum of *in vitro*, *in vivo*, and other data commonly still has to be assessed.

As noted above in Part B of the European requirements, whilst a cosmetic product must comply with all regulatory stipulations, it is necessary also that the safety of each cosmetic product (including that of the ingredients that have not been restricted) be assessed independently by an individual with appropriate expertise, formally designated "the assessor". This individual must be able to demonstrate the necessary qualifications and experience. Furthermore, each product must have an associated dossier containing technical details of the ingredients and a safety (toxicological) assessment of every ingredient as well as the final formulation placed on the consumer market.

21.3 REACH and the Classification, Labelling, and Packaging (CLP) Regulations

REACH (Registration, Evaluation, Authorisation and Restriction of Chemicals) details can be found on the EU website. Although comprehensive in some respects, the large burden of responsibility for completion of a REACH assessment of chemicals and any subsequent risk assessment has, deliberately, been placed on industry, since it is their responsibility to place chemicals on the market in a safe manner. Only a small minority of assessments will be cross checked by members of the European Chemicals Agency (ECHA). To assist with the assessments, extensive guidance has been published and is updated from time to time to take account of toxicological progress, including in non-animal methods–see the website for detailed information.

Test methods for the identification of skin sensitizers are clearly set out within the legislation, but it also encourages a wide range of other evidence to be taken into account, including from chemical structure and human evidence. Toxicological evaluations follow the principles set out in the GHS (Globally Harmonised System), which endeavours to harmonise testing and assessment internationally, based largely on test guidelines developed under the auspices of the Organisation for Economic Cooperation and Development.

21.4 Specific Restrictions

As already mentioned above with respect to fragrance allergens, where the general regulation of cosmetic products is deemed to have been insufficiently effective, *ad hoc* specific restrictions may be applied. Thus, although there is, for example, a positive list for preservatives permitted in cosmetic products (Annex Ⅴ), it has long been recognised that this category is associated with the causation of contact allergy. Therefore, from time to time, despite prospective limits being applied, clinical evidence demonstrates the need for further action. By this route, certain preservative substances previously allowed in cosmetics have been subject to additional restriction, for example being discouraged in leave-on products. It is not appropriate here to catalogue such restrictions (not least since all the information is readily available on the EU websites). However, what is evident is that, at least in principle, failure to assess the (contact allergy) risk to human health invites the parallel risk that the substance may become heavily restricted or banned entirely from cosmetics.

Reference

Basketter D, White I R. Legislative Aspects of Cosmetic Safety in the European Union: The Case of Contact Allergy. Cosmetics, 2016, 3: 17.

Key Words & Phrases

recasting 重铸；重塑；改正
verbatim *adj./adv.* 逐字的（地）；一字不差的（地）
foreseeable *adj.* 可预料的；可预见的；可预知的
inhalation *n.* 吸入；吸入剂；吸入药
duly *adv.* 适当地；适时地；恰当地；按时地；准时地
intimate *adj.* 亲密的；密切的；个人隐私的
credential *n.* 资格证书；证件；资格；资历
non-exhaustive 非详尽无遗的
imminent *adj.* 迫在眉睫的；即将发生的；临近的
in silico 生物信息学；预测基因
stipulation *n.* 规定；约定；合同
dossier *n.* 材料汇编；卷宗；档案
auspice *n.* 支持；赞助；资助；主办；保护
ad hoc *adj.* 临时安排的；特别的
causation *n.* 诱因；起因；原因

Chapter 22 The United States' Regulatory Approach to the Safety and Efficacy of Cosmetics

In United States, the law does not require cosmetic products and ingredients, other than color additives, to have the approval of Food and Drug Administration (FDA) before they go on the market, but there are laws and regulations that apply to cosmetics on the market in interstate commerce. The two most important laws pertaining to cosmetics marketed in the United States are the Federal Food, Drug, and Cosmetic Act (FD&C Act) dating back to 1938 and the Fair Packaging and Labeling Act (FPLA) dated 1967. FDA regulates cosmetics under the authority of these laws.

22.1 What Kinds of Products Are "Cosmetics" under the Law?

The FD&C Act defines cosmetics by their intended use, as "articles intended to be rubbed, poured, sprinkled, or sprayed on, introduced into, or otherwise applied to the human body...for cleansing, beautifying, promoting attractiveness, or altering the appearance" [FD&C Act, sec. 201(i)]. Among the products included in this definition are skin moisturizers, perfumes, lipsticks, fingernail polishes, eye and facial makeup, cleansing shampoos, permanent waves, hair colors, and deodorants, as well as any substance intended for use as a component of a cosmetic product. It does not include soap.

But, if the product is intended for a therapeutic use, such as treating or preventing disease, or to affect the structure or function of the body, it's a drug [FD&C Act, 201(g)], or in some cases a medical device [FD&C Act, 201(h)], even if it affects the appearance. Other "personal care products" may be regulated as dietary supplements or as consumer products.

People often use the term "personal care products" to refer to a wide variety of items that we commonly find in the health and beauty departments of drug and department stores. These products may fall into a number of different categories under the law.

- Products intended to cleanse or beautify are generally regulated as cosmetics. Some examples are skin moisturizers, perfumes, lipsticks, fingernail polishes, makeup, shampoos, permanent waves, hair colors, toothpastes, and deodorants. These products and their ingredients are not subject to FDA premarket approval, except color additives (other than coal tar hair dyes). Cosmetic companies have a legal responsibility for the safety of their products and ingredients.
- Products intended to treat or prevent disease, or affect the structure or function of the body, are drugs. This is true even if a product affects how you look. Some examples are treatments for dandruff or acne, sunscreen products, antiperspirants, and diaper ointments. Generally, drugs must receive premarket approval by FDA or, if they are

nonprescription drugs, conform to special regulations, called "monographs", for their category.

- Some are both cosmetics and drugs. Examples include anti-dandruff shampoos and antiperspirant-deodorants, as well as moisturizers and makeup with SPF (sun protection factor) numbers. They must meet the requirements for both cosmetics and drugs.
- Some may belong to other categories, including medical devices (such as certain hair removal and microdermabrasion devices), dietary supplements (such as vitamin or mineral tablets or capsules), or other consumer products (such as manicure sets).

Under the FD&C Act, cosmetic products and ingredients, with the exception of color additives, do not require FDA approval before they go on the market. Drugs, however, must generally either receive premarket approval by FDA through the New Drug Application (NDA) process or conform to a "monograph" for a particular drug category, as established by FDA's Over-the-Counter (OTC) Drug Review. These monographs specify conditions whereby OTC drug ingredients are generally recognized as safe and effective, and not misbranded. Certain OTC drugs may remain on the market without an NDA approval until a monograph for its class of drugs is finalized as a regulation. However, once FDA has made a final determination on the status of an OTC drug category, such products must either be the subject of an approved NDA [FD&C Act, sec. 505(a) and (b)], or comply with the appropriate monograph for an OTC drug.

22.2 What Does the Law Say about the Safety and Labeling of Cosmetics?

The FD&C Act prohibits the marketing of adulterated or misbranded cosmetics in interstate commerce.

"Adulteration" refers to violations involving product composition–whether they result from ingredients, contaminants, processing, packaging, or shipping and handling. Under the FD&C Act, a cosmetic is adulterated if–

- "it bears or contains any poisonous or deleterious substance which may render it injurious to users under the conditions of use prescribed in the labeling thereof, or under conditions of use as are customary and usual" (with an exception made for coal-tar hair dyes);
- "it consists in whole or in part of any filthy, putrid, or decomposed substance";
- "it has been prepared, packed, or held under insanitary conditions whereby it may have become contaminated with filth, or whereby it may have been rendered injurious to health";
- "its container is composed, in whole or in part, of any poisonous or deleterious substance which may render the contents injurious to health"; or
- except for coal-tar hair dyes, "it is, or it bears or contains, a color additive which is unsafe within the meaning of section 721(a)" of the FD&C Act. (FD&C Act, sec. 601)

"Misbranding" refers to violations involving improperly labeled or deceptively packaged products. Under the FD&C Act, a cosmetic is misbranded if–

- "its labeling is false or misleading in any particular";
- its label does not include all required information. (An exemption may apply to cosmetics that are to be processed, labeled, or repacked at an establishment other than where they were originally processed or packed; see Title 21, Code of Federal Regulations, section 701.9)
- the required information is not adequately prominent and conspicuous;
- "its container is so made, formed, or filled as to be misleading";
- it is a color additive, other than a hair dye, that does not conform to applicable regulations issued under section 721 of the FD&C Act; and
- "its packaging or labeling is in violation of an applicable regulation issued pursuant to section 3 or 4 of the Poison Prevention Packaging Act of 1970". (FD&C Act, sec. 602)

Under the FD&C Act, a product also may be misbranded due to failure to provide material facts. This means, for example, any directions for safe use and warning statements needed to ensure a product's safe use.

In addition, under the authority of the FPLA, FDA requires a list of ingredients for cosmetics marketed on a retail basis to consumers [Title 21, Code of Federal Regulations (CFR), section 701.3]. Cosmetics that fail to comply with the FPLA are considered misbranded under the FD&C Act (FPLA, section 1456). This requirement does not apply to cosmetics distributed solely for professional use, institutional use (such as in schools or the workplace), or as free samples or hotel amenities.

FDA can take action against cosmetics on the market that are in violation of these laws, as well as companies and individuals who market such products.

22.3 Does FDA Approve Cosmetics Before They Go on the Market?

FDA's legal authority over cosmetics is different from our authority over other products we regulate, such as drugs, biologics, and medical devices. Under the law, cosmetic products and ingredients do not need FDA premarket approval, with the exception of color additives. However, FDA can pursue enforcement action against products on the market that are not in compliance with the law, or against firms or individuals who violate the law.

In general, except for color additives and those ingredients that are prohibited or restricted by regulation, a manufacturer may use any ingredient in the formulation of a cosmetic, provided that–

- the ingredient and the finished cosmetic are safe under labeled or customary conditions of use,
- the product is properly labeled, and
- the use of the ingredient does not otherwise cause the cosmetic to be adulterated or misbranded under the laws that FDA enforces.

22.4 Who Is Responsible for Substantiating the Safety of Cosmetics?

Companies and individuals who manufacture or market cosmetics have a legal responsibility to ensure the safety of their products. Neither the law nor FDA regulations require specific tests to demonstrate the safety of individual products or ingredients. The law also does not require cosmetic companies to share their safety information with FDA.

FDA has consistently advised manufacturers to use whatever testing is necessary to ensure the safety of their products and ingredients. Firms may substantiate safety in a number of ways. FDA has stated that "the safety of a product can be adequately substantiated through (a) reliance on already available toxicological test data on individual ingredients and on product formulations that are similar in composition to the particular cosmetic, and (b) performance of any additional toxicological and other tests that are appropriate in light of such existing data and information". (Federal Register, March 3, 1975, page 8916).

In addition, regulations prohibit or restrict the use of several ingredients in cosmetic products and require warning statements on the labels of certain types of cosmetics.

22.5 Can FDA Order the Recall of a Hazardous Cosmetic from the Market?

Recalls of cosmetics are voluntary actions taken by manufacturers or distributors to remove from the marketplace products that represent a hazard or gross deception, or that are somehow defective [21 CFR 7.40(a)]. FDA is not authorized to order recalls of cosmetics, but we do monitor companies that conduct a product recall and may request a product recall if the firm is not willing to remove dangerous products from the market without FDA's written request.

22.6 What Actions Can FDA Take Against Companies or Individuals Who Market Adulterated or Misbranded Cosmetics?

FDA may take regulatory action if we have reliable information indicating that a cosmetic is adulterated or misbranded. For example, FDA can pursue action through the Department of Justice in the federal court system to remove adulterated and misbranded cosmetics from the market. To prevent further shipment of an adulterated or misbranded product, FDA may request a federal district court to issue a restraining order against the manufacturer or distributor of the violative cosmetic. Cosmetics that are not in compliance with the law may be subject to seizure. "Seizure" means that the government takes possession of property from someone who has violated the law, or is suspected of doing so. FDA also may initiate criminal action against a person violating the law.

In addition, FDA works closely with U.S. Customs and Border Protection to monitor imports. Under section 801(a) of the FD&C Act, imported cosmetics are subject to review by

FDA at the time of entry through U.S. Customs. Products that do not comply with FDA laws and regulations are subject to refusal of admission into the United States. They must be brought into compliance (if possible), destroyed, or re-exported. FDA does not inspect every shipment of cosmetics that comes into this country, but imported cosmetics are still subject to the laws we enforce, even if they are not inspected upon entry.

FDA takes regulatory action based upon agency priorities, consistent with public health concerns and available resources.

22.7 Can FDA Inspect Cosmetics Manufacturers?

FDA can and does inspect cosmetic manufacturing facilities to assure cosmetic product safety and determine whether cosmetics are adulterated or misbranded under the FD&C Act or FPLA.

22.8 Does FDA Test Cosmetics or Recommend Testing Labs?

Although FD&C Act does not subject cosmetics to premarket approval by FDA, we do collect samples for examination and analysis as part of cosmetic facility inspections, import inspections, and follow-up to complaints of adverse events associated with their use. FDA may also conduct research on cosmetic products and ingredients to address safety concerns.

FDA does not function as a private testing laboratory, and in order to avoid even the perception of conflict of interest, we do not recommend private laboratories to consumers or manufacturers for sample analysis.

22.9 Do Cosmetics Firms Need to Register with FDA or Get an FDA License to Operate?

Under the law, manufacturers are not required to register their cosmetic establishments or file their product formulations with FDA, and no registration number is required to import cosmetics into the United States.

However, we encourage cosmetic firms to participate in FDA's Voluntary Cosmetic Registration Program (VCRP) using the online registration system. Cosmetic manufacturers, distributors, and packers can file information on their products that are currently being marketed to consumers in the United States and register their manufacturing and/or packaging facility locations in the VCRP database. To learn more and access this program, see Voluntary Cosmetic Registration Program (VCRP).

Reference

US Food and Drug Administration. Cosmetics Laws & Regulations.[EB/OL][2023-03-24]. https://www.fda.gov/cosmetics/cosmetics-guidance-regulation/cosmetics-laws-regulations.

Key Words & Phrases

pertaining to 关于；有关；适合；属于
coal tar 煤焦油
diaper *n.* 尿布；（婴儿的）尿片
nonprescription *n.* 非处方药
microdermabrasion *n.* 微晶磨皮术；微晶磨皮；微晶换肤术；微晶换肤仪
adulterate *v.* 掺杂，掺假
violation *n.* 违反；违法；违章；越轨；侵犯；破坏；违例；犯规
filthy *adj.* 肮脏的；污秽的
putrid *adj.* 腐烂的；腐臭的
insanitary *adj.* 不卫生的；不洁的
seizure *n.* 没收；起获；充公；没收的财产

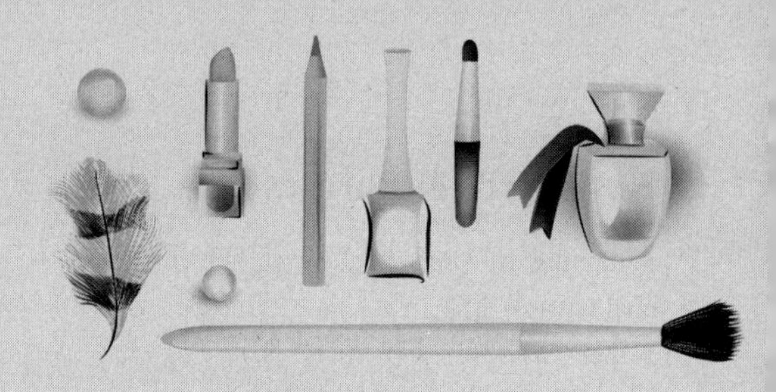

Unit 6

Formulations, Technologies, Analyses and Packaging of Cosmetics

In the manufacture of cosmetics, formulations, technologies, analyses and packaging are very important.

In this unit, you will see:

① Some examples on formulations, technologies and analyses;

② Some materials on packaging of cosmetics.

Chapter 23 Formulations, Technologies and Analyses of Cosmetics

There are a large number of different cosmetics used for a variety of different purposes. And different cosmetics have different formulations, production technologies and analysis methods. Following are four examples for the manufacture of some kinds of cosmetics.

23.1 Example 1: The Manufacture and Analyses of a Cream Formulation

Only the experimental section is shown here. For the research background and results, please see the other parts of the paper in reference.

23.1.1 Materials

The active ingredient (X), cholesterol, span 65 and solutolHS-15 were obtained from Sigma-Aldrich, Inc. (Gillingham, UK). Baobab oil was purchased from Aromatic Natural Skin Care (Forres, UK), jojoba and coconut oil from Southern Cross Botanicals (Knockrow, Australia). The emulsifying wax was obtained from CRODA International Plc (Goole, East Yorkshire, UK). Other excipients of the cream and tris buffer solutions were of analytical grade.

23.1.2 Methods

(1) Preparation of niosomes Five niosome formulations, labelled A to E, were prepared using the thin-film hydration technique, with cholesterol (45%), span 65 (45%), solutol HS-15 (10%) each dissolved in 4 mL organic solvent (chloroform) in a 250 mL round bottom flask, for the manufacture of 300 μmol of vesicles. The chloroform was removed using a rotary evaporator at 60℃, 40r/min and a vacuum of (464±10)mbar (1bar=10^5Pa). After placing the 250 mL flask at the interface of the H_2O in the bath, the pressure was allowed to drop until no chloroform was left and a thin film of the mix formed on the flask wall. A total of 5 mL of tris buffer pH 7.4 with 0.01 mL or 10 uL of the active, X (*i.e.*, total active concentration added was 0.002 in volume ratio) was added to hydrate the lipid films, followed by gentle agitation–enabling the formation of multi-lamellar vesicles and the entrapment of the active in the vesicles. The mix was intermittently incubated at 60℃ for a period of 10 minutes while shaking to allow for the complete detachment of the lipid film, encouraging more entrapment. After this, the newly prepared niosomes were separated *via* sephadex G-50 column chromatography and characterized. According to results obtained from characterization studies on all five niosome formulations, models C and D were proven to be of excellent quality (*i.e.*, sizes of 592 and 601 nm and –49.2 and –34.5 mV surface charge, respectively) and were therefore considered lead formulations and incorporated into the cream base *via* manual mixing.

(2) Preparation of creams Four 100g O/W model creams containing active-loaded niosomes (labelled as model ⅠA-ⅣA), and their baselines without niosomes (labelled as model ⅠB-ⅣB), were prepared with the formulas stated in Table 23.1, according to the following method: the oil phase and water phase ingredients were weighed in two separate beakers. After heating the oil phase and water phase to 75℃, both phases were mixed together for 18 minutes at 9500r/min using the Silverson L5M electric homogenizer to obtain a uniform mix. At a cool down temperature of 40℃, 5% of the active niosomes suspended in water were added to each cream model in batch "A" and further mixed manually with a glass stirrer for two minutes, to

Table 23.1 Ingredient and amount variables in 100 g of each cream formulation.

Phase	INCI	Composition/%	ⅠA	ⅠB	ⅡA	ⅡB	ⅢA	ⅢB	ⅣA	ⅣB
Oil	Stearyl alcohol	Stearyl alcohol					1	1	1	1
	Simmondsia chinensis seed oil	Jojoba oil	4	4	5	5	5	5		
	Adansonia digitata seed oil	Baobab oil	4	4	5	5			5	5
	Cocos nucifera	Coconut oil					5	5	5	5
Water	Glycerine	Glycerine	5	5			5	5	5	5
	Propylene glycol	Propylene glycol			5	5				
	Aqua	Water	73.7	78.7	71.7	76.7	71.7	76.7	71.7	76.7
Active		Entrapped active	5		5		5		5	

avoid disruption of the vesicles. The newly formulated products were collected into eight separate 100 g glass jars with plastic caps, labelled ⅠA-ⅣA and ⅠB-ⅣB, with and without actives, respectively. The first, labelled model Ⅰ (1∶1 of jojoba and baobab oil) contained a water phase of 85%, oil phase (10%) and emulsifier (5%) while the remaining three had an equal percent composition of water phase (83%), oil phase (12%) and emulsifier (5%), labelled Ⅱ (1∶1 of jojoba and baobab oil), Ⅲ (1∶1 of jojoba and coconut oil) and Ⅳ (1∶1 of baobab and coconut oil).

(3) Sensory lexicons and definitions A sensory lexicon was devised in three different stages, for all formulated oil-in-water products: (a) appearance–pourability (b) pick-up–firmness and elasticity/stretchability (c) rub-out–spreadability and stickiness. Each stage was correlated with rheological parameters, as shown in Table 23.2, to help provide information on the identity and quality of the test products.

Table 23.2 Proposed protocol of rheological parameters–sensory attribute pairs, and their description.

Stage of usage	Sensorial attribute	Description	Rheological parameter
Appearance	Pourability	Ability of a product to flow or be pumped out of the container when a force is applied	Viscometry; Yield stress
Pick-up	Firmness	The degree to which the product is able to hold its shape or structure in the presence of force	Oscillatory; Amplitude sweep
	Elasticity/ Stretchability	It is the ability of the product to deform or expand (strain) by resisting an external force (stress)	Oscillatory; Frequency sweep
Rub-out	Spreadability	The force required to cause flow of the product	Viscometry; Yield stress
	Stickiness	Ability of product to attach to the skin, yielding a sticky skin feel	Oscillatory; Frequency sweep

(4) Instrumental rheology and sensory characterization To obtain the rheological measurements of the cream models, a Kinexus lab+ Rotational Rheometer (Malvern Panalytical Instruments, Malvern, UK) was used with a stainless-steel parallel plate of 20 mm diameter at a constant temperature of (32±1)℃, a gap size of 0.25 mm, and a humidity of 33%. All measurements were performed in triplicate (n=3).

Yield stress: pourability and spreadability–a stress range of 0.001 Pa to 10000 Pa at a ramp time of 2 min and a decade of 10 was applied.

Strain amplitude sweep with LVR determination: firmness–the samples were oscillated over a shear stress range of 0.001 Pa to 10000 Pa, at a frequency of 1 Hz and a decade of 10.

Frequency sweep: stickiness and elasticity or stretchability–the samples were oscillated over a frequency range of 50 to 0.05 Hz, at a percent strain within the LVR.

(5) Statistical analysis Statistical evaluation of results obtained for the formulated creams was achieved using the SPSS software (SPSS UK Ltd, IBM, Woking, UK). To indicate whether any significant correlations ($p < 0.05$) exist between the rheological data obtained on all eight O/W creams, Pearson's Chi-square test was conducted.

23.2 Example 2: The Manufacture and Analyses of a Suncream Formulated with Thermal Spring Waters from Ourense (NW Spain) and *Sargassum muticum* Extracts

Only the experimental section is shown here. For the research background and results, please see the other parts of the paper in reference.

23.2.1 Materials

(1) Water Thermal spring water was collected from four different sources of Ourense city (Galicia, Spain): As Burgas (B), A Chavasqueira (C), O Tinteiro (T), and Muíño da Veiga (M). It was carried in glass-capped bottles and stored at 4 ℃ until use. Double-distilled water (W) (MilliQ, Merck Millipore, Germany) was used to prepare a suncream control.

(2) Extracts The *Sargassum muticum* extract (SR) used as antioxidant ingredient was obtained by the extraction and fractionation of brown algae collected in Praia Mourisca (Alcabre, Spain) during Summer. The algal biomass was subjected to autohydrolysis (LSR 1 ∶ 30, lg R_0=3.56) in a stainless-steel reactor PARR 4843 (Parr Instr. Co., Moline, IL, USA). The autohydrolysis liquors, separated by filtration with a Büchner filter, were subjected to microfiltration and nanofiltration in a 200 Da membrane (both 60.7cm×101.6cm in length, Iberlact, Spain) in 6 steps. The retentate liquor was freeze-dried to obtain the SR extract.

A commercial solid extract of *Fucus* sp. (FE) (Guinama, Valencia, Spain) and α-tocopherol (TF) (Sigma-Aldrich, Munich, Germany) were used to prepare the extract control emulsions. A further suncream was prepared without any extract as a blank sample (Bl).

23.2.2 Methods

(1) Suncream formula The oil-in-water suncream was prepared by homogenisation of 600 g of oil phase with 65 g of water phase in a mixer (constant stirring, 70℃). An oil phase (337.81 g cream basis, 112.64 g dimethicone 350, 56.32 g avocado oil, 150.15 g sunscreen, 337.81 g titanium dioxide, 6.6 g Fenonip) was melted in a water bath until the temperature reached 70℃ . The water phase was prepared by adding 6 g propyleneglycol and 1.5 g carbopol ultrez 10 to 427 g of each tested thermal water with further mixing. In order to form a gel, 1.5 g triethanolamine was added to the emulsion and mixed. All ingredients were purchased from Guinama (Alboraya, Spain). When the temperature decreased to 40℃ , 0.45 mL rose oil and 3 mL tetramer cyclomethicone were added.

Four parts of 140 g from this batch were disposed in different beakers to prepare four suncreams by adding one of the three selected extracts at 0.15% (by mass) as the antioxidant ingredient in each of them. One cream without any extract was prepared as a control. Every mixture was carefully homogenised and then bottled in three clean glass-capped flasks of 50 mL.

Five suncreams were elaborated, in triplicate, with four thermal spring waters (B, C, T, and M), with bidistilled water (W), and a blank with W. Each formula was combined with *Sargassum muticum* extract (SR), *Fucus* sp. commercial extract (FE), and commercial α-tocopherol (TF) as a control emulsion, to obtain 15 different formulations. A further formulation was prepared without antioxidant extract (Bl).

(2) Analytical methods

① Determination of the water composition. The content of the dissolved anions fluoride (F^-), chloride (Cl^-), nitrate (NO_3^-), nitrite (NO_2^-), and sulfate (SO_4^{2-}) in the thermal spring water samples was determined by ionic chromatography (Dionex, ICS-3000, Sunnyvale, CA, USA). Briefly, water was filtered using 0.22μm regenerated cellulose filters and injected into a chromatograph equipped with a conductivity detector (eluent 4.5mmol/L Na_2CO_3/0.8mmol/L $NaHCO_3$, flow rate 1mL/min, working temperature 30℃). A multielement standard (100×10^{-6}, CPAchem, Stara Zagora, Bulgaria) was used for identification.

② Emulsion pH and color determination. The pH of the thermal water and the suncream samples was measured using a digital pH meter (GLP 21, Crison, Barcelona, Spain). A portable colorimeter (Konica Minolta CR-600d, Osaka, Japan) was used to measure the color of the suncreams. The measurements were tested three times.

The CIELab color space parameters lightness (L*), redness-greenness (a*) and yellowness-blueness (b*) were determined from month 1 to month 9 every two months. The instrument was calibrated with a white ceramic tile.

③ Sensory analysis. A sensory test was designed to evaluate the influence of using thermal spring water and natural extracts as ingredients of the suncream preparations. Samples were presented at room temperature in 50-mL flasks, randomly coded with three digits according to UNE-EN ISO 8587:2010.

A total of 18 untrained healthy judges (20-55 years old) agreed to participate as volunteers and completed the questionnaires from each cream sample at time 0 (4 thermal waters, 1 distilled water, and 4 extracts). Only one panelist was discarded, because the questionnaire was incomplete.

Hedonic judgements concerning five appearance attributes (gloss, consistency, color, smell intensity, smell preference), seven skin parameters (spreadability, softness, skin gloss, skin feel, skin smell intensity, skin smell persistence, skin smell preference), and the global preference for each suncream were collected. The intensity of each attribute was rated on a 0-10 scale (0, not detected; 10, high intensity). The sensory attributes were assessed one after the other by the panelists, which was done on the inner side of the forearms. Paper tissues and water were provided for careful skin cleaning and drying before the measurements.

④ Oxidation experiments. The emulsions were stored in triplicate in 50-mL glass-capped flasks (room temperature, darkness) during a period of 9 months. The experimental conditions tried to simulate and evaluate the normal oxidation process of a conventional commercial suncream, which, once opened, is used by the consumer during a long period under non-refrigerated storage.

Aliquots of each emulsion were removed periodically to determine their peroxide value and *p*-anisidine value. The spectrophotometric determination of the peroxide value assay (pV) was performed by mixing one aliquot of emulsion with isooctane ∶ 2-propanol (3 ∶ 2, volume ratio). Peroxide quantification was carried out in accordance with Díaz *et al.* using a cumene hydroperoxide standard curve. The spectrophotometric determination of the *p*-anisidine value (pA) was carried out in accordance with AOCS Cd 18-90. The total oxidative value (TOTOX) was calculated to assess the oxidative deterioration of the lipids as the sum of the peroxide and *p*-anisidine values defined above ($TOTOX=2\times pV+pA$).

⑤ Statistical analysis. The effect of the use of thermal water and natural antioxidants in the preparation of suncream formulations on hedonic judgements concerning five appearance and seven skin attributes and on global preference were tested using an analysis of variance (ANOVA).

Fisher's test at a significance level of $p < 0.05$ was used to determine the statistical differences between the cream samples, based on the characteristics described by the panelists. ANOVA was applied to the descriptive analysis, samples, and the panelists data, followed by a means separation using the Fisher's test at a 95% significance level. Hierarchical cluster analysis (HCA) and principal component analysis (PCA) were carried out using a Statistica 8.0 program (Stat-Soft, Tulsa, OK, USA).

23.3 Example 3: The Manufacture and Analyses of Several Hair Conditioner Products

Only the experimental section is shown here. For the research background and results, please see the other parts of the paper in reference.

23.3.1 Materials

Hydroxyethyl-behenamidopropyl-dimonium chloride (HBD-C1) and behentrimonium-methosulfate were provided by CRODA (Snaith, United Kingdom), whereas cetrimonium chloride was supplied by BASF (Ludwigshafen, Germany). The remaining raw materials were provided by the cosmetic company Belleza Express S.A. (Cali, Colombia) and used as received. Water Type Ⅱ (ultra-pure water) was obtained from a purification system called Millipore Elix Essential (Merck KGaA, Darmstadt, Germany).

23.3.2 Surface tension and contact angle measurements

Surface tension and static contact angle measurements were carried out using the pendant and sessile drop methodology. Here, a video-based optical contact angle instrument (OCA15EC Dataphysics Instruments, Filderstadt, Germany) with version 4.5.14 SCA20 and SCA22 software were used. Data were recorded on an IDS video camera, where information was gathered from approximately 400 to 800 frames for reference as the static angle. Moreover, the point of capture was defined where the reflected incident light completely disappeared (about 1s from leaving the dispensing system). Drop volumes ranged from 5×10^{-3} to 15×10^{-3}mL, whereas the liquid deposition was fixed to 1cm for all assays. Each measurement was carried out at approximately (22±1)℃ and 60%±5% relative humidity.

23.3.3 Determination of the required HLB for the oil phase

To determine the required HLB (hydrophile-lipophile balance) in the oil phase for the emulsified prototypes, 11 formulations were synthesized (in the oil phase only) using a range of preservatives, ultra-pure water, and an emulsifying system. In these formulations, the same proportions were maintained as described in Table 23.3. Only the emulsifying system was changed to a Span 80 and Tween 80 blend, with the mixture combined using different proportions to obtain mixed HLB values, corresponding to 6, 7, 8, 9, 10, 11, 12, 13, 14, 15, and 16. Subsequently, each prototype was packed inside a 15 mL Falcon™ tube and exposed to a thermal stress assay for three weeks. The storage conditions were changed from (40±1) ℃ to (4±0.5)℃ every four days.

23.3.4 Preparation of emulsions

The emulsified systems were developed according to the formulation shown in Table 23.3.

Table 23.3 Hair conditioner product prototype formulations

<table>
<tr><th>Individual component</th><th>Ingredients</th><th>Mass ratio/%</th></tr>
<tr><td rowspan="3">Oil phase</td><td>Cetearyl-alcohol</td><td rowspan="3">4.8</td></tr>
<tr><td>Coco-caprylate</td></tr>
<tr><td>Shea butter</td></tr>
<tr><td>viscosity agent</td><td>Hydroxy-ethyl-cellulose</td><td>0.5</td></tr>
<tr><td>Wetting agent</td><td>Glycerin</td><td>0.2</td></tr>
<tr><td rowspan="3">Preservative</td><td>Methyl-isothiazolinone</td><td rowspan="3">0.2</td></tr>
<tr><td>Phenethyl-alcohol</td></tr>
<tr><td>Propylene-glycol-2-methyl ether</td></tr>
<tr><td>pH modifier</td><td>Citric acid</td><td>0.05</td></tr>
<tr><td rowspan="3">Emulsifier system</td><td>Lauryl-glucoside (neutral)</td><td>0.2</td></tr>
<tr><td>Cationic surfactant 1</td><td rowspan="2">1</td></tr>
<tr><td>Cationic surfactant 2</td></tr>
<tr><td>Dispersing phase</td><td>Water</td><td>q.s. (quantum sufficit)</td></tr>
</table>

All prototypes were synthesized in triplicate using different binary cationic surfactant blends, as shown in Table 23.4. At first, the aqueous phases (composed of hydroxy-ethyl-cellulose, glycerin, and ultra-pure water) were mixed using a homo-mixer (Ultra Turrax® T-25, IKA®, Staufen, DE-BW, Germany) at 6000r/min for 10 min and subsequently heated to 80℃ (mixture 1). At the same time, the oily ingredients were weighed together with the surfactants (mixture 2) and heated to 75℃ in a separate vessel. The oil phase (mixture 2) was then added to the aqueous phase (mixture 1) with a 9000r/min agitation for 10 min until a white emulsion was formed with a viscous consistency. The emulsions were continuously stirred at 400r/min speed using a propeller-type stirrer (IKA® RW 20, IKA®, Staufen, DE-BW, Germany) and heated to 40℃ , whereas the remaining ingredients (*i.e.*, preservative and pH modifier) were incorporated into the emulsion. The resulting mixture was cooled to room temperature. Subsequently, the prototypes were selected according to the viscosity criteria, which must have a minimum viscosity of 15000cP(1cP=1mPa · s). These conditions are fundamental to both intrinsic organoleptic characteristics and the stabilization of these products.

Table 23.4 Designing the cationic surfactant mixture for the hair conditioner prototypes

Emulsified system prototype	Cationic surfactant mixture	Ratio
1	HBD-Cl and BT-MS	1 ∶ 03
2	HBD-Cl and BT-MS	1 ∶ 01
3	HBD-Cl and BT-MS	3 ∶ 01
4	HBD-Cl and CT-Cl	1 ∶ 03
5	HBD-Cl and CT-Cl	1 ∶ 01
6	HBD-Cl and CT-Cl	3 ∶ 01
7	HBD-Cl and PQ-70	1 ∶ 03
8	HBD-Cl and PQ-70	1 ∶ 01
9	HBD-Cl and PQ-70	3 ∶ 01
10	BT-MS and CT-Cl	1 ∶ 03
11	BT-MS and CT-Cl	1 ∶ 01
12	BT-MS and CT-Cl	3 ∶ 01
13	BT-MS and PQ-70	1 ∶ 03
14	BT-MS and PQ-70	1 ∶ 01
15	BT-MS and PQ-70	3 ∶ 01
16	CT-Cl and PQ-70	1 ∶ 03
17	CT-Cl and PQ70	1 ∶ 01
18	CT-Cl and PQ-70	3 ∶ 01

23.3.5 Accelerated stability tests

One hundred milliliters of each prototype conditioner were packed in a polyethylene-terephthalate container and stored at accelerated stability conditions [40℃ ±2℃ and (75±5)% relative humidity] for 12 weeks.

The ratio is given in parts that are 1% of the total cationic surfactant in each formulation. HBD-Cl: hydroxyethyl-behenamidopropyl-diammonium chloride, BT-MS: behentrimonium methosulphate, CT-Cl: cetrimonium chloride, and PQ-70: polyquaternium-70.

23.3.6 Zeta potential, pH, and conductivity measurements

Zeta potential measurements were carried out using a zetasizer nano ZSP (Malvern Instruments, UK) at (25±2)℃ temperature, with equilibration times of 120 s in a DTS 1070 capillary cell. Here, the attenuator position and intensity were set automatically. To prepare a sample, 130 mg of the emulsifier was diluted in 20 mL of ultra-pure water and manually stirred. From this, a 50 μL aliquot was taken and diluted with 1 mL of ultra-pure water before each zeta potential measurement was taken. The electrical conductivity and the pH of the emulsions were determined using a CR-30 conductivity meter and a Starter-2100 pH meter, respectively.

23.3.7 Particle size analysis

The particle size distribution of the emulsions was obtained using a Mastersizer 3000 (Malvern Instruments, UK), equipped with a helium/neon laser at a wavelength of 632.8 nm. Previously, 0.6 g of the emulsion was diluted with 10 mL of ultra-pure water at (25±2)℃ and stirred at 400r/min. The appropriate amount sample was determined when the dark level was reached (between 2% and 8%).

23.3.8 Rheological profile and viscosity

The rheological profile was obtained using different shear rates (20 s^{-1} to 150 s^{-1}). For this, a viscometer (micro-visc, RheoSense Inc., San Ramon, CA, USA) was used. The viscosity was measured using a Brookfield® viscometer with a No. 4 needle (Brookfield®, Middleboro, MA, USA) at 10r/min.

23.3.9 Creaming index (CI)

Creaming index values were determined from the ratio of sediment volume and the total emulsion volume. Here, 15 mL of the emulsion was centrifuged in a Wincom 80-2 centrifuge at 3000r/min for 4 h. The results were expressed as the creaming index (CI) according to

$$CI=HS/HE\times 100$$

where HS is the sediment height and HE is the sample height before centrifugation.

23.4 Example 4: The Manufacture and Analyses of Lipsticks, Lip Balms and Skin Creams Containing Alkenones

Only the experimental section is shown here. For the research background and results, please see the other parts of the paper in reference.

23.4.1 Materials

The marine microalgae *Isochrysis* was purchased from Necton S.A. (Olhão, Portugal). Alkenones were isolated and purified from the *Isochrysis* biomass as previously described.

Microcrystalline wax, ozokerite, castor oil, triglyceride, isoeicosane, meadowfoam seed oil, lanolin alcohol, candelilla wax, carnauba wax, Red 7, mica pearl white, tocopherol, petrolatum, jojoba oil, almond oil, avocado butter, and xanthan gum were purchased from Making Cosmetics (Snoqualmie, WA, USA). The following ingredients were received as gifts: hexamethyldisiloxane (Dow Corning, Midland, MI, USA); isohexadecane (Presperse, Somerset, NJ, USA); isopropyl stearate (Lubrizol, Wickliffe, OH, USA); C12-15 alkyl benzoate (Phoenix Chemical, Calhoun, GA, USA); ethylhexyl methoxycinnamate and Germaben II (INCI: propylene glycol, diazolidinyl urea, methylparaben, and propylparaben; Ashland, Covington, KY, USA); propanediol (DuPont Tate Lyle Bio Products, Loudon, TN, USA); sorbitan oleate, polysorbate 80, and cetyl alcohol (Croda, Edison, NJ, USA), stearic acid (Acme Hardesty, Blue Bell, PA, USA), and glyceryl monostearate (Corbion, Lenexa, KS, USA). All ingredients were of cosmetic grade.

23.4.2 Methods

(1) Determination of solubility Formulating for Efficacy™ (ACT Solutions Corp, Kirkwood, DE, USA), hereinafter referred to as 'FFE', is an *in-silico* modeling software used in the cosmetic and personal care and the pharmaceutical industry in formulation design. In this project, FFE was utilized to predict the solubility of alkenones based on Hansen Solubility Parameters (HSPs) in a variety of solvents. Solubility was necessary information for later steps of this study. In order to add ingredients to FFE, we created their simplified molecular-input line-entry system (SMILES), as this information was not available in the published literature or chemical databases (*e.g.*, PubChem). ChemDraw (Release 15.0, CambridgeSoft, Waltham, MA, USA), a molecule editing software, was used to create the SMILES.

(2) Melting point determination Melting point was determined using differential scanning calorimetry (DSC) analysis. Using a Mettler MT 5 microbalance (Mettler Toledo, Columbus, OH, USA), a 6-mg sample was sealed in an aluminum crucible. A pinhole was created on the lid of the crucible to vent gas buildup. DSC was performed at a 10℃/min ramp from 0-400℃ using a DSC 822e instrument (Mettler Toledo, Columbus, OH, USA) attached to a F25-ME refrigerated/heating circulator (Julabo, Allentown, PA, USA). Nitrogen gas was purged at a rate of 10 mL/min. TA Universal analysis software was used to obtain the scans.

(3) Thickening capability test This test was used to determine how alkenones thicken emollients commonly used in personal care and makeup product formulations.

In order to select comparators for alkenones, a literature search was done on the melting point of various, commonly used waxes. The ones with melting points close to that of the alkenones were tested with the above-mentioned DSC method to confirm the melting range. Two waxes with similar melting ranges, microcrystalline wax and ozokerite, were identified as good comparators. The melting point peak was 69℃ for microcrystalline wax, and 74℃ for ozokerite.

Solvents identified by FFE were used in this test. To evaluate the thickening of the waxes, 9.0 g of solvent was weighed on an analytical balance with an accuracy of 0.001 g into a 25-mL glass beaker. The solvent, while covered with aluminum foil to avoid evaporation, was heated approximately 5℃ above the melting point of the particular wax. When the solvent reached the desired temperature, it was removed from the heat, and 1.0 g of the wax was added to the solvent. Mixing continued until room temperature was reached. The mixture was then reweighed and the evaporated solvent was replaced. The beaker was left covered with a piece of aluminum foil overnight. The stability of the mixture was checked the following day, and if it remained stable, viscosity was measured. Stability was qualitatively assessed based on signs of separation and any visible change in color.

A Brookfield viscometer DV-I (Brookfield Engineering Laboratories, Middleboro, MA, USA) was used with a concentric cylinder spindle (21#) and a small sample adapter to determine the viscosity of the different wax-solvent mixtures. The tests were performed at 21℃ . The spindle was rotated from 0-100r/min. All measurements were done in triplicate.

(4) Lipstick and lip balm formulation　Based on the melting-point determination, microcrystalline wax and ozokerite were selected to be used as comparators for the alkenones. A lipstick formula (Table 23.5) and a lip balm formula (Table 23.6) containing both microcrystalline wax and ozokerite were selected with the purpose that the alkenones would be tested as an alternative for either microcrystalline wax or ozokerite. This way, both the compatibility of the various waxes and the suitability of the alkenones in a lipstick/lip balm formula could be tested.

Table 23.5　Lipstick formulas

Lipstick formula	INCI name	Lipstick 1/% (by mass)	Lipstick 2/% (by mass)	Lipstick 3/% (by mass)
Phase A	Castor oil	25.8	25.8	25.8
	Triglyceride	16	16	16
	Isoeicosane	17	17	17
	Meadowfoam seed oil	5	5	5
	Lanolin alcohol	5	5	5
	Microcrystalline wax	2	2	—
	Ozokerite wax	5	—	5
	Alkenone wax	—	5	2
	Candelilla wax	7	7	7
	Carnauba wax	3	3	3
Phase B	Red 7 (and) castor oil	2	2	2
	Mica pearl white	11	11	11
Phase C	Tocopherol	0.2	0.2	0.2
	Propylene glycol (and) diazolidinyl urea (and) methylparaben (and) propylparaben	1	1	1

For the lipstick, phase A was added to a glass beaker and heated to 80℃ to melt the waxes. The pigment dispersion in phase B was prepared using an EXACT three-roll mill (EXACT

Technologies, Inc, Oklahoma City, OK, USA). Ingredients of phase B were added to melted phase A one-by-one and mixed until the color was uniform. Then the mixture was removed from heat. Phase C was added to phase A/B, and the mixture was poured into a metal lipstick mold while still hot. When settled, the sticks were topped off. The mold then was put into the refrigerator for 15 min. The lipsticks were then removed from the mold and inserted into plastic lipstick cases.

Table 23.6 Lip balm formulas

Lipstick formula	INCI name	Lip balm 1/% (by mass)	Lip balm 2/% (by mass)	Lip balm 3/% (by mass)
Phase A	Petrolatum	45.5	45.5	45.5
	Jojoba oil	22	22	22
	Almond oil	13	13	13
	Microcrystalline wax	8	8	—
	Ozokerite	8	—	8
	Alkenones	—	8	8
	Avocado butter	3	3	3
Phase B	Tocopherol	0.5	0.5	0.5

For the lip balm, phase A was heated to 80℃ while mixing until all ingredients were melted. Then the mixture was removed from the heat and phase B was added with mixing until a homogenous mixture was obtained. The mixture was then poured into lip balm tubes.

(5) Cream formulation Four oil-in-water (O/W) emulsions were formulated. One contained the alkenones, while the other three were formulated with commonly used waxes as comparators for the alkenones. The comparator waxes were cetyl alcohol, stearic acid, and glyceryl monostearate. The waxes acted as thickeners in the oil phase. The composition of the creams can be found in Table 23.7.

Table 23.7 Cream formulas

Cream formula	INCI name	Cream 1/% (by mass)	Cream 2/% (by mass)	Cream 3/% (by mass)	Cream 4/% (by mass)
Phase A	Water	68.8	68.8	68.8	68.8
	Propanediol	5	5	5	5
	Xanthan gum	0.2	0.2	0.2	0.2
Phase B	Isopropyl isostearate	15	15	15	15
	Alkenone	5	—	—	—
	Cetyl alcohol	—	5	—	—
	Stearic acid	—	—	5	—
	Glyceryl monostearate	—	—	—	5
	Sorbitan oleate	1.5	1.5	1.5	1.5
	Polysorbate 80	3.5	3.5	3.5	3.5
Phase C	Propylene glycol (and) Diazolidinyl urea (and) Methyl paraben (and) Propyl paraben	1	1	1	1

All four creams were formulated identically. First, water was heated to 80℃. Xanthan gum was mixed with propanediol to form a slurry. This slurry was added to the water. The oil phase (*i.e.*, phase B) was combined in a separate beaker and heated to 80℃. When both phases reached the same temperature, the oil phase was added to the water phase with propeller mixing. The emulsion was briefly homogenized, then it was removed from the heat. The emulsion was allowed to cool with continuous propeller stirring, and phase C was added when the emulsion reached 45℃. The emulsion was allowed to cool to room temperature under continuous mixing. Water loss was checked by weighing the creams and evaporated water was replaced. The emulsion was mixed again and was then homogenized for 10 s. Then, each cream was stored in plastic jars.

(6) Viscosity, thixotropy, and rheology Rheological properties and viscosity of the emulsions were evaluated using a Discovery hybrid rheometer DHR-3 (TA Instruments, New Castle, DE, USA). A 40 mm 2° cone and plate geometry at (25.0±0.1)℃ was used to test samples of ~0.8 mL. The shear rates ranged from 0.1-100 s^{-1} in steady state flow for viscosity and thixotropy measurements. Dynamic viscoelasticity was measured as a function of frequency in the linear viscoelastic region (LVE).

First, the LVE range was determined with an amplitude sweep. The amplitude sweep helps establish the extent of the material's linearity. Below the critical stress level, the structure is intact, and G' (storage modulus) and G" (loss modulus) remain constant. Increasing the stress above the critical stress level disrupts the network structure. Above the critical stress, the material's behavior is non-linear and G' usually declines. After the LVE was defined, a frequency sweep at an oscillation stress below the critical value was performed. The frequency sweep provides more information about the effect of colloidal forces, type of network structure in the cream and the interactions among droplets.

(7) Liquid crystalline structure Oleosomes are liquid crystal regions that form around the oil drops in an emulsion. These are multilayers of lamellar liquid crystals surrounding the oil droplets that become randomly distributed as they progress into the continuous phase. The rest of the liquid crystals produce the "gel" phase that is viscoelastic.

The creams were observed with a polarized light microscope (Accu-Scope® 3000 LED, Accu-Scope, Commack, NY, USA). For the microscopic evaluation, a small amount of each emulsion was placed on a microscopic slide and quickly covered with a cover slip. The sample was finger pressed to spread and create a cream layer as thin as possible. A 40× objective lens was used with cross polarizers in a bright field to detect birefringence. Detecting optical linear birefringence has been a standard tool in studying the anisotropic properties of materials for nearly two centuries. Birefringence can be detected using a polarized light microscope. The sample is placed between two polarizers oriented with their planes of vibration being mutually perpendicular. When an isotropic sample is placed between crossed polars, the state of the polarization of light is unchanged and in theory, no light is transmitted through the optical system. The light can only be transmitted if the state of polarization is changed, *i.e.*, the sample is birefringent. Oleosomes appear under optical microscopy as "Maltese crosses".

(8) Stability testing Stability of the lipsticks, lip balms and creams were monitored at two temperatures, room temperature (25℃) and an elevated temperature (45℃) in stability cabinets for 10 weeks. Samples were checked visually at weeks 1, 2, 4, 6, 8, and 10. Lipsticks and lip balms were assessed in their final containers. Samples of creams were placed into 1.5-mL centrifuge tubes, and those tubes were placed into stability cabinets.

References

[1] Adejokun D A, Dodou K. Quantitative Sensory Interpretation of Rheological Parameters of a Cream Formulation. Cosmetics, 2020, 7: 2.

[2] Balboa E, Conde E, Constenla A, Falqué E, Domínguez H. Sensory Evaluation and Oxidative Stability of a Suncream Formulated with Thermal Spring Waters from Ourense (NW Spain) and *Sargassum muticum* Extracts. Cosmetics, 2017, 4: 19.

[3] Agredo P, Rave M C, Echeverri J D, Romero D, Salamanca C H. An Evaluation of the Physicochemical Properties of Stabilized Oil-in-Water Emulsions Using Different Cationic Surfactant Blends for Potential Use in the Cosmetic Industry. Cosmetics, 2019, 6: 12.

[4] McIntosh K, Smith A, Young L K, Leitch M A, Tiwari A K, Reddy C M, O'Neil G W, Liberatore M W, Chandler M, Baki G. Alkenones as a Promising Green Alternative for Waxes in Cosmetics and Personal Care Products. Cosmetics, 2018, 5: 34.

Key Words & Phrases

baobab oil 猴面包树油
baseline *n*. 基线；基础；起点
homogenizer *n*. 均化（匀浆）器；均质机
lexicon *n*. 全部词汇；词汇表；字典
pourability *n*. 倾倒性；可浇注性；流动性
stretchability *n*. 可拉伸性；可延性；拉伸性
spreadability *n*. 铺展性；蔬菜汁的延展性；覆盖性；涂抹性
stickiness *n*. 黏稠度；黏性
rheological *adj*. 流变的
triplicate *n*. 一式三份
chi-square 卡方分布；卡方；卡方检验
suncream *n*. 防晒霜
Sargassum muticum extract 马尾藻提取物
brown algae 褐藻门；褐藻
microfiltration *n*. 微滤
nanofiltration *n*. 纳米过滤
freeze-dried 冷冻干燥
Fucus *n*. 墨角藻属；墨角藻；褐藻；黑角菜属之海藻；岩藻
homogenisation *n*. 均化；同质化；均一化；均质化
elaborate *v*. 详尽阐述；详细制订；精心制作
ionic chromatography 离子色谱
multielement *n*. 多元素；多元件
colorimeter *n*. 色量计，色度计；比色计
calibrate *v*. 标定；校准（刻度，以使测量准确）
panelist *n*. 小组成员
hedonic *adj*. 享乐的；享乐主义的
aliquot *n*. 试样；整除数；可分量
p-anisidine *n*. 对氨基苯甲醚；对甲氧基苯胺
cumene hydroperoxide 过氧化氢异丙苯
variance *n*. 方差；变化幅度；差额
hydroxyethyl-behenamidopropyl-dimonium chloride 羟乙基苯胺丙基二氯化铵
behentrimonium-methosulfate 山嵛基三甲基铵甲基硫酸盐
ultra-pure water 超纯水
pendant and sessile drop methodology 悬挂式和固定式跌落法
organoleptic *adj*. 影响（或涉及）器官（尤指味觉、嗅觉或视觉器官）的；感官的
zeta potential Zeta 电位
viscometer *n*. 黏度计
alkenone *n*. 烯酮；烯酮类
Isochrysis *n*. 等鞭金藻属
hexamethyldisiloxane *n*. 六甲基二硅醚；六甲基二硅氧烷
alkyl benzoate 烷基苯甲酸酯
propylene glycol *n*. 丙二醇
sorbitan oleate 山梨醇油酸酯
crucible *n*. 坩埚；熔炉
pinhole *n*. 针刺的孔；针孔

comparator *n.* 坐标量测仪；比较仪；比较器
concentric cylinder spindle 同心圆柱主轴
slurry *n.*（土、煤末或水泥混合而成的）泥浆；稀泥
thixotropy *n.* 触变性；摇溶性；触变剂；搅溶性
viscoelasticity *n.* 黏弹性
viscoelastic region 黏弹性区
amplitude sweep 振幅扫描
critical stress 临界应力
frequency sweep 频率扫描；扫频
oscillation *n.* 摆动；摇摆；振动；浮动；振幅
colloidal force 胶体力
oleosome *n.* 油质体
lamellar *adj.* 薄片状的；层式的；成薄层的
birefringence *n.* 双（光）折射；二次光折射；折射率
anisotropic property 各向异性特性
Maltese cross 黑十字花样

Chapter 24 Cosmetic Packaging

24.1 Description

The term cosmetic packaging includes primary and secondary packaging. Primary packaging, also called cosmetic containers, is housing the cosmetic product. It is in direct contact with the cosmetic product. Secondary packaging is the outer wrapping of one or several cosmetic containers. An important difference between primary and secondary packaging is that any information that is necessary to clarify the safety of the product must appear on the primary package. Otherwise, much of the required information can appear on just the secondary packaging. The cosmetic container shall carry the name of the distributor, the ingredients, define storage, nominal content, product identification (*e.g.*, batch number), warning notices, and directions for use. The secondary packaging shall, in addition, carry the address of the distributor and information on the cosmetic's mode of action. The secondary packaging does not need to carry any product identification notice. In cases where the cosmetic product is only wrapped by one single container, this container needs to carry all the information.

24.2 Purpose of Cosmetic Packaging

There are multiple reasons why care must be put into cosmetic containers. Not only must they protect the product, they need to provide conveniences for vendors and ultimately consumers.

(1) Protection The main purpose of a container is to store the product so that it is not degraded through storage, shipping and handling. Degradation and damage can be caused by various causes. These causes can be categorized into biological, chemical, thermal causes, damage caused by radiation and damage caused by human interaction, by electric sources or by pressure. In addition to protecting the product, packaging also plays a big role marketing cosmetic products. While product quality is a major factor in the product's success, its packaging must be attractive since that's the essence of beauty marketing. Package design must capture the imagination and be associated with enhancing appearance. One of the keys to attractive packaging is the artistic use of colors. Most relevant for the marketer is the outer secondary packaging. However, there are cosmetics which are distributed in one single cosmetic container.

(2) Creation of brand awareness Cosmetic packages must not only convey beauty, they must equate to brand awareness. Since the package is what the consumer initially sees, it is very influential in shaping perceptions about the product. Part of building brand awareness for a

cosmetic product is associating it with emotion. Since it is not a survival product it is marketed to appeal to the desire to enhance appearance. The packaging must stimulate this emotion.

(3) Labelling　Labels tell consumers what they need to know about the product, as far as how to use it and where it comes from. Companies must list the ingredients and the function of the product, especially when it is unclear. The label must contain the contact information of the entity responsible for putting the product on the market. Labels also provide product tracking information. The label must be easy to read, particularly for a customer where the product is being displayed. Certain compositions, such as perfumes, can be listed as one ingredient. Secondary packages are what the consumer sees as the outermost package. Primary packages are within the secondary package. Certain information can appear just on secondary packages. The most important information, particularly if the product is prone to misuse, must be displayed on both the primary and secondary packaging.

(4) Information accuracy　One of the most important aspects of regulations on labeling is that the information is accurate. Although the FDA does not have the resources to inspect all cosmetic products on the market, it can issue penalties for various violations involving packaging and labeling. It is the manufacturer's responsibility to make sure that its product is safe for public consumption.

(5) Avoidance of misleading information　None of the information, including name and address, may be misleading. Words can be abbreviated only if it is clear what they represent. All text must be printed clearly on the packaging. Smaller packages in which text is too difficult to read should include tags with legible text.

(6) Listing of ingredients　Ingredients must be listed in a certain order with priority given to ingredients that represent 1% or more of the volume. These ingredients must be listed in descending order, based on mass. This group of ingredients is then followed by those that represent 1% or less of the product and listed in any order. Colorants may also be listed in any order.

24.3 Packaging in Multiple Layers

Many times cosmetic products are packaged in multiple layers. Whenever it is difficult to detect for the consumer, the number of units should be listed on the outer package, which should contain details about how to use the product and warnings on what to do if it is misused. It is essential that the product is protected from environmental elements such as mould and bacteria.

The packaging must be sufficient enough to protect the mechanical, thermal, biological, and chemical properties of the product. It should also be strong enough to withstand human tampering and radiation damage.

24.4 Standards and Regulations

The FDA oversees cosmetic packaging but does not test products. It leaves testing for safety up to manufacturers. It still provides regulations and can issue recalls when a product is associated with safety hazards. While the FDA does not have many restrictions on ingredients for cosmetic products, it does require that certain chemicals and colorants be listed.

As far as EU regulations regarding packaging, manufacturers must be compliant with EC No. 1223/2009. One of these requirements involves the manufacturer issuing a safety report before putting the product on the market. The manufacturer must also disclose any serious undesirable effects (SUE) to the EU. Marketers are required to list nano-materials. The EU's definition of "ingredients" does not include raw or technical materials used in production that do not end up in the final product. In some cases when durability is an issue, the manufacturer must list an expiration date after the product has been opened. The words "best used before" are common for identifying the product expiration date.

Standard ISO 22715 provides specifications for the packaging and labeling of all cosmetic products that are sold or distributed at no charge; *i.e.*, free samples. National regulations dictate what products are to be regarded as cosmetics. While ISO 22715 is not legally binding, national regulations regarding cosmetic products can be even stricter than those laid out in ISO 22715. The link between standards and regulations is that a standard often represents the common denominator of national law, as the standardization committee consists of members of most countries.

24.5 Environmental Aspects

The container must be made of materials resistant to hot or cold temperatures. It must also protect the product from ultraviolet rays, which can potentially damage the product. The container also cannot absorb product substances. Traditionally, plastic material or glass have been used to house cosmetics. Aluminum has become a popular type of container due to its lightweight yet sturdy quality, flexibility, durability and recyclability. A key factor in what type of material can be used for containers is how compatible the material is with the product.

Plastic packaging materials are controversially discussed because of their polluting effects in particular to the marine environment. In 2014, a scientific study estimates the amount of floating plastics in the world's oceans to 5 trillion pieces with an accumulated mass of 250,000 tons.

Nowadays there is high and increasing demand for packaging made from bioplastics. Indeed, biodegradable and bio-based polymer matrices will be an added value *versus* the petrochemical-based polymers that are not bio-recyclable (compostable or biodegradable, following UNI EN ISO 13432). In this case, however, polymer crystallinity, structural conformation, and molecular weight must be strictly controlled in order to assess polymer degradability. Several biodegradable polymer materials have been widely studied: poly(lactic acid) (PLA), polyhydroxyalkanoates (PHAs) and polysaccharides.

Reference

Cinelli P, Coltelli M B, Signori F, Morganti P, Lazzeri A. Cosmetic Packaging to Save the Environment: Future Perspectives. Cosmetics, 2019, 6: 2.

Key Words & Phrases

brand awareness 品牌知名度
prone to 易于
legible *adj*. 易读的；可读的
sturdy *adj*. 结实的；坚固的
poly(lactic acid) 聚乳酸
polyhydroxyalkanoate 聚羟基乙醇酸酯

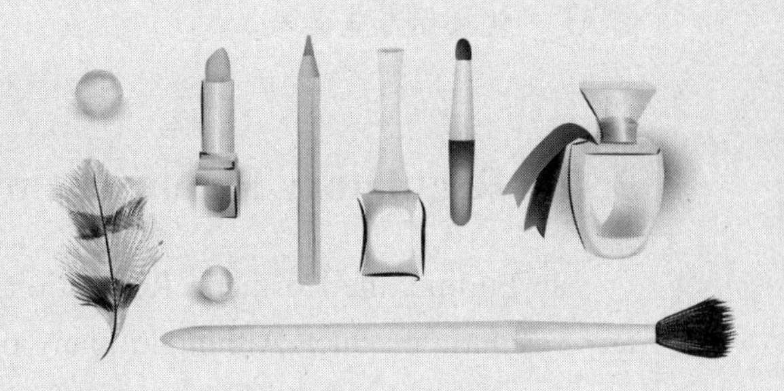

Unit 7
Safety Testing, Efficacy and Sensory Evaluation of Cosmetic Products

To ensure safety, cosmetic products must be controlled by a series of safety tests. On the other hand, to attract users, cosmetics manufacturers usually use original combinations of ingredients that give efficacies, effects, and concepts. Therefore, in this unit, you will see:

① Safety testing of cosmetic products;

② Efficacy of cosmetics;

③ Sensory evaluation.

Chapter 25 Safety Testing of Cosmetic Products

25.1 Introduction

Cosmetic products need to have a proven efficacy combined with a comprehensive toxicological assessment. The 7th Amendment to the European Cosmetics Directive has banned animal testing for cosmetic products and for cosmetic ingredients in 2004 and 2009, respectively. Then, the European Cosmetic Regulation No. 1223/2009 and the specific Regulation No. 655/2013 specify the required data to proof the safety and support the claims. Largely driven by regulatory authorities, a wide range of alternatives to animal testing have been developed and validated for safety testing of cosmetic products and adopted as test guidelines. This chapter discusses the main *in vitro* alternative models used in safety testing of cosmetic products and cosmetic ingredients with a focus on regulatory requirements, genotoxicity potential, skin sensitization potential, skin and eye irritation, endocrine properties, and dermal absorption. Advantages and limitations of each model in safety testing of cosmetic products are discussed and novel technologies capable of addressing these limitations are presented.

25.2 Regulatory Requirements for Cosmetics Safety Assessments

In Europe, the Cosmetic Regulation No.1223/2009 sets the framework for the safety of any cosmetic product. Although, many other geographical areas do not specify the detailed documentation to establish their own frameworks, their regulations share the common goal of ensuring safety of the final consumers.

Some ingredients must be included in so-called "positive" lists, for the ones having specific functions (Annex Ⅵ for colorants, Annex Ⅴ for preservatives, Annex Ⅴ for UV filters). An ingredient with such a function should then comply to the requirements of the given Annex. Some ingredients are prohibited (Annex Ⅱ) or restricted to particular uses (Annex Ⅲ).

The origin of those regulatory limitations is mainly safety. In Europe, some of the ingredients are evaluated by the SCCS (Scientific Committee on Consumer Safety), which publishes its opinion with safe conditions of use, before the ingredient is listed in an annex. The SCCS publishes opinions based on the evidence presented to it, combined with guidance. That is helpful, rather than spelling out the prescriptive demand for strict adherence to precise regulatory "guidelines". The European committee regularly provides a guidance for the evaluation of the safety of ingredients. In the USA, the CIR (Cosmetic Ingredient Review), established from a trade association (currently the PCPC) with the support of the FDA prioritizes and assesses cosmetic ingredients, generally consider groups of similar substances based on chemical families or plant-derived ingredients. The CIR's report does not include the risk assessment.

All regulated ingredients must have a favorable opinion of the SCCS, such as the recent ones on resorcinol, for its use in hair dyes, propylparaben as preservative (updated opinion discarding any concern related to endocrine disruption) or octocrylene as UV filter (other update related to endocrine disruption).

However, the committee can also give its opinion on substances for non-regulated uses (titanium dioxide in inhaled products or aluminum in lipsticks).

Some publications can also be available from national authorities, related to particular concern for a country (example of phenoxyethanol in France), or specific investigations allowing a better management of the risk, as in the case of "technically unavoidable concentrations" of heavy metals, studied in Germany.

Transversal regulations can have consequences on the safety of the substances used in cosmetic products: the CLP Regulation (classification, labelling and packaging of substances and mixtures) of major importance for CMR (carcinogenic, mutagenic and reprotoxic) substances. The carcinogenic, mutagenic, and reprotoxic substances are considered as the most dangerous substances; their harmonized classification in Europe is rarely based on epidemiological information (asbestos, benzene, *etc.*) and more generally based on experimental results in animals (musk xylene, Disperse Yellow 3, *etc.*). The Annex ⅩⅦ of REACH can be of major importance for a very limited number of substances: D4 (cyclopentasiloxane) and D5 (cyclotetrasiloxane) are prohibited silicones in rinsed products above 0.1% (under entry 70 of the Annex XVII of REACH for restrictions). This decision is not triggered by toxicological properties but by their fate in the environment: these are the PBT and vPvB effects (for Persistent, Bioaccumulative and Toxic, very Persistent, very Bioaccumulative).

The list of SVHC (substances of very high concern) includes substances based on concern regarding reprotoxicity, carcinogenicity, endocrine disruption or effects for the environment, PBT or vPvB.

Those programs are somehow linked to each other (the general concern of endocrine disruption justified a call for data from the European Commission to revise the opinions of the SCCS (*e.g.*, benzophenone-3, octocrylene, benzyl salicylate…) in the past two years.

(1) Substances restricted by an annex When the SCCS receives a mandate from the European Commission to assess the safety of a substance for a regulated function, the opinion is based on the analysis of the scientific dossier submitted by the industry.

The scientific opinion considers each endpoint, including local tolerance (skin irritation, phototoxicity when relevant), genotoxicity, systemic toxicity including reprotoxicity and sub-chronic/chronic toxicity. Characterization of dermal absorption is essential to calculate the SED (systemic exposure dose).

The exposure of the substance is considered as its expected concentration in cosmetic products, either in one given product or in several products, when a broad use is expected, as it would be for a preservative.

(2) Substances not restricted by an annex Any other substance, ingredient, or impurity has the obligation of being safe for the consumer, based on the toxicological profile, as required by the Annex I and Guidelines, using regularly updated data from supplier or literature.

There are two points of view: the one of the suppliers of the ingredient and the one of the Responsible Person for a cosmetic product using the ingredient (the Responsible Person being the legal entity in Europe responsible for the product, generally the manufacturer). They do not have the same regulatory obligations. However, they should have the same purpose: consumer safety.

Any supplier of a cosmetic ingredient, such as any company which manufactures and markets a substance in the European Union, must register its substance according to the annual tonnage.

Even if the intrinsic toxicity of a substance is independent from its production, the number of toxicological results required in a REACH registration dossier depend on the annual tonnage. Highly toxic substances and substances of low toxicity have the same requirements (but important concern should be taken into account among the program of SVHC: substances of very high concern). No toxicological data are requested for substances registered below 1 to 10 tpa (ton per annum) and increasing information is required to be submitted with increasing tonnage bands.

For tonnage of 10-100 tpa (Annex VII): toxicological requirements include data for *in vitro* skin irritation/corrosion, *in vitro* eye irritation, skin sensitization, *in vitro* gene mutation in bacteria, acute toxicity, and short-term toxicity (28 days).

At 10 to 1000 tpa (Annex Ⅷ): toxicological requirements include data for *in vitro* mutagenicity study in mammalian cells or *in vitro* micronucleus study, *in vitro* gene mutation in mammalian cells, *in vivo* skin irritation, *in vivo* eye irritation, possibly testing proposal for *in vivo* genotoxicity, acute toxicity, and screening for reproductive/developmental toxicity.

At 100 to 1000 tpa (Annex Ⅸ) following endpoints are added: the sub-chronic toxicity (90 days), prenatal developmental toxicity in one species, and extended one-generation reproductive toxicity.

Finally, above 1000 tpa (Annex Ⅹ) a long-term repeated dose toxicity (≥12 months) if triggered, with developmental toxicity in a second species, extended one-generation reproductive toxicity, and carcinogenicity.

It is then important to realize that for ingredients produced below 10 to 1000 tpa, no information is available about the DNA damage (micronucleus test), and below 100 tpa, neither any sub-chronic toxicity nor any information on the full cycle of reproduction is known. A supplier of cosmetic ingredients should then think about the need of the cosmetic brands (Responsible Persons in general) who need to prove the safety of each ingredient.

The cosmetic brand (the Responsible Person) is the one responsible of the product. Studies can be made on the product, to confirm a good acceptability in humans. It is mostly to confirm the absence of eye and skin irritation, by *in vitro* test and other complementary tests in humans (the grail being the use test in normal conditions of use, to confirm the absence of objective irritation and absence of signs of discomfort). The tests for photo-toxicity or skin sensitization are rarely performed. It should be reminded that the Human Repeat Insult Patch Test (HRIPT) is non ethical and usually the historical data are significantly poor from a statistical point of view using a small size panel. However, the new *in vitro* tests for skin sensitization are quite promising, particularly if they can cover multiple Key Events of the Adverse Outcome Pathway, and if they can be applied to the finished product. Both the SENS-IS and Genomic Allergen Rapid Detection (GARD) assays analyze the genomic response of the cells to the exposure of the substance or the product to predict sensitization, including its potency, with GARD assay being able to quantify the dose-effects relationship, thus providing a good perspective for its use in quantitative risk assessment. Any test done on the finished product, as those two last ones, and the tests made on eleuthero-embryo from fish or amphibians discussed here are of particular relevance, since a large part of the risk assessment on the product in based on individual data of substances.

The major part of the safety then relies on the toxicological data of the substances. The toxicological results can come from the supplier, when they have a REACH registration dossier, or when they voluntary produce additional *in vitro* data. It can also be existing data from literature or *in silico* predictions Quantitative Structure-Activity Relationship (QSAR) or read-across. The safety assessor, working with the Responsible Person, makes a comprehensive search of existing toxicological information to write the toxicological profile of the ingredient, and possibly identify any data gaps. Pragmatically, toxicological profiles of ingredients often lack some information. Among the most current data gaps includes following endpoints: skin sensitization, DNA damage, chronic toxicity, and dermal absorption. With one exception, *in vitro* assays exist for all these endpoints, most of them with OECD guidelines, or with good results of validation. When it is chosen not to perform the test (data waiving), a rationale is absolutely needed as justification. *In vitro* micronucleus test is one of the missing tests which has no reason to be lacking, since an *in vitro* OECD test exists for a long time. Probably there is a misunderstanding of the Responsible Person who might not realize that it is absolutely complementary to the *in vitro* mutagenicity test in bacteria, since both tests investigate two independent types of abnormalities of DNA, both predictive of cancer.

In some cases, a reliable *in silico* prediction, with one, or even better, consensus from several complementary software, can waive or replace such tests. This solution can be cheaper than testing and the rationale can be robust. *In silico* predictions are also a good strategy when associated to partially concluded results, such as the *in vitro* mutagenicity test. This test is not sufficient to investigate genotoxicity, but a QSAR prediction can provide a good orientation before performing the *in vitro* micronucleus assay, to better understand the potential of a substance to induce DNA damage. Such approaches are widely accepted for the regulatory assessment of pharmaceutical impurities under ICH M7 guideline.

Currently, with other methods gathered in the so-called NAMs (New Approach Methodologies), read-across is a major tool to predict the systemic toxicity of a substance in the absence of any animal testing. Finding structural analogues, selecting them based on relevant criteria, and predicting an endpoint-specific toxicity based on the results formerly obtained with those analogues is both a very ethical way to use existing data, and provides a relevant and reliable solution for predicting sub-chronic/chronic toxicity and reprotoxicity. This parameter is one of the criteria of toxicokinetic (absorption, distribution, metabolism, and excretion; ADME) which should be better used in the future to enhance the application of NAMs.

Last but not least, although dermal absorption could help calculating a precise margin of safety, it is hardly investigated. This rare information is of equal importance in the calculation of the MoS (Margin of Safety) as the systemic NOAEL (no observed adverse effect level) and the exposure. Generally unknown, it is, by default, estimated to 50% according to the Notes of Guidance from the SCCS. For some substances, a "very low rate" can justify to avoid investigating systemic toxicity. The mathematical modeling of dermal absorption is an important field of research but no robust model is currently available. Some models identified good predictivity but were limited to small substances below 300 Da. A recent preliminary retrospective analysis of the ingredients with opinions of the SCCS showed that physicochemical properties of the substance can differentiate the ones with low and high dermal absorption (the threshold being at 2%).

25.3 Genotoxicity Assessment of Cosmetic Products

In the second part of the 20th century, many research teams have developed different kind of tests based on different mechanisms showing direct DNA damages, to detect direct DNA reactive substances that alter DNA and therefore the genetic code. In the 70s, Bruce Ames developed the most famous bacterial Reverse Mutation test, the "Ames test". The most relevant mutagen tests were quickly taken into account by regulatory authorities to identify genotoxic substances in cosmetics and also by cosmetics companies for optimization of the methods and refined cosmetics ingredients. Test battery strategies for genotoxicity evaluation have been issued by regulatory agencies and guidelines are published by OECD.

In the safety assessment of cosmetic ingredients, the assessment of genotoxic potential is crucial. The SCCS 10th Revision recommended to use an *in vitro* battery of two tests. One test for the evaluation of the potential for mutagenicity: bacterial reverse mutation test (OECD 471) Ames test and a second *in vitro* micronucleus test (OECD 487) for the evaluation of chromosome damage (clastogen and aneuploidy). The combination of both tests allowed the detection of all relevant genotoxic carcinogens. The test system should be exposed to the test item both in the absence and in the presence of a metabolic activation system (S9-fraction from the livers of rats treated with Aroclor 1254 or a combination of phenobarbital and β-naphthoflavone).

25.4 Skin Sensitization Assessment of Cosmetic Products

Skin sensitizers are chemicals that have the intrinsic potential to induce a state of hypersensitivity in humans, that upon repeated topical exposure may result in the development of allergic contact dermatitis (ACD). Sensitization involves the activation of an adaptive immune response and the priming of immunological memory, and once acquired, it is often a chronical condition, and elicitation of clinical symptoms can only be prevented by avoiding exposure to the inducing agent. Proactive identification and evaluation of skin sensitization potential is therefore of central importance for safety evaluation of chemicals and represents a key toxicological endpoint among regulatory authorities across multiple industries, and not least for cosmetics, where the intended route of exposure often is *via* dermal application.

Before a new cosmetic ingredient is placed on the European market, evaluation of its safety profile, including the assessment of skin sensitization hazards and potency is mandatory. Following the revision of Annex Ⅶ of the REACH regulation, as well as the transformation of the cosmetic directive into a regulation (EC1223/2009), traditional animal models, such as the Guinea Pig based assays (GPMT or the Buehler test) or the murine Local Lymph Node Assay (LLNA), are no longer allowed to meet the information requirements for substances exclusively intended for use in cosmetic products. To this end, a plethora of New Approach Methods (NAMs), such as in chemico and *in vitro* methods, have been validated and incorporated into official test guidelines by the OECD, serving as viable replacements for animal studies. These methods are designed to target individual Key Events (KE) in the Adverse Outcome Pathway (AOP) for skin sensitization, which recapitulates the most important key mechanistic events that are required for the development of skin sensitization. Currently, three technical Test Guidelines (OECD TG 442 C, D and E) describe a total of seven such methods, including the KE1 based Direct

Peptide Reactivity Assay (DPRA) and the Amino acid Derivative Reactivity Assay (ADRA), the KE2 based assays KeratinoSens and LuSens, and the KE3 based assays h-CLAT, U-SENS, and the IL-8 Luc assay. According to the current testing paradigm, these methods should not be considered as stand-alone assays, but rather in the context of a tiered testing strategy, a so-called defined approach (DA), where a fixed data integration procedure is used to arrive at a final classification, based on the readout from several NAMs. Importantly, based on the empirical data, accuracies of the proposed DAs, ranging between 75.6% to 85.0%, were superior to that of the LLNA (74.2%) for predicting human skin sensitization hazard. In addition to the current OECD adopted assays, several alternative and innovative assays are in the process of being validated and adapted as official TGs, some showing predictive performances similar to the proposed DAs, also when considered as stand-alone assays. Thus, skin sensitization testing is an ever-moving target, and to provide guidance to testing and safety evaluation to the cosmetic industry, the Scientific Committee on Consumer Safety (SCCS) publishes the "Notes of Guidance for the Testing of Cosmetic Ingredients and Their Safety Evaluation", ensuring that testing can be performed in compliance with EU cosmetic legislations.

25.5 Endocrine Properties Assessment of Cosmetic Products

On the 13 December 2017 the European Parliament adopted scientific criteria to define endocrine disruptors which came into force for plant protection products and biocides in 2018. This has been a major step towards the future implementation of similar criteria for regulation of cosmetics in Europe. Despite the discrepancies due to the particular context of cosmetics, a few lessons relating to endocrine assessment strategies have been learnt from experience.

Adopted criteria for endocrine disruptors are closely related to the WHO definition of 2012. An endocrine disruptor is defined by three main criteria: its endocrine mode of action, its capacity to cause an adverse effect, and the plausible link between this endocrine activity and the related adverse outcome.

Regulatory authorities require datasets to permit a conclusive assessment on the disruptive capacity of an endocrine active sample. However, for cosmetic ingredients this will be difficult as availability of comprehensive endocrine test systems is very limited without accessing animal experimentation. Therefore, alternative models will be required to overcome this difficulty that can provide data which will contribute to safety of cosmetics for the endocrine system in an ethical manner.

Since 2002, experts representing OECD member countries have published test guidelines dedicated to endocrine assessment of chemicals. These internationally acknowledged methods are listed, and their proper usage is described within the OECD Guidance Document 150. According to this document, adversity should be assessed (using laboratory animals) to achieve a conclusive assessment of an endocrine disruptor. OECD validated methods cover so far EATS (Estrogen, Androgen, Thyroid, and Steroidogenic) endocrine pathways, for which specific adverse physiological outcomes have been characterized.

25.6 Assessment of Dermal Absorption of Cosmetic Products

Assessment of dermal absorption is a crucial aspect of cosmetic product and ingredient safety, as opposed to drugs, which almost always enter the body in other ways. *In vitro* dermal absorption studies are the gold standard method for skin pharmacokinetic evaluation and are suitable to predict the expected dermal absorption by humans.

The purpose of the dermal absorption testing, also known as dermal penetration or percutaneous penetration, is to provide a measurement of the absorption or penetration of a substance through the skin barrier and into the skin.

Detailed guidance on the performance of *in vitro* skin absorption studies is available (OECD 2004, 2011, 2019). In addition, the SCCNFP (Scientific Committee on Cosmetics and NonFood Products) adopted a first set of "Basic Criteria" for the *in vitro* assessment of dermal absorption of cosmetic ingredients back in 1999 and updated in 2003 (SCCNFP/0750/03). The SCCS updated this Opinion in 2010 (SCCS/1358/10). A combination of OECD 428 guideline with the SCCS "Basic Criteria" (SCCS/1358/10) is considered to be essential for performing appropriate *in vitro* dermal absorption studies for cosmetic ingredients.

Dermal absorption studies are conducted to determine how much of a chemical penetrates the skin, and thereby whether it has the potential to be absorbed into the systemic circulation. Therefore, knowledge of dermal absorption phenomena is essential for:

Safety issues—the presence of systemic test item may lead to systemic adverse effects, the quantities absorbed is taken into consideration in toxicological risk assessment to extrapolate human exposure and calculate the margin of safety (MoS); and

Therapeutic aspects-the quantities penetrated can be taken into consideration to predict the therapeutic concentration at the target sites in skin tissue.

In vitro dermal absorption studies are applied in different sectors and for different purposes:

Formulation screening-for selection of lead candidate formulation;

Bioequivalence-to determine if the new product has the same degree of dermal absorption as reference product. *In vitro* dermal absorption assay was recently used to demonstrate bioequivalence, and the results of the comparison were accepted by the FDA in connection with the marketing authorization for Lotrimin Ultra cream;

Cosmetics and consumer products-Dermal absorption rate is part of the toxicological profile of any ingredient. Almost always provided for any submission to the SCCS, the *in vitro* dermal absorption studies can then be part of the safety assessment of a cosmetic product;

Pharmaceutical products-*in vitro* dermal absorption studies are part of safety and efficacy assessment of topical products;

Chemical/agrochemical-*in vitro* dermal absorption studies are part of safety assessment purposes. With respect to pesticides, the results of the *in vitro* dermal absorption studies alone are accepted for pesticides risk assessment purposes in the European Union and other countries.

Different types of formulations can be assessed through *in vitro* dermal absorption studies: creams, gels, ointments, suspensions, foam, patches, aqueous, solvent, hair dyes, shampoo, foundation, moisturizer, cleansers, soaps, sunscreen, *etc*.

25.7 Skin and Eye Irritation Assessment of Cosmetic Products

Assessment of skin and eye irritation potential of an ingredient or formulation is an important part in cosmetic ingredient safety assessments.

Dermal irritation is defined as the production of reversible damage of the skin, following the application of a test substance for up to 4 h (OECD 404). Eye irritation is defined as the occurrence of changes in the eye following the application of a test substance to the anterior surface of the eye, which are fully reversible within 21 days of application (OECD 405).

Skin and eye irritation are assessed using reconstructed human tissue-based test methods. Commercially available 3D-models based on reconstructed human epidermis (RhE) are used for skin irritation testing (OECD test Method 439) and 3D-model based on reconstructed human cornea-like epithelium (RhCE) is used for eye irritation testing (OECD Test Method 492). It should be noted that there are different *in vitro* models that address serious eye damage and/or identification of chemicals not triggering classification for eye irritation or serious eye damage, but we will only focus on RhCE model.

Reference

Barthe M, Bavoux C, Finot F, Mouche I, Cuceu-Petrenci C, Forreryd A, Chérouvrier Hansson A, Johansson H, Lemkine G F, Thénot J-P, Osman-Ponchet H. Safety Testing of Cosmetic Products: Overview of Established Methods and New Approach Methodologies (NAMs). Cosmetics, 2021, 8: 50.

Key Words & Phrases

endocrine *n.* 内分泌；内分泌系统
propylparaben *n.* 尼泊金丙酯；对羟基苯甲酸丙酯
octocrylene *n.* 氰双苯丙烯酸辛酯；奥克立林
inhale *v.* 吸入；吸气
reprotoxic *adj.* 对生殖系统有毒性的
asbestos *n.* 石棉
musk xylene 二甲苯麝香；二麝香
persistent *adj.* 持久的；持续的；坚持不懈的
mandate *n.* 授权；命令
local tolerance 局部耐受性
obligation *n.* 职责；责任；（已承诺的或法律等规定的）义务
tonnage *n.* 吨位；吨数
mutagenicity *n.* 诱变性
sub-chronic 亚慢性的
carcinogenicity *n.* 致癌性
ethical *adj.*（有关）道德的；伦理的；合乎道德的
potency *n.* 影响力；支配力；效力
eleuthero-embryo 刺五加胚
amphibian *n.* 两栖动物
pragmatically *adv.* 实用主义地；讲究实效地
consensus *n.* 一致的意见；共识
toxicokinetic *adj.* 毒代动力学的；毒物动力学的
metabolism *n.* 新陈代谢；代谢
excretion *n.* 排泄；排泄物；分泌；分泌物
dermal *adj.* 皮肤的；真皮的
mutagen *n.* 诱变剂
clastogen *n.* 断裂剂
aneuploidy *n.* 非整倍性
phenobarbital *n.* 苯巴比妥
naphthoflavone *n.* 萘黄酮
mandatory *adj.* 强制性的；强制的
plethora *n.* 过多；过量；过剩
discrepancy *n.* 差异
pharmacokinetic *n.* 药代动力学
percutaneous *adj.* 经皮的；通过皮肤的
extrapolate *v.* 推断；推知；外推
agrochemical *n.* 农用化学品
epithelium *n.* 上皮细胞

Chapter 26 Efficacy of Cosmetics

To attract users, cosmetics manufacturers usually use original combinations of ingredients that give efficacies, effects, and concepts such as whitening, anti-aging, wrinkle prevention, anti-inflammatory, acne prevention, moisturizing, slimming, scalp care, hair growth, and damage hair treatment *etc*. The following is an example on efficacy evaluation of cosmetics.

Example: Anti-aging Efficacy of the Cosmetics Containing Red Palm Fruit Extract

Only the evaluation section is shown here. For the other experiments and results, please see the other parts of the paper in Reference.

26.1 Evaluation of the Day and Night Creams

26.1.1 Physical properties

The appearance of the day and night creams was recorded for the organoleptic properties color, smell, and product clarity. The pH value was measured using a digital calibrated pH meter (SevenEasy™ pH, Mettler Toledo, Switzerland). The pH values were determined three times and reported as the mean±standard deviation. The viscosity was measured using a rheometer (HAAKE MARS 60) equipped with a Peltier temperature control system (ThermoFisher Scientific, Karlsruhe, Germany).

26.1.2 *In vitro* sun protection factor (SPF) evaluation

The photoprotective efficacy was assessed *in vitro* by determining the SPF value of the formulations, which was calculated using an SPF analyzer (Optometrics LLC/SPF-290F, Miami, FL, USA) and performed according to a previous study. Each day cream and night cream were spread (2 mg/cm^2) over a polymethyl methacrylate (PMMA) plate (70.7mm^2) using a fingertip. The plate samples were then dried for 15 min, protected from light, and exposed to a xenon arc solar simulator. Measurements were made under the following laboratory conditions: (24.0±0.5)℃ and 50%-60% RH. The measurement was performed by scanning nine spots on each sample at 2 nm interval between 290 and 400 nm. The UVB protection efficacy was recorded as SPF. An intact, non-coated, PMMA plate was used as a blank reference, representing 100% light transmission. All measurements were performed in triplicates.

26.1.3 Stability study

The day and night creams were separately packaged in clear polyethylene (PE) plastic jar containers (50 g) and stored at a temperature of (30±0.5)℃ and a relative humidity of (75±5)% for 6 months to determine the stability under actual storage conditions in Thailand (hot and humid zone). In addition, the accelerated stability study was performed by freeze-thaw cycling at 2-8℃ for 48 h and 45℃ for 48 h and repeated for six cycles. The content of vitamin E, β-carotene, and physical properties, including appearance, pH, and viscosity, were recorded.

26.2 Clinical Study of Safety and Skin Efficacy of the Day and Night Creams

26.2.1 Ethics consideration

The current study is a prospective, open-label, randomized clinical trial. The study protocol was approved by the Human Research Ethics Committee of Walailak University (WUEC-19-186-01) on 18 October 2019.

26.2.2 Subjects

All participants were female, aged 25-50 years, and lived in Nakhon Si Thammarat Province, Thailand. There are no similar clinical trials with *E. guineensis* oil extract. The clinical study of *E. guineensis* extract, as an anti-aging cream, reported in the literature was used to calculate the sample size, but it was not a similar study. The number of subjects used for skin efficacy testing in this study was determined as per Gaspar *et al.*, who studied the effect of skincare products on skin water content (stratum corneum water content and transepidermal water loss) and skin roughness in 14 human volunteers per group. In addition, Lademann *et al.* studied the effect of skin creams on hydration and skin elasticity (stratum corneum and elasticity) by testing 15 volunteers per group. Therefore, we set the number of volunteers to a larger sample size of 35 participants.

26.2.3 Skin irritation protocol

This test is used to determine whether the product used for testing causes irritation to the skin to confirm the safety of the products. A single-patch test protocol was used for the evaluation. General information of the participants who used facial cream products, either day or night creams, was recorded. The principal investigator described the information of the products and study methods to all volunteers. Each participant was tested with the two products (day and night creams) for the skin irritation study by applying 0.5 g of each cream separately into the aluminum well of the Finn chamber® occlusive patch. Each well was 0.8 cm in diameter or 0.5cm^2 in area. After cleaning the experimental area with 0.9% sodium chloride solution, the Finn chamber® occlusive patch containing the tested cream was applied to the forearm of the participants and covered with waterproof tape (Nexcare, 3M, Thailand). After 48 h of application, the Finn chamber® was removed, and the surface-mounted product was wiped off the skin with a cotton swab moistened with 0.9% sodium chloride. The principal investigator or skin specialist observed and evaluated the skin reactions and recorded the relevant scores.

26.2.4 Clinical efficacy study protocol

After completing the skin irritation study and excluding the participants that did not compile the results, the participants were divided into two groups by computer randomization. Group 1 received 50 g of day cream packed in a PE plastic jar. Group 2 received a night cream with the same package. The principal investigator advised the instructions for the cream use. Briefly, after washing and wiping the face to dry, the cream was applied daily on the face in the morning for day cream or before bedtime for night cream users, for consecutive 30 days. The participants had to avoid applying other facial care products to their faces during the clinical study period. The participants in both groups were investigated five times. At the first meeting, the skin parameters of the participants were measured before applying the cream as the baseline value (1st measurement). The 2nd measurement was performed 30 min after the first application of the cream. The 3rd, 4th, and 5th measurements were performed after applying the cream for 7, 14, and 30 days, respectively. Before the measurements, the volunteers were housed in an air-conditioned room at (25±2)℃ for 30 min to adjust their skin. Skin was measured in the same area every time. The skin condition measurement included the following tests:

(1) The stratum corneum water content and the amount of water accumulated in the epidermis were measured using Corneometer® CM 852 (Courage + Khazaka Electronic Co., Ltd., Köln, Germany).

(2) TEWL or the amount of water lost from the skin was measured using Tewameter® TM 300 (Courage + Khazaka Electronic Co., Ltd., Germany).

(3) The skin elasticity was measured using Cutometer® MAP 580 (Courage + Khazaka Electronic Co., Ltd., Germany).

(4) The melanin index representing pigment amount and skin redness was measured using Mexameter® MX 18 (Courage + Khazaka Electronic Co., Ltd., Germany).

The skin was photographed with a DSLR camera (Nikon D70, Nikon Corp. Tokyo, Japan) and the depth of wrinkles on the skin was recorded with the Visioscan SV 600 (Courage + Khazaka Electronic Co., Ltd., Germany) on Day 1 before cream application and on Day 30 before skin measurement. Some volunteers were photographed with permission, and publications of those photographs do not identify the subjects and are already acknowledged. The subjects were photographed with the DSLR camera from the front, and either side, at an angle of 45° , without editing the picture program.

All experiments were performed at the left cheek (L), right cheek (R), and forehead (N) of the participants.

26.2.5 Satisfactory survey by questionnaires

Besides the clinical study to evaluate the efficacy of the products, a satisfaction survey of the participants was conducted to confirm the improvement in moisture, elasticity, and skin color change. The questionnaire to survey the efficacy of the anti-aging day and night products were answered by the volunteers of the clinical efficacy study (35 participants per group). After completing the clinical study, the questionnaire was sent to the participants to gauge their satisfaction with the products. The questionnaire for satisfaction assessment of day and night creams was developed based on product appearance, performance, and how the user felt about the change in skin conditions when applied. The topic assessment details included color, texture, smell, skin performance improvement including hydration, radiance, firmness, reduction of wrinkles, and skin irritation. Satisfaction scores were divided into five levels, with a score of 5 indicating the highest, and a score of 1 indicating the lowest satisfaction level.

26.3 Statistical Analysis

Paired t-test results were analyzed using the SPSS program (statistical significance level, $p < 0.05$) to determine the results of skin improvement (before and after treatment). The results of the questionnaire were collected and analyzed by considering the mean and standard deviation. All data are presented as percentages. The difference in satisfaction, before and after using the product, was analyzed using paired t-test at a 95% confidence level ($p < 0.05$).

26.4 Results and Discussion

26.4.1 Skin irritation study of day and night creams

The active ingredients of the day and night creams included *E. guineensis* fruit extract, enriched with tocopherol, tocotrienol, and β-carotene, which showed antioxidant activity. A clinical study was conducted on volunteers to determine the anti-aging efficacy of the day and night creams. A total of 71 subjects, who participated in the irritation test, were enrolled in the study. After 48 h of the irritation test, skin changes were assessed. It was found that all 71 participants did not develop redness or erythema, skin swelling, blisters, or pustules (edema) after 48 h of exposure to either product. The erythema score and the edema score were 0 for both products. However, one volunteer developed a small area of red skin at the night cream application site (erythema score=1 and edema score=1) approximately 24 h after patch removal. However, the skin reaction occurred at the waterproof plaster patch outside the aluminum well of the Finn chamber®. To ensure the safety of this participant, she was excluded from the clinical efficacy study in the next step. With respect to percentage of irritation or abnormalities on the volunteers' skin, the day and night creams did not cause any irritation. The calculated MII of the day cream after the 48 h patch test and 24 h after patch removal was 0.00. The MII value of the night cream was 0.00 after the 48 h patch test, and it was 0.028 24 h after patch removal (n=1). The MII results indicated that both day and night facial creams were non-irritating (NI) or had good cutaneous compatibility. Both products can be applied safely to the skin because their ingredients are safe for use in cosmetics. Although the night cream had a higher incidence of irritation than the day one, it was still considered a safe product based on its MII result.

26.4.2 Clinical efficacy by skin assessment

The volunteers participating in the clinical efficacy study were the same as in the skin irritation study (35 participants per group). However, one volunteer per group did not show up for clinical study appointments, so the participants remained 34 persons per group. The cumulative water content in the epidermis or stratum corneum was measured using Corneometer® CM 825. After 30 days of application of the day and night creams, the accumulation of water content in the subject's epidermis significantly increased by 8.73%±2.05% and 3.54%±1.85%, respectively, compared to the first use on Day 1. The first application of the day and night creams showed a significant increase in water content up to 19.05%±1.70% and 4.79%±2.11%, respectively, compared to the baseline. This indicated that the day and night creams have high efficacy in increasing the water content of the skin. In addition, the skin of all volunteers had an average dermal water content between 53.57 and 109.24 AU. A value greater than 45 AU indicates that the skin is adequately hydrated.

TEWL was measured using Tewameter® TM 300 in 34 subjects from each group using the day and night creams. The mean water loss from the skin of the subjects in the two groups decreased after 7, 14, and 30 days of cream application, compared to the baseline values. If the water loss from the skin is between 10 and 25g/(h · m^2), the skin is healthy because there is little evaporation of water from the skin, which allows the skin to retain adequate moisture. Day and night creams were found to decrease water loss from the skin epidermis, after application for 7 days, and increased after 14 and 30 days with statistical difference ($p < 0.05$).

The day cream seems to slightly increase the water content and reduce water loss from the skin epidermis compared to the night cream. However, both day and night creams improved skin health with a statistically significant difference ($p < 0.05$) after application for 30 days, compared with baseline values. This is likely the result of various components in the formulations containing the important active substances from red palm fruit extracts, such as tocopherol, tocotrienol, β-carotene, several fatty acids, and other moisturizing agents, in the cream formula, which acts as a barrier to protect the skin from losing water. Thus, the skin remained moist and healthy.

Skin elasticity is one of the key qualities that helps the skin appear young and healthy. Skin elasticity can be assessed with the Cutometer® MPA 580, which measures the elasticity of the epidermis using suction pressure to mechanically deform the skin. The probe aperture of Cutometer® was sucked onto the skin for a while, and when the force was withdrawn, the skin would try to return to its original state. This value indicated the elasticity of the epidermis. The skin elasticity of the volunteers applying the day and night creams for 30 days significantly increased to (45.25±9.14)% and (45.52±10.97)%, respectively. The facial creams containing *E. guineensis* fruit extract loaded SLNs that contained active substances such as tocopherol, tocotrienol, β-carotene, and fatty acid, and the process of preparing creams with SLN technology allowed substances to be absorbed and permeated into the skin cells quickly. Antioxidant-rich formulations improve skin radiance and enhance skin elasticity in women.

The difference in skin color between individuals results from genetic diversity, depending on the amount of melanin and hemoglobin. Skin color can be assessed using a Mexameter® MX 18 probe, which measures the melanin index. For the 34 volunteers applying the day and night creams, the average value of melanin pigment decreased in both experimental groups, indicating that both skin creams could increase the radiance of the skin. The night cream resulted in reduced melanin less than that of the day cream, on average, (18.29±4.12)% and (16.18±3.83)%, respectively. Likely because of the sunscreen in the day cream, the reduction in melanin was caused by the physical and chemical sunscreens in the day cream. The reduction in melanin pigment results from important antioxidants such as tocopherol, tocotrienol, and *β*-carotene. Antioxidants inhibit the process of oxidation and melanogenesis of melanin pigment formation, so they help the skin appear bright and clear.

Photographs of the skin texture of volunteers, applying the day and night creams were obtained by Visioscan VC98. The subjects were photographed before the application of cream (Day 0), and after 7, 14, and 30 days from top to bottom. The results show that the volunteers had smoother and more hydrated skin, and deep grooves in the skin looked shallower when compared to the images before using the day and night products.

Reference

Plyduang T, Atipairin A, Sae Yoon A, Sermkaew N, Sakdiset P, Sawatdee S. Formula Development of Red Palm (*Elaeis guineensis*) Fruit Extract Loaded with Solid Lipid Nanoparticles Containing Creams and Its Anti-aging Efficacy in Healthy Volunteers. Cosmetics, 2022, 9: 3.

Key Words & Phrases

viscosity *n.* 黏度；（液体的）黏性

rheometer *n.* 流变仪

polymethyl methacrylate 有机玻璃；聚甲基异丁烯酸；聚甲基丙烯酸甲酯

xenon arc 氙弧

solar simulator 太阳模拟器

freeze-thaw 冻融；冷冻-解冻；解冻；冻融法；冻融作用

E. guineensis 油棕

consecutive *adj.* 连续的；连续不断的

gauge *v.*（用仪器）测量；判定；判断（尤指人的感情或态度）；估计

radiance *n.* 容光焕发；红光满面；（散发出来的）光辉

swelling *n.* 膨胀；肿胀；肿胀处；浮肿处

pustule *n.* 脓疱；脓包

aperture *n.* 光圈；缝隙；小孔；（尤指摄影机等的光圈）孔径

skin radiance 皮肤光泽

Chapter 27 Sensory Evaluation of Cosmetic Products

Sensory evaluation (or sensory analysis) is the science involved with the assessment of the organoleptic attributes of a product by human senses (sight, smell, taste, touch and hearing) . It is also a test method in which people is the "measuring instrument".

Sensory evaluation is a very important test method for cosmetics. In a sensory evaluation of cosmetic products, visual sense is often used to evaluate the shape, color, uniformity, brightness, and transparency. Tactile sense is often used to evaluate the smearing, absorption, viscosity, residue, rheology, and hardness. Finally, olfactory sense is often used to evaluate the aroma, purity, intensity, and odor.

Cosmetic sensory evaluation can be roughly divided into objective evaluation and subjective evaluation. The objective evaluation also includes discrimination tests and descriptive tests. In the descriptive analysis, qualitative descriptors and quantitative descriptors are used to describe the sensory attributes accurately, unambiguously and judgmentally. The evaluation results are generally presented with data and charts instead of words. This kind of method requires that the evaluator must be professional. In this kind of sensory evaluation some product characteristics or differences, which cannot be find by consumers, can be find. Subjective evaluation, also known as affective tests, includes preference tests and acceptability tests. The purpose is to test consumers' preference for different sensory characteristics. Generally, some consumers (such as more than 30) serve as evaluators. No professional training is need for them. And there are no strict requirements for the judgment of sensory characteristics of samples, which is only determined by personal preferences.

The following are two examples for sensory evaluation of cosmetic products.

27.1 Example 1: Sensory Evaluations of Personal-care Products Formulated with Natural Antioxidant Extracts

Only the sensory evaluation is shown here. For the other experiments and results, please see the other parts of the paper in reference.

Many natural products, botanicals, or waste materials–derived from agricultural products, foods and beverages–can be used in cosmetics products. Sensory test was carried out to evaluate the possibility of using natural extracts as ingredients of some cosmetic preparations and their acceptability by consumers. Independently of the antioxidant activity, the ethanolic extracts were added to provide their odoriferous characteristics to the different cosmetic formulations and to evaluate their acceptance by the consumers. Likewise, the color differed with raw material too and the volunteers scored also this difference.

Sensory analysis was performed to determine the preference for the natural extracts, because the composition of the individual formulations was the same, except in this ingredient.

A control sample which did not contain any extract from the studied vegetal raw materials was used for reference. The most preferred and valued attributes in all cosmetics were spreadability, softness, consistence/texture, and skin feel, but the ANOVA (analysis of variance) results showed that these attributes in each cosmetic product were comparable and the use of the different extracts caused only a significant effect ($p < 0.05$) on two parameters: color and odor.

Participants from the two genres accomplished the sensory analysis (26 women and 29 men), and, as can be seen in Figure 27.1, no great differences between them were found. Only four samples demonstrated significant differences: women evaluated better than men the *Acacia* flower extract when it was included in the hand cream and exfoliating preparations, and the control-exfoliating and the shampoo with grape pomace extract obtained better scores with men than with women. Female participants valued body oil and clay masks better than males, and, on the contrary, men valued the shampoo more. Hand cream elaborated with acacia flowers or shiitake attained the best scores, with 7.20 and 7.04 points, respectively.

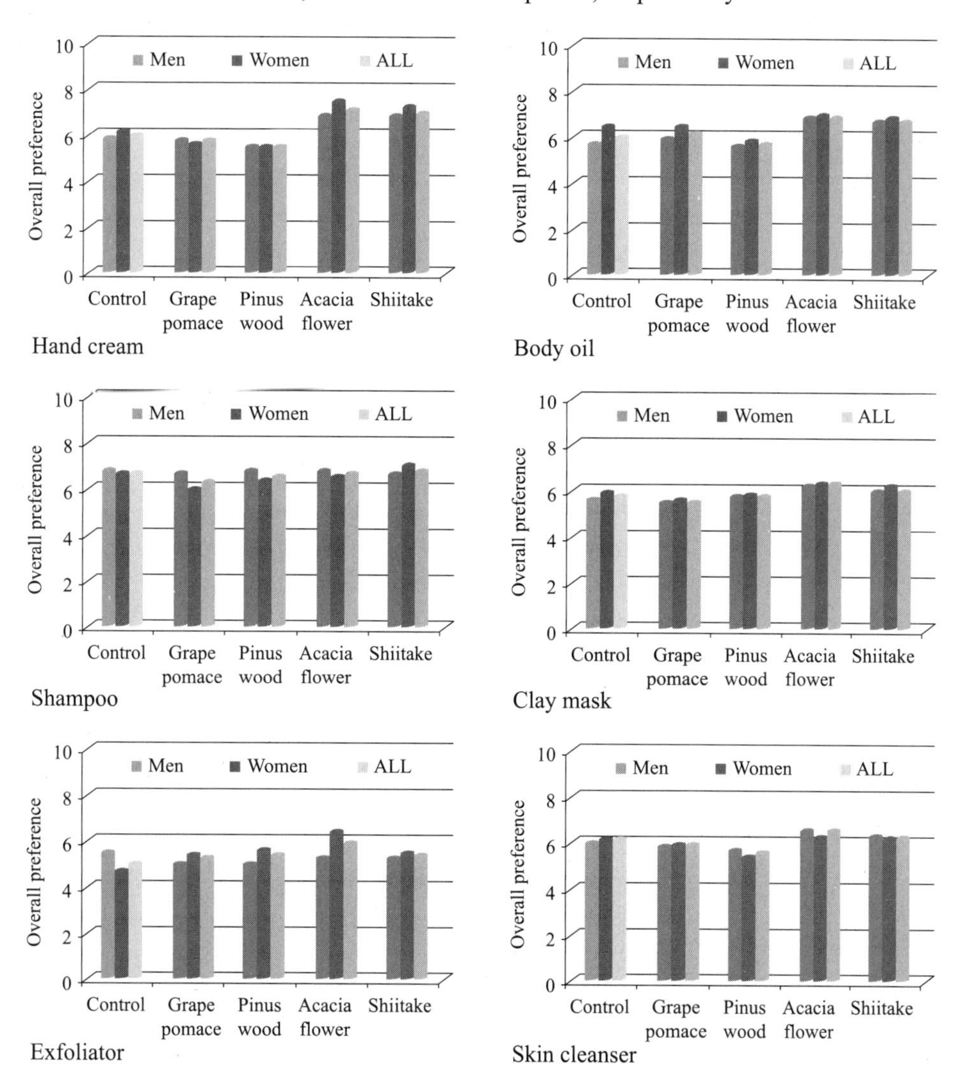

Figure 27.1 Overall appreciation for tested personal-care products according to the volunteers' gender

From the four assayed raw materials, the least preferred extracts in all cosmetics were red grape pomace and *Pinus* wood ethanolic extracts. The average overall preference showed that these extracts were best valued as aromatic additives in the shampoo, and the worst score was attained by the exfoliating body. The cosmetic most valued with the acacia or shiitake extracts were the hand creams, and the worst was the body exfoliating cream. Control samples (without any added extracts) achieved similar scores than when GPE and PWE were added into all personal-care products, except in shampoo, where it was obtained the best punctuation along with the shiitake extract.

The participants were also divided into three age-segments: <20, 20-30, and 30-45 years, with 20, 19, and 16 persons, respectively. This last group perceived acacia and shiitake extracts with the significant highest values for all formulations, revealing the influence of color and odor of these natural extracts (Figure 27.2). In opposition, the participants under 20 years of age preferred the extracts obtained from *Pinus* wood and grape pomace or the control sample (without any extract), except in hand cream. These two extracts received lower score in the group of older participants. Consumers in the 20-30 years old segment evaluated samples with the highest overall preference values in all personal-care products, and the preferred extracts were acacia flowers and shiitake extracts.

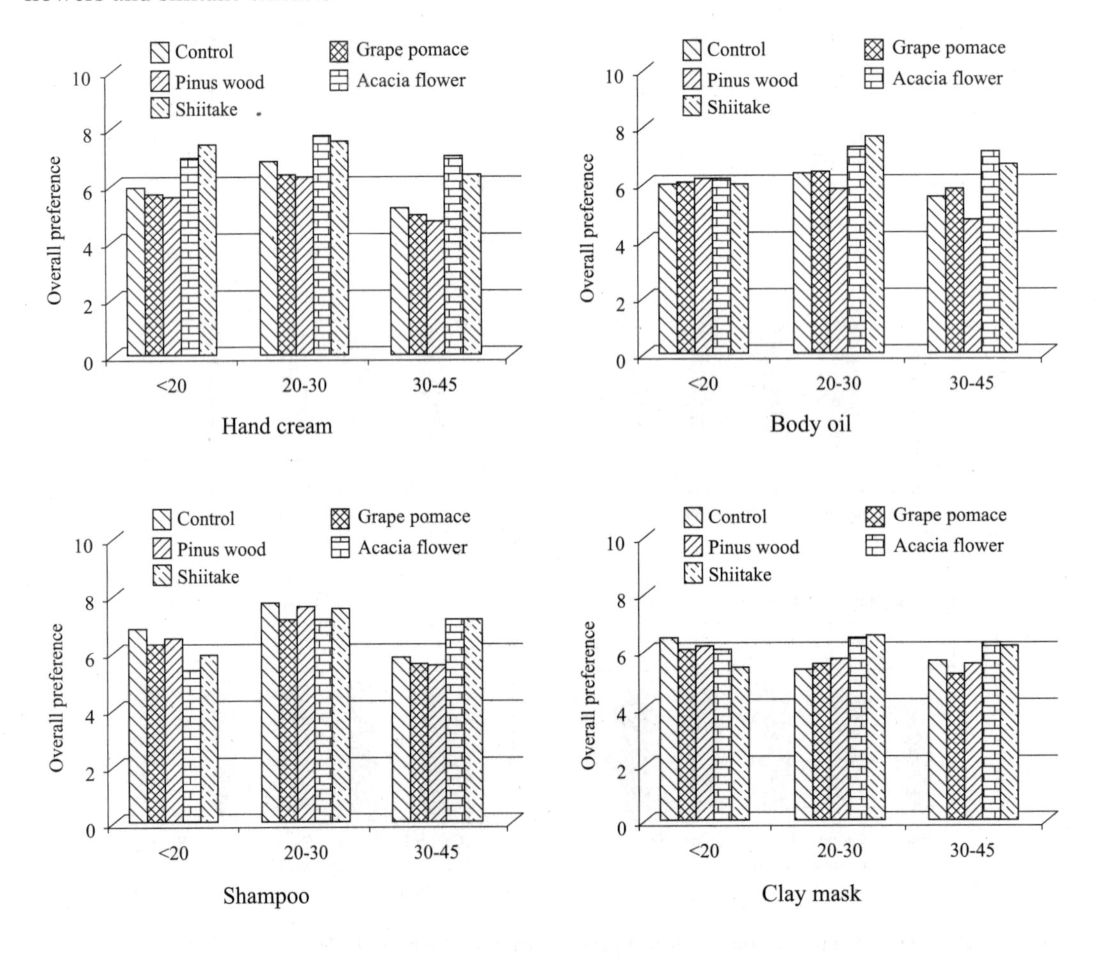

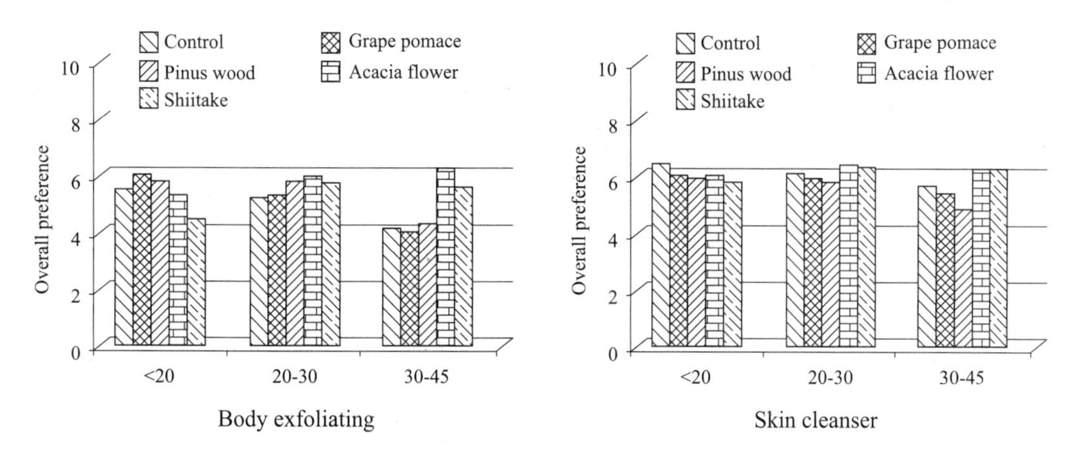

Figure 27.2 Overall appreciation for tested personal-care products according to the volunteers' age-interval

The sensory characterization confirms which properties mostly influence consumer acceptance. All volunteers considered that the different tested personal-care products have a good spreadability, softness, and good skin penetration ability and posterior skin feeling, but also a pleasant color and fragrance. These results have shown that a variety of plant materials can be used as additives in cosmetic products to supply color and aroma.

Principal component analysis (Figure 27.3) was performed to know how consumer acceptance is based on sensory attributes. The interrelationships among the extracts from natural

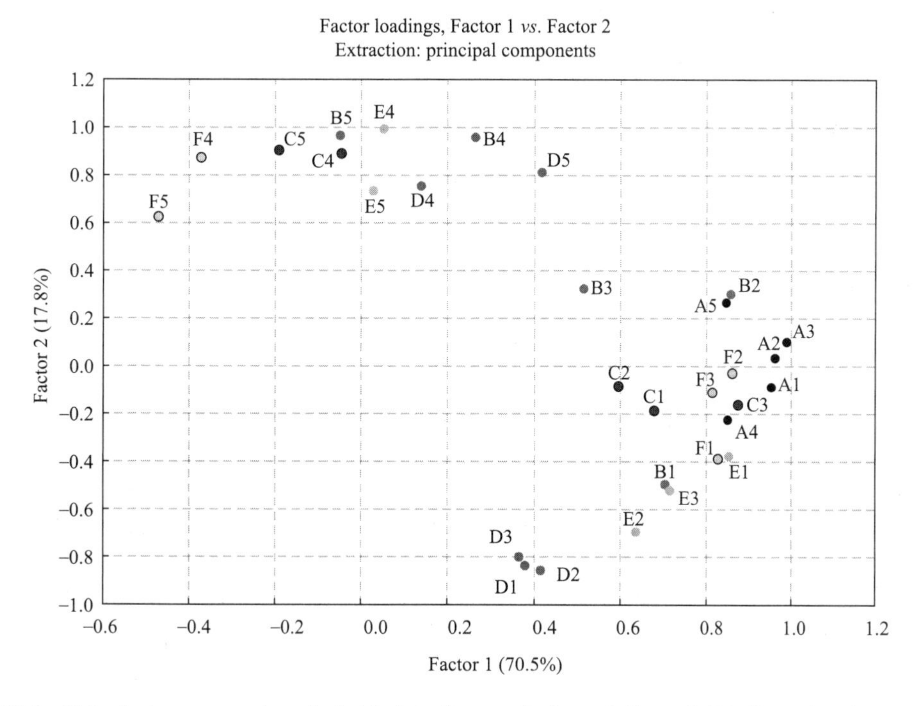

Figure 27.3 Principal component analysis bi-plot of cosmetic formulations. A-F indicate hand cream (HC), body oil (BO), shampoo (S), clay mask (CM), body exfoliating cream (BE), and a skin cleanser formulation (SC) respectively. 1-5 indicate control (without extract), grape pomace extract (GPE), pine wood extract (PWE), acacia flower extract (AFE), and shiitake extract (SE), respectively

raw materials used in personal care-products and sensory descriptors showed that the first two factors explained 88.3% of the total variance among the extracts. The first component accounted for 70.5% of the data variability and the second for 17.8%. In the PCA the extract samples are clearly grouped into two clusters. Cluster 1, located in the upper half, was associated with the cosmetics elaborated from acacia flowers and shiitake extracts, which had the greatest acceptance. Cluster 2, in the positive part of Factor 1 and in the negative part of Factor 2, was characterized by the other two extracts (grape pomace and pine wood) and the control (without added extract). No differences between different cosmetic formulations were found.

All products were well-tolerated because any visible skin irritation or erythema was not observed. Besides expected appearance, spreadability, softness, and skin feeling, the color and odor also play an important role on overall preference and, consequently, on purchase intent. According to the obtained results, fragrance and color were two important attributes for consumer preference and they are essential additives to make personal-care products, even, cosmetic companies use colors in packaging design to communicate the properties of their fragrances. Between the four assayed vegetal extracts, the floral aroma and yellow color provided by the acacia flower extract were evaluated higher by all consumers, independently of genre or age; likewise, this ethanolic extract also presented the highest *in vitro* antioxidant activity.

27.2 Example 2: Sensory Evaluation of a Suncream Formulated with Thermal Spring Waters from Ourense (NW Spain) and *Sargassum muticum* Extracts

Only the sensory evaluation is shown here. For the other experiments and results, please see the other parts of the paper in reference.

Sensory analysis

The ANOVA results indicated significantly large differences regarding the perception of the cream's appearance and skin attributes, consistency, color, spreadability, softness, and the preference for its smell on the skin ($p < 0.05$). Individually, the use of a specific thermal water in the formulations has a significant effect on six sensory attributes: consistency, color preference, spreadability, softness, skin feel, and smell preference on the skin. The presence of a specific extract implied significant differences in gloss and spreadability. However, combined factors (waters × extracts) only supposed significant differences in gloss and in smell preference on the skin.

Hierarchical cluster analysis (HCA) allows a preliminary study to be carried out in order to search for natural groupings among the samples based on their similarities. As shown in the dendrogram (Figure 27.4), it is possible to distinguish clearly between cream samples elaborated with *S. muticum* (SR) extract and those obtained with other extracts. *S. muticum* extract creams showed the highest values for gloss, spreadability, and softness. They showed the highest rating for skin feel, with the same value as for those prepared with α-tocopherol, and the lowest value in consistency, smell preference in glass/on skin, color, and intensity and persistence of smell on skin.

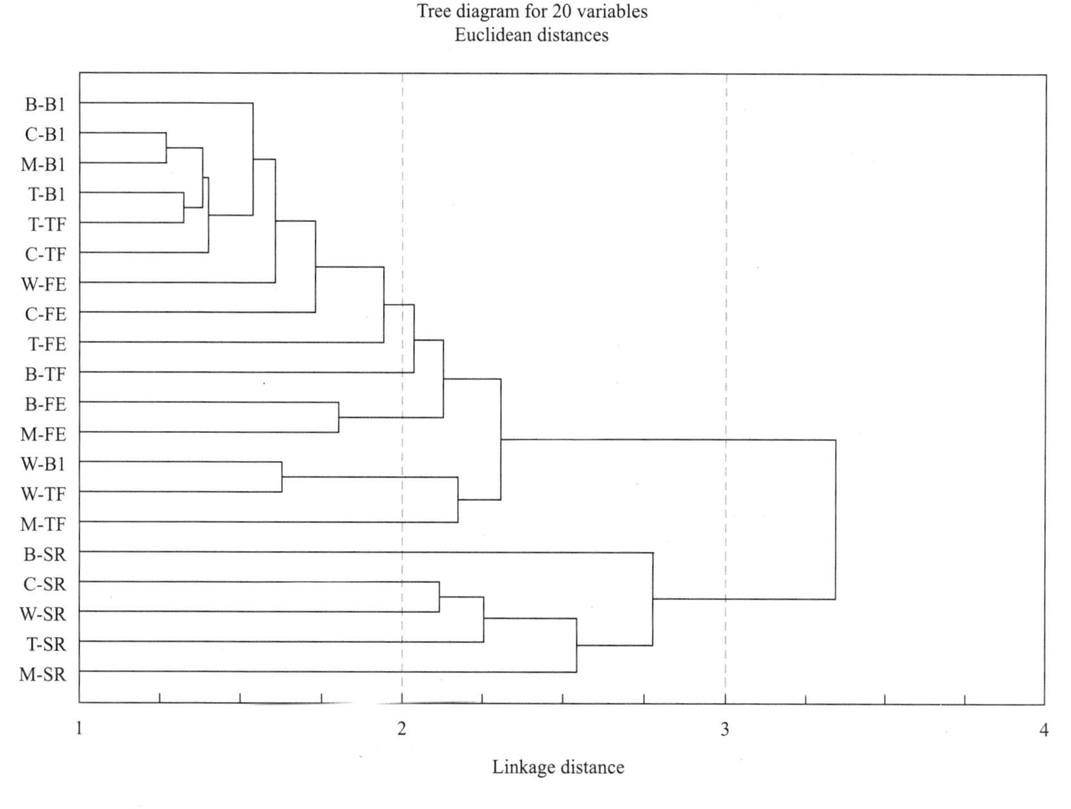

Figure 27.4 Dendrogram for cream samples obtained from the hierarchical cluster analysis

A principal component analysis was applied to sensory descriptive data to study the relation between global preference of cream samples and sensory attributes. Results explained 89.33% of the total data variation with 55.63% and 33.71% contributions from the first and second principal components, respectively (Figure 27.5). Results indicated that overall acceptability (upper left quadrant) was strongly correlated with attributes related to hedonic feeling on the skin after spreadability. Furthermore, variables associated with the appearance in the glass (lower right quadrant) exerted less influence on the global preference of cream.

The preference of water employed to elaborate creams differed with the extract used (Figure 27.6). When no extract was used (Bl), the preferred thermal water was As Burgas. A Chavasqueira and O Tinteiro thermal waters were better evaluated when included in the preparations with α-tocopherol (TF) and with *S. muticum* (SR) extracts. Similar results were obtained for creams formulated with A Chavasqueira or the bidistilled water and *Fucus* sp. extract (FE).

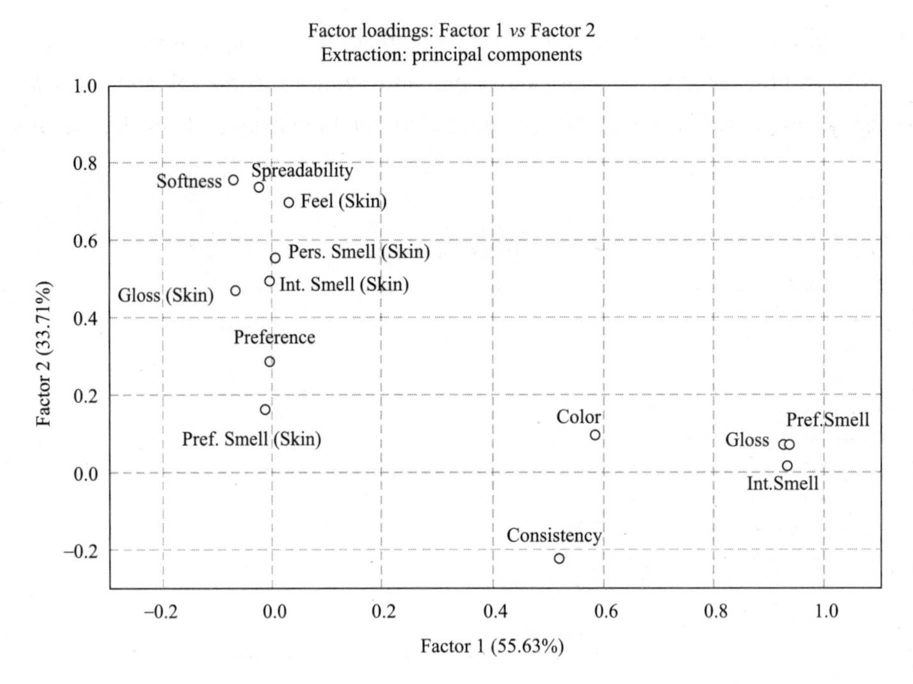

Figure 27.5 Principal component analysis showing the relation among sensory attributes and the global preference of cream

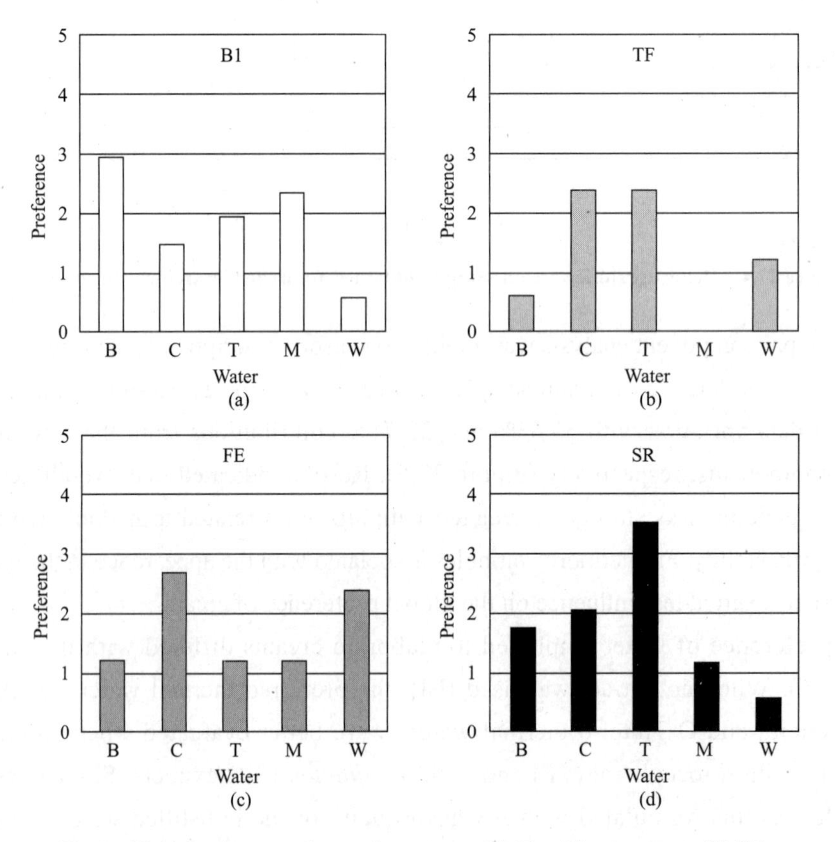

Figure 27.6 Preference of water by extract: (a) Bl, without extract; (b) TF, tocopherol; (c) FE, *Fucus*; (d) SR, *Sargassum muticum*

According to the average score obtained from 17 judges, among the creams prepared with extracts, there was a global preference for those prepared with the *S. muticum* extract (9.1 points), followed by the *Fucus* sp. extract (8.5 points) and tocopherol (6.5 points).

In general, *S. muticum* preparations presented higher sensory acceptance than those prepared with *Fucus* sp. extract or α-tocopherol, but it was almost the same as for creams prepared without antioxidant. The total score calculated as the sum of the individual scores from 0 to 50 for these creams was 155, 145, 110, and 158, respectively.

The presence of thermal water in cosmetic formulations was also a very well-considered parameter. According to the average score obtained, the most preferred thermal spring water was O Tinteiro (9.0 points), followed by A Chavasqueira (8.5 points) and As Burgas (6.5 points). Formulas with thermal spring water from O Tinteiro and A Chavasqueira and with *S. muticum* algal-derived extract (SR) were preferred.

For SR emulsions, compared to the emulsions prepared with other extracts, the most valued attribute was gloss, and the least valued attributes were consistency and odour. They also presented slightly better spreadability and softness.

References

[1] Soto M L, Parada M, Falqué E, Domínguez H. Personal-care Products Formulated with Natural Antioxidant Extracts. Cosmetics, 2018, 5: 13.

[2] Balboa E, Conde E, Constenla A, Falqué E, Domínguez H. Sensory Evaluation and Oxidative Stability of a Suncream Formulated with Thermal Spring Waters from Ourense (NW Spain) and *Sargassum muticum* Extracts. Cosmetics, 2017, 4: 19.

Key Words & Phrases

tactile sense 触觉；触感
olfactory sense 嗅觉
objective evaluation 客观评价
subjective evaluation 主观评价
odoriferous *adj.* 有气味的；散发气味的
Acacia flower 金合欢花
pomace *n.*（水果榨汁后的）果渣；油渣
shiitake *n.* 香菇
Pinus *n.* 松属；松树；松科松属；松柏科松属
posterior *adj.* 在后面的；在后部的
genre *n.* 体裁；类型
hierarchical cluster analysis 系统聚类分析；层次聚类分析
exert *v.* 发挥；施加；行使；运用
S. muticum 海黍子；马尾藻

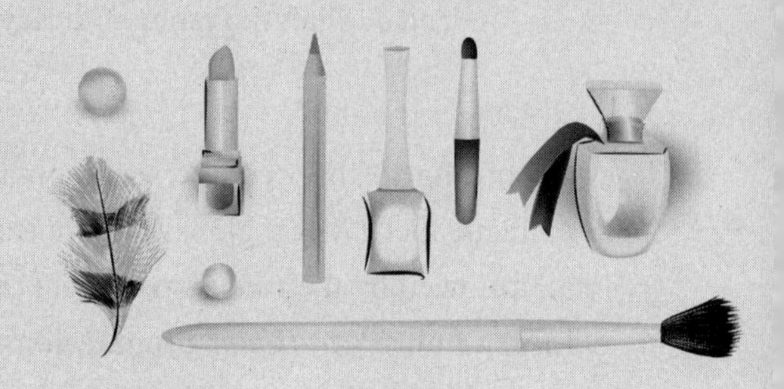

Unit 8

Biotechnology Applied to Cosmetics

Biotechnology is defined as the application of knowledge in life sciences to create products or services that are beneficial to humans, being used to improve the quality and efficiency of food production, or even the production of cosmetic active ingredients, drugs, and vaccines.

Since biotechnology became a growing area, there are currently more than 250 products obtained by biotechnological processes, including therapeutic proteins and monoclonal antibodies.

Biotechnology has had an impact on cosmetics in several ways. Cosmetics companies use biotechnology to discover, develop, and produce components of cosmetic formulations and to evaluate the activity of these components on the skin, in particular, how they can affect the changes associated with aging. Thus, biotechnology represents a good alternative tool for developing active ingredients that are able to slow down the aging process.

In this unit, you will see a description of some active ingredients obtained through biotechnological processes, found in the available literature.

Chapter 28 Active Ingredients Obtained by Biotechnological Processes

There are several chemical compounds that have been used in cosmetics and that may have potential adverse effects on their users. Because of these effects and the environmental impact of chemical compounds, herbal extract cosmetics have attracted consumer's interest and have become very important in the cosmetics industry since the early 1990s. Currently, some certain compounds resulting from biotechnological processes have shown interesting skin care properties and may come to be considered effective ingredients. Some of them, such as kojic acid, hyaluronic acid and peptides, were introduced in the previous chapters. The following are some brief introductions on growth factor, enzymes and stem cell.

28.1 Growth Factors

A growth factor is defined as a biologically active molecule that is secreted and can affect cell growth. Growth factors may act on specific cell surface receptors that subsequently transmit these cell signals to other intracellular components. The ability of growth factors to promote growth, differentiation, and/or cell division has attracted the attention of not only the pharmaceutical industry, but also the cosmetics industry.

Human growth factors are considered extraordinary molecules in the cosmetics industry, thanks to their important role in skin care. The use of these molecules for skin rejuvenation is thought to be an emerging and promising strategy. Advances in knowledge of the role of growth factors in wound healing and regeneration have aroused great interest in the role that these molecules may play in the repair of skin structures. As the endogenous functionalities of the growth factors decrease as a result of the smaller reduction of skin cells during skin death, an exogenous supplementation of growth factors can promote the repair of aging skin and revitalize it. A combination of growth factors together with antioxidants, matrix building agents, and skin conditioning agents can be effective in treating skin anti-aging. A cosmetic formulation based on this combination of ingredients was launched on the market, called TNS (Tissue Nutrient Solution) Recovery Complex System with NouriCel-MD™ (USA). This product promoted the disappearance of wrinkles and pigmentation and, at the same time, improved skin firmness.

Growth factors can be applied topically or injected. Several clinical studies have shown that topical application of animal growth factors, or the injection of autologous growth factors, can also increase collagen synthesis in the dermis. The purpose of administering topical or injectable growth factors is to increase the activity of the cells responsible for the remodeling of the dermis, in order to delay or reverse the aging of the skin. The clinical applications of topical and injectable growth factors are promising and still need to be studied in terms of their safety, efficacy, tolerability, and stability.

Human epidermal growth factor (hEGF) can speed up the healing process and was also found to be effective in the treatments of wrinkles, age spots, and freckles. Pure hEGF can potentially be produced on a large scale through genetic engineering. HEGF had been produced in several hosted systems, including *E. coli* and *S. cerevisiae*. When *E. coli* is used as host, the yield is not appropriate for industrial requirements, as the hEGF cytoplasm tends to form inclusion bodies, which can be rapidly degraded by proteases. Thus, the total production cost ends up increasing owing to the additional production steps required to release hEGF from inclusion bodies. In addition, the hEGF produced by prokaryotic systems is lower compared with that produced by eukaryotic systems. Therefore, the use of eukaryotic systems, such as *P. pastoris*, can produce the growth factor on a large scale. Skin Actives (USA) markets heterologously expressed EGF in *E. coli* to use as a skin conditioning agent.

28.2 Enzymes

Enzymes are proteins that are present in living organisms and catalyze several biochemical reactions that are necessary for life.

Isolated microorganisms from various environments represent a source of enzymes that can be used in industrial processes. Using recombinant DNA technology, it is possible to clone the genes encoding these enzymes, and thus express them heterologously in strains commonly used in the pharmaceutical and cosmetics industries.

In the cosmetics industry, various types of enzymes are used to develop formulations that facilitate the course of biochemical skin reactions, protecting the skin from aging. These enzymes are also responsible for protecting the skin against some external agents (such as UV radiation) and against free radicals.

The use of enzymes in cosmetics provides a specific biochemical pathway that is more beneficial and leads to a better performance of the skin. One of such enzymes is the superoxide dismutase (SOD), which through its mechanism of action, prevents damage caused by free radicals and other harmful pollutants. SOD enzymes control the levels of a variety of reactive oxygen species (ROS) and reactive nitrogen species (formed through UV exposure and other radiation, as well as from normal cellular metabolism), limiting the potential toxicity of these molecules and controlling cellular aspects that are regulated by their signaling functions. ROS produced in the metabolic pathways have been shown to lead to skin deterioration and, therefore, SOD is considered an anti-aging enzyme as it helps to remove these ROS in humans. In addition, SOD maintains the integral keratin structure, promotes skin elasticity, and provides a smooth feeling to the skin. SOD can be obtained through genetic modification of *S. cerevisiae.*

Proteases are enzymes that break down proteins into peptides and later into amino acids. In cosmetics, proteases are primarily aimed at promoting skin exfoliation, which corresponds to the scaling of the keratinized superficial corneal layer, and to increase the absorption of water and other ingredients present in cosmetics. By promoting exfoliation, these proteases will improve the appearance of the skin. Bromelain, papain, and chymotrypsin are the examples of herbal proteases used in cosmetics, but cannot be used by most individuals, owing to the risk of allergy. Seki *et al.* reported that subtilisin, a serine protease produced by *Bacillus licheniformis*, is an effective skin exfoliator. Commercial proteases for cosmetic use can be obtained by the recombinant DNA technology.

Another type of enzyme that has gained interest is DNA repair enzymes such as photolyases. When DNA repair is deficient and the melanin present in the skin cannot protect the skin from the damage caused by solar radiation, the risk of accumulation of cancer-induced mutations induced by UV radiation may increase. DNA photolyases can reverse these lesions by eliminating thymine dimers that are formed and play a critical functional role in DNA repair. Navarrete-Dechent and Molgó performed a clinical study to evaluate the usefulness of a new topical sunscreen containing DNA photolyase for the treatment of actinic keratoses. The cream used was applied twice a day for three months. They concluded that DNA photolyase decreased the number of lesions, supporting a role for photolyase as a treatment to reverse UV damage. These results have encouraged the research for new highly active photolyases and the development of photolyase-containing products. Marizcurrena *et al.* produced a bacterial recombinant photolyase from *Hymenobacter sp*. UV11, and did the characterization of its DNA repair ability. So, an enzyme was easily produced in a host cell and showed potential UV-damaged DNA repair activity *in vitro*. This work showed that the results obtained could contribute to the development of cosmetic products containing photolyases.

28.3 Stem Cells

A stem cell is an undifferentiated cell that can self-renew to replicate or can originate several specialized cell types. For example, hematopoietic stem cells can differentiate into red blood cells, white blood cells, and platelets.

Owing to the limited life cycle of most somatic cells, the ability of stem cells to replace damaged somatic cells is crucial for the tissue homeostasis of many organisms. Therefore, there is a tremendous interest in understanding the mechanisms of stem cells' self-renewal and differentiation, given their potential applications in regenerative medicine and aging studies.

With aging, the functional capacities of stem cells diminish, resulting in reduced organ function and delayed tissue regeneration. Thus, the decrease in stem cell function results in changes in the physiology of the tissue itself, which may affect the organism's health and viability.

Unlike human cells, plant cells are not only capable of regenerating tissues, but can also originate a new plant. For cosmetics, plant stem cells are reproduced in cell culture by a micropropagation method, which is involved *in vitro* cell culture. Many companies scaled up the production of stem cells from *in vitro* cultures to bioreactors that are used on a large scale. Plant stem cell extracts are sources of many active ingredients that are safe for the human body.

Plant stem cells are responsible for several effects, such as the following:

- Prolongation of fibroblasts life;
- Increased epidermis flexibility;
- Regulation of cell division;
- Reconstruction of damaged epidermis;
- Activation of cell DNA repair;
- Protection against UV radiation.

The pioneer company in the production of plant stem cells for the cosmetic industry is Mibelle AG Biochemistry (Switzerland), which produced liposomes containing apple stem cells from Uttwiler Spätlauber, a rare variety of Swiss apple (PhytoCellTec™ *Malus domestica*), in 2008. The clinical study that was conducted confirmed the efficacy of these stem cell extracts in reducing wrinkles after 28 days. This company has also introduced plant stem cell extracts from *Vitis vinifera* (PhytoCellTec™ Solar *Vitis*, Switzerland), *Saponaria pumila* (PhytoCellTec™ nunatak®, Buchs, Switzerland), and *Argania spinosa* (PhytoCellTec™ Argan, Switzerland) in the cosmetics market. These extracts are presented in the form of a suspension and also show great effects in the treatment of wrinkles and improve the activity of epidermal stem cells.

Stem cells from other plants have also been tested. For example, *Syringa vulgaris* contains verbascosides with skin anti-inflammatory and anti-aging effects; *Lycopersicon esculentum* from tomato has antioxidant properties, protecting skin cells from oxidative stress; *Coffea bengalensis* and *Nicotiana sylvestris* stimulate fibroblasts collagen production, promoting skin regeneration; orange stem cells (Citrustem™) improve skin elasticity and smoothness; and stem cells from ginger leaves (*Zingiber officinale*) reduce skin pores and sebum production, which leads to a smooth texture.

According to some studies, one of the strongest inhibitors of the human cell aging process is kinetin, which is a cytokinin found in high concentrations in stem cells of, for example, *Citrus* fruits and raspberries. Kinetin is a natural antioxidant that protects proteins and nucleic acids from oxidation processes and other types of damage. This compound allows cells to remove the excess of free radicals to protect them from oxidative stress and can be responsible for reducing the protein glycation. This cytokinin has been found to contribute to the prevention of skin aging. Kinetin, which can be designated as a natural growth hormone, is important for stimulating skin stem cells as it improves the epidermis barrier function, stimulates keratinocytes production, reduces trans-epidermal water loss, and reduces superficial wrinkles. According to the literature, kinetin has little or no photoprotection effect compared with other compounds and, therefore, should not be used in sunscreens.

Reference

Gomes C, Silva A C, Marques A C, Sousa Lobo J, Amaral M H. Biotechnology Applied to Cosmetics and Aesthetic Medicines. Cosmetics, 2020, 7: 33.

Key Words & Phrases

rejuvenation *n.* 更新；复苏
exogenous *adj.* 外源性的
revitalize *v.* 使恢复；使更强壮
autologous *adj.* 先天的；固有的；遗传的；同源（种）的
freckle *n.* 雀斑；斑点；黑斑
S. cerevisiae 酿酒酵母
cytoplasm *n.* 细胞质
prokaryotic *adj.* 原核生物的
eukaryotic *adj.* 真核的；真核生物的
P. pastoris 酵母菌
heterologously *adv.* 异源地
superoxide dismutase 超氧化物歧化酶
metabolic *adj.* 代谢的；新陈代谢的
protease *n.* 蛋白酶；蛋白[水解]酶
bromelain *n.* 菠萝蛋白酶
papain *n.* 木瓜蛋白酶
chymotrypsin *n.* 胰凝乳蛋白酶；糜蛋白酶
subtilisin *n.* 枯草杆菌蛋白酶
Bacillus licheniformis 地衣芽孢杆菌
actinic keratosis 光线性角化病
Hymenobacter *n.* 薄层菌属
stem cell 干细胞
hematopoietic *adj.* 造血的
platelet *n.* 血小板
somatic *adj.* 体细胞的；躯体的；体壁的
homeostasis *n.* 稳态；体内平衡；内环境稳定；内稳态；稳定性
micropropagation *n.* 微体繁殖
liposome *n.* 脂质体
Syringa vulgaris 紫丁香；欧丁香
verbascoside *n.* 毛蕊花糖苷
Lycopersicon esculentunm 番茄
Coffea bengalensis 孟加拉咖啡；本伽兰西斯种咖啡
Nicotiana sylvestris 美花烟草
cytokinin *n.* 细胞分裂素；细胞激动素
raspberry *n.* 树莓

Appendix Ⅰ: Names and Symbols of Selected Elements

Name	Symbol	Name	Symbol	Name	Symbol
aluminum	Al	holmium	Ho	rhenium	Re
antimony	Sb	hydrogen	H	rhodium	Rh
argon	Ar	indium	In	rubidium	Rb
arsenic	As	iodine	I	ruthenium	Ru
barium	Ba	iridium	Ir	samarium	Sm
beryllium	Be	iron	Fe	scandium	Sc
bismuth	Bi	krypton	Kr	selenium	Se
boron	B	lanthanum	La	silicon	Si
bromine	Br	lead	Pb	silver	Ag
cadmium	Cd	lithium	Li	sodium	Na
cesium	Cs	lutetium	Lu	strontium	Sr
calcium	Ca	magnesium	Mg	sulfur	S
carbon	C	manganese	Mn	tantalum	Ta
cerium	Ce	mercury	Hg	technetium	Tc
chlorine	Cl	molybdenum	Mo	tellurium	Te
chromium	Cr	neodymium	Nd	terbium	Tb
cobalt	Co	neon	Ne	thallium	Tl
copper	Cu	nickel	Ni	thulium	Tm
dysprosium	Dy	niobium	Nb	tin	Sn
erbium	Er	nitrogen	N	titanium	Ti
europium	Eu	osmium	Os	tungsten	W
fluorine	F	oxygen	O	vanadium	V
gadolinium	Gd	palladium	Pd	xenon	Xe
gallium	Ga	phosphorus	P	ytterbium	Yb
germanium	Ge	platinum	Pt	yttrium	Y
gold	Au	potassium	K	zinc	Zn
hafnium	Hf	praseodymium	Pr	zirconium	Zr
helium	He	promethium	Pm		

Appendix Ⅱ: Multiplicative Prefixes

No.	prefixes	No.	prefixes
1	mono	20	icosa
2	di(bis)	21	henicosa
3	tri(tris)	22	docosa
4	tetra (tetrakis)	23	tricosa
5	penta (pentakis)	30	triaconta
6	hexa (hexakis)	31	hentriaconta
7	hepta (heptakis)	35	pentatriaconta
8	octa (octakis)	40	tetraconta
9	nona (nonakis)	50	pentaconta
10	deca (decakis)	60	hexaconta
11	undeca	70	heptaconta
12	dodeca	80	octaconta
13	trideca	90	nonaconta
14	tetradeca	100	hecta
15	pentadeca	200	dicta
16	hexadeca	500	pentacta
17	heptadeca	1000	kilia
18	octadeca	2000	dilia
19	nonadeca		

Vocabulary

A

abrasive action　磨削作用
absolute configuration　绝对构型
acacia flower　金合欢花
acacia tree　刺槐；金合欢树
acetaldehyde　*n.* 乙醛
acetylated lanolin alcohol　乙酰化羊毛脂醇
acetylcholine　*n.* 乙酰胆碱
acetylcholinesterase　*n.* 乙酰胆碱酯酶
acid halide　酰基卤；酸性卤化物
acne　*n.* 痤疮；粉刺
acne ointment　痤疮膏
acne treatment lotion　痤疮治疗乳液
acrylic nail　水晶指甲
actinic keratosis　光线性角化病
acute poisoning　急性中毒
ad hoc　*adj.* 临时安排的；特别的
adenovirus　*n.* 腺病毒；腺病毒科
adequate　*adj.* 充足的；足够的；合格的；合乎需要的
adulterate　*v.* 掺杂；掺假
aerobe　*n.* 需氧菌；好氧微生物
aerosol　*n.* 气溶胶；气雾剂
aesthetic　*adj.* 审美的；美学的；美的；有审美观点的；艺术的
agar　*n.* 琼脂
agarobiose　*n.* 琼脂二糖；琼胶二糖
agaropectin　*n.* 硫琼胶；琼脂胶
agarose　*n.* 琼脂糖
age spot　老年斑
agglomeration　*n.* 集聚；聚集；(杂乱聚集的) 团
agrochemical　*n.* 农用化学品
Albizia　*n.* 合欢；合欢属
alboctalol　*n.* 白桑八醇
alembic　*n.* 蒸馏器 (釜，罐)；净化器具
aliquot　*n.* 试样；整除数；可分量
alkanet　*n.* 朱草；紫草根；由朱草提制的红色染料
alkenone　*n.* 烯酮；烯酮类
alkoxylation　*n.* 烷氧基化
alkyl benzoate　烷基苯甲酸酯
alkylamidopropylamine *N*-oxide　烷基酰胺丙基胺 *N*-氧化物
alkylamidopropylbetaine　*n.* 烷基酰胺丙基甜菜碱
alkylbetaine　*n.* 烷基甜菜碱
alkyldimethylamine *N*-oxide　烷基二甲胺 *N*-氧化物
allantoin　*n.* 尿囊素
allergic reaction　过敏反应
allowed colorant　允许用的着色剂
allowed preservative　允许用的防腐剂
aloe　*n.* 芦荟
Aloe ferox　开普芦荟
Aloe vera　芦荟汁 (用于生产护肤霜等)；芦荟
Aloe vera leaf gel　芦荟叶凝胶
aloesin　*n.* 芦荟素
alopecia areata　斑秃；局限性脱发
amaranth　*n.* 紫红色；苋属植物；苋菜
amber　*n.* 琥珀色；琥珀；黄褐色
amino-modified silicone oil　氨基改性硅油
ammonium lauryl sulfate　十二烷基硫酸铵；月桂基硫酸铵
ammonium thioglycolate　巯基乙酸铵；硫代乙醇酸铵
amodimethicone　*n.* 氨端聚二甲基硅氧烷
amphibian　*n.* 两栖动物
amphiphilic　*adj.* 两亲的；两性分子的
amphoteric　*adj.* 两性的
amplitude sweep　振幅扫描
analog　*n.* 模拟；[结构]类似物
anaphylaxis　*n.* 过敏反应
anesthetic　*n.* 麻醉药；麻醉剂
aneuploidy　*n.* 非整倍性
3,6-anhydro-L-galactopyranose　3,6-脱水-L-半乳吡喃糖
anion　*n.* 阴离子
anisotropic property　各向异性特性
annatto　*n.* 胭脂树红；胭脂树
anthranilate　*n.* 邻氨基苯甲酸盐
anthraquinone dye　蒽醌染料
anti-aging product　抗衰老产品

anti-collagenase 抗胶原酶
anti-elastase 抗弹性蛋白酶
antifungal *n.* 抗真菌剂；抗霉剂
anti-inflammatory *adj./n.* 抗炎的（药）
anti-keratinizing 抗角质化
antioxidant *n.* 抗氧化剂；抗氧化物
antiperspirant *n.* 止汗剂
antiproliferative activity 抗增殖活性；抗增殖
antiseptic *n.* 防腐剂；抗菌防腐药
antistatic *adj.* 抗静电的
anxiety *n.* 焦虑；忧虑；担心；害怕；渴望
aperture *n.* 光圈；缝隙；小孔；（尤指摄影机等的光圈）孔径
apiaceae *n.* 伞形科
apoptosis *n.* 凋亡；细胞凋亡
arabic gum 阿拉伯树胶
arabinose *n.* 阿拉伯糖；果胶糖
arazine *n.* *N*-乙酰基-*S*-法尼基-L-半胱氨酸
arbutin *n.* 熊果苷
armpit *n.* 腋窝
aromatic dye 芳香染料
arsenate *n.* 砷酸盐
arsenic acid 砷酸
arsenite *n.* 亚砷酸盐
arsenous acid 亚砷酸
artocarpanone *n.* 桂木二氢黄素
Artocarpus gomezianus 长圆叶菠萝蜜
Artocarpus heterophyllus 木菠萝，菠萝蜜
Artocarpus xanthocarpus 黄果菠萝蜜
asbestos *n.* 石棉
ascorbate *n.* 抗坏血酸盐
ascorbic acid 抗坏血酸；维生素C
asteraceae *n.* 菊科；菊科植物
asthma *n.* 气喘；哮喘
Astragalus *n.* 距骨；紫云英属
astringent *n.*（用于化妆品或药物中的）收敛剂；止血剂
atopic *adj.* 特应性的；异位的
auspice *n.* 支持；赞助；资助；主办；保护
autologous *adj.* 先天的；固有的；遗传的；同源（种）的
avobenzone *n.* 阿伏苯宗；亚佛苯酮
avocado oil 鳄梨油
azelaic acid 杜鹃酸；壬二酸
azide *n.* 叠氮化物
azo dye 偶氮染料

B

Bacillus licheniformis 地衣芽孢杆菌
Bacillus subtilis 枯草杆菌
bacteriostatic activity 抑菌活性
baobab oil 猴面包树油
barium sulfate 硫酸钡；重晶石
barrier cream 护肤霜；隔离霜
basecoat *n.* 底漆；最下面的一层
baseline *n.* 基线；基础；起点
bassorin *n.* 西黄蓍胶；黄蓍胶糖
bearberry *n.* 熊果；熊莓
beeswax *n.* 蜂蜡
behentrimonium chloride 山嵛基三甲基氯化铵
behentrimonium-methosulfate 山嵛基三甲基铵甲基硫酸盐
behenyl alcohol 山嵛醇；二十二醇
benzalkonium chloride 苯扎氯铵
benzethonium chloride 氯化苄乙氧铵；苄索氯铵
benzophenone *n.* 二苯甲酮
benzoyl peroxide 过氧化苯甲酰
benzylhemiformal *n.* 苄基半缩甲醛
Betula pendula 欧洲白桦；垂枝桦
beverage *n.* 饮料
binary compound 二元化合物
binder *n.* 结合剂；黏合剂
biocide *n.* 抗微生物剂；杀虫（菌）剂
biopsies *n.* 活组织检查
birefringence *n.* 双（光）折射；二次光折射；折射率
bismuth oxychloride 氯氧化铋
bleaching agent 漂白剂；增白剂
blemish balm 修护霜；伤痕保养霜；遮瑕霜
blend *v.*（和某物）混合；融合；掺和；调制
blusher *n.* 胭脂；脸红的人
botulinum toxin 肉毒杆菌毒素；肉毒毒素
bract *n.* 苞；苞片；托叶
brand awareness 品牌知名度
breast augmentation 隆胸术；丰胸；隆胸；隆乳；隆乳手术
bridged polycyclic 桥接多环
bridgehead *n.* 桥头；桥塔
brittle *adj.* 易碎的；脆性的

bromelain *n.* 菠萝蛋白酶
bronopol *n.* 溴硝丙二醇；溴硝醇
bronzer *n.* 古铜色化妆品
brown algae 褐藻门；褐藻
butylated hydroxyanisole 丁基羟基茴香醚
butylated hydroxytoluene 丁基羟基甲苯

C

cactus *n.* 仙人掌；仙人掌科植物
caffeine *n.* 咖啡碱
calamine *n.* 炉甘石
calibrate *v.* 标定；校准（刻度，以使测量准确）
Camellia oleifera 油茶
camouflage makeup 迷彩妆
camphor *n.* 樟脑
candelilla shrub 小烛树灌木
candelilla wax 小烛树蜡；堪地里拉蜡
capric triglyceride 癸酸甘油三酯
carbene *n.* 卡宾；碳烯
carbonyl *n.* 羰基
carboxylic acid 羧酸
carboxymethyl cellulose 羧甲基纤维素
carbyne *n.* 卡拜；碳炔
carcinogen *n.* 致癌物
carcinogenic *adj.* 致癌的
carcinogenicity *n.* 致癌性
cardiovascular *adj.* 心血管的
carnauba wax 巴西棕榈蜡；棕榈蜡
carnosine *n.* 肌肽
carotene *n.* 胡萝卜素
carotenoid *n.* 类胡萝卜素；类叶红素；脂色素
carrageenan *n.* 卡拉胶
carrageenan *n.* 角叉聚糖；角叉菜胶；卡拉胶
carrageenin *n.* 角叉聚糖；角叉菜胶；卡拉胶
carrier peptide 载体肽
castor oil 蓖麻子油
catechin gallocatechin 儿茶素-棓儿茶素
catecholamine secretion 儿茶酚胺分泌
cation *n.* 阳离子
cationic polymer 阳离子聚合物
Caucasian *n.* 高加索人
causation *n.* 诱因；起因；原因
cedarwood *n.* 杉木；雪松属木料
cell migration 细胞迁移
ceramide *n.* 神经酰胺
cetearyl alcohol 鲸蜡醇；棕榈醇
cetrimonium bromide 十六烷基三甲基溴化铵；西曲溴铵
cetrimonium chloride 西曲氯铵；十六烷基三甲基氯化铵
cetyl alcohol 十六醇
cetyltrimethylammonium chloride 十六烷基三甲基氯化铵
chalcogenide *n.* 硫属化物
characteristic (functional) group 特征基团；官能团
charge number 电荷数
chelating agent 螯合剂；络合剂
chelator *n.* 螯合剂
chemical peel 化学换肤；化学脱皮术；果酸换肤；化学剥脱
chirality *n.* 手性
chi-square 卡方分布；卡方；卡方检验
chloral hydrate 水合氯醛；水合三氯乙醛
chlorhexidine *n.* 洗必泰；双氯苯双胍己烷
chlorhexidine digluconate 氯己定二葡糖酸盐
chlorhexidine gluconate 葡萄糖酸洗必泰
chloroacetate *n.* 氯乙酸盐
4-chloroaniline 4-氯苯胺
chlorobutanol *n.* 三氯叔丁醇；氯丁醇；氯代丁醇
4-chloro-*m*-cresol 4-氯间甲酚
chlorophyll *n.* 叶绿素
cholesterol *n.* 胆固醇
chondrus crispus 角叉菜；鹿角菜；爱尔兰海藻
chrome hydroxide green 氢氧化铬绿
chrome oxide green 氧化铬绿；铬绿
chymotrypsin *n.* 胰凝乳蛋白酶；糜蛋白酶
ciclopirox olamine 环吡司胺；环吡酮胺
cinnabar *n.* 辰砂；朱砂（可用作颜料）
cinnamate *n.* 肉桂酸酯；肉桂酸；肉桂酸盐
cinnamon *n.* 肉桂皮；桂皮香料
cinnamon leaf oil 肉桂叶油
cis and *trans* isomer 顺反异构体
citrate *n.* 柠檬酸盐
citric acid 柠檬酸
citrulline *n.* 瓜氨酸
citrus *n.* 柑橘属
clastogen *n.* 断裂剂

cleanser *n.* 洁肤液；洁肤霜；清洁剂
cleansing balm 卸妆膏；洁面膏
climbazole *n.* 甘宝素；氯咪巴唑；咪菌酮；苯咪丁酮
clinical test 临床试验
coal tar 煤焦油
cocamide DEA 椰油酸二乙醇酰胺
cocamide MEA 椰油酸单乙醇酰胺
cocamidopropyl betaine 椰油酰胺丙基甜菜碱
cochineal *n.* 胭脂虫红颜料
cocoa butter 可可脂
cocoamphoacetate *n.* 椰油乙酸盐
cocoamphodiacetate *n.* 椰油二乙酸盐
coconut oil 椰子油
coenzyme Q10 辅酶Q10
Coffea bengalensis 孟加拉咖啡；本伽兰西斯种咖啡
cold cream 洁面乳；润肤膏
collagen *n.* 胶原蛋白；胶原
collagen stimulation 胶原蛋白刺激
collagenase *n.* 胶原酶
colloidal force 胶体力
colloidal oatmeal 天然胶体燕麦；胶状燕麦精华；胶原燕麦；燕麦凝胶
colloidal sulfur 胶体硫黄；硫黄胶；胶态硫
color cosmetic 彩妆
color lake 色淀
colorimeter *n.* 色量计；色度计；比色计
com starch 玉米淀粉
Combretum *n.* 风车子属；风车藤属；风车子
comedogenic *adj.* 产生粉刺的
comparator *n.* 比测（值）器；比长仪；比色计；比较器（块）
compensate *v.* 补偿；弥补
compulsory *adj.* 强制性的；（因法律或规则而）必须做的；强制的；强迫的
concealer *n.* 遮瑕膏
concentric cylinder spindle 同心圆柱主轴
conditioner *n.* 护发剂；护发素；（洗衣后用的）柔顺剂
conditioning agent 调节剂
conjugated bond 共轭键
consecutive *adj.* 连续的；连续不断的
consensus *n.* 一致的意见；共识
constitutional repeating unit 组成重复单元；重复结构单元
contact urticarial 接触性荨麻疹
contaminant *n.* 致污物；污染物
contiguous *adj.* 相接的；相邻的
contour *n.* 外形
coordination *n.* 协作；协调；配合；配合物
Copernicia prunifera 巴西棕榈树
copolymer *n.* 共聚物
coriander *n.* 芫荽；香菜
corneocyte *n.* 角化细胞；角质层细胞
cortex *n.* 皮层；皮质
cortex fragment 皮质片段；皮层片段
cosmeceutical *n.* 药妆品；药妆；药用化妆品
cosmetic *n.* 化妆品；美容品
cosmetic surgery 整容手术
creatine *n.* 肌酸
credential *n.* 资格证书；证件；资格；资历
crimson *n.* 深红色
criterion *n.* 标准
critical stress 临界应力
crow’s feet 鱼尾纹；眼角的鱼尾纹
crucible *n.* 坩埚；熔炉
cumene hydroperoxide 过氧化氢异丙苯
Curcuma xanthorrhiza 印尼莪术
cutaneous *adj.* 皮肤（上）的
cutaneous penetration 皮肤穿透
cuticle *n.* 角质层
cyanine dye 花青染料；菁染料
cyclobutane pyrimidine 环丁烷嘧啶
cyclomethicone *n.* 环甲硅脂；环聚二甲基硅氧烷
cyclooxygenase *n.* 环加氧酶；环氧合酶
cytokine *n.* 细胞因子；细胞激素
cytokinin *n.* 细胞分裂素；细胞激动素
cytologic *adj.* 细胞学的
cytoplasm *n.* 细胞质

D

dandruff *n.* 头皮屑
dark circle 黑眼圈；黑圈
day cream 日霜
daywear *n.* 便服；日常衣服
decorative cosmetic 装饰性化妆品
decorin-like 类饰胶蛋白聚糖
decyl glucoside 癸基葡糖苷

delineate *v.* 勾画；描述；描画；解释
Dendropanax morbifera 黄漆木
deodorant *n.* 除臭剂；解臭剂
deoxyarbutin *n.* 脱氧熊果苷
depigmentation *n.* 脱色；褪色
depilatory *n.* 脱毛剂
derivative *n.* 衍生物
dermal *adj.* 皮肤的；真皮的
dermal density 真皮密度
dermatitis *n.* 皮炎
dermatologically *adv.* 皮肤病学地
dermatologist *n.* 皮肤科医生；皮肤科医师
dermis *n.* 真皮
dermoscopic *n.* 皮肤镜
desquamation *n.* 脱屑；剥离
detergent agent 洗洁精
D-galactose D-半乳糖
diabete *n.* 糖尿病
diaper *n.* 尿布；（婴儿的）尿片
diatragacanth *n.* 泛棘
dibenzoylmethane *n.* 二苯甲酰甲烷
dibutyl phthalate 邻苯二甲酸二丁酯
diester *n.* 双酯
dihydroxyacetone *n.* 二羟基丙酮
dimeric stilbene 二聚二苯乙烯
dimethicone *n.* 聚二甲基硅氧烷；二甲基硅油
dimethiconol *n.* 聚二甲基硅氧烷醇
dipropylene glycol 一缩二丙二醇；二丙二醇
disaccharide *n.* 双糖；二糖
discrepancy *n.* 差异
disinfectant *n.* 消毒剂；杀菌剂
disodium lauryl sulfosuccinate 月桂醇磺基琥珀酸酯二钠
dispersity *n.* 色散度；弥散度
dissipate *v.* 消散；驱散；（使）消失
diureide *n.* 二酰脲
divalent group 二价基
DMDM hydantoin 二羟甲基二甲基乙内酰脲
dodecyl *n.* 十二烷基
dossier *n.* 材料汇编；卷宗；档案
double-blind 双盲的
duly *adv.* 适当地；适时地；恰当地；按时地；准时地
dye *n.* 染料

E

E. coli 大肠杆菌
E. guineensis 油棕
eau de Cologne 科隆香水；古龙水
eau de parfum 淡香精；香水
eau de toilette 淡香水
eau fraiche 清香水；淡香水；幽香水
eczema *n.* 湿疹
eczematous *adj.* 湿疹性的
edema *n.* 浮肿；水肿
efficacy *n.* 功效；（尤指药物或治疗方法的）效力
elaborate *v.* 详尽阐述；详细制订；精心制作
elastase *n.* 弹性蛋白酶
elastin *n.* 弹性蛋白
electronegative *adj.* 电负性的
electropositive *adj.* 阳性的；带正电的；正电性的
element sequence 元素序列
eleuthero-embryo 刺五加胚
embalmment *n.* 尸体防腐法；（尸体的）防腐处理；薰香
emergency *n.* 突发事件；紧急情况
emollient *n.* 润肤剂；润肤霜
end-group 端基
endocrine *n.* 内分泌；内分泌系统
endocuticle *n.* 内表皮；内角皮
endogenous *adj.* 内源；内生的
endogenous plasma 内源性血浆
endonuclease *n.* 核酸内切酶；内切核酸酶
endotoxin *n.* 内毒素
enkephalin *n.* 脑啡肽
entrust *v.* 委托；交托；托付
enzyme inhibitor peptide 酶抑制剂肽
eosin *n.* 曙红（一种红色荧光染料）
eosine Y 曙红 Y
epidermis *n.* 表皮
epidermolysis *n.* 表皮松解
epigallocatechin *n.* 表没食子儿茶素
epigallocatechin-3-gallate 表没食子儿茶素-3-没食子酸酯
epithelial *adj.* 上皮的；皮膜的
epithelium *n.* 上皮细胞
ergothionine *n.* 麦硫因
ericaceae *n.* 杜鹃花科

erosion *n.* 侵蚀；糜烂；腐蚀
erythema *n.* 红斑
erythemogenic *adj.* 引起红斑的
essential oil 精油
ester *n.* 酯
esterification *n.* 酯化反应
estrogenic *adj.* 动情（激素）的；雌激素的
estrogenicity *n.* 雌激素活性
ethical *adj.*（有关）道德的；伦理的；合乎道德的
ethnicity *n.* 种族渊源；种族特点
ethoxylated fatty alcohol 乙氧基脂肪醇
ethyl cellulose 乙基纤维素
ethylenediaminetetraacetic acid 乙二胺四乙酸
ethylhexul methoxycinnamate 甲氧基肉桂酸乙基己酯
eukaryotic *adj.* 真核的；真核生物的
eupafolin *n.* 楔叶泽兰素；泽兰叶黄素
excipient *n.* 赋形剂
excretion *n.* 排泄；排泄物；分泌；分泌物
exempt *v.* 豁免；免除
exert *v.* 发挥；施加；行使；运用
exfoliant scrub cleanser 去角质/磨砂洁面乳
exogenous *adj.* 外源性的
explicit *adj.* 明确的；清楚明白的；易于理解的；（说话）清晰的；直言的；坦率的；不隐晦的；不含糊的
exponentially *adv.* 以指数方式
expression *n.* 压出法
extracellular matrix 细胞外基质
extrapolate *v.* 推断；推知；外推
eye shadow 眼影膏
eyebrow pencil 眉笔
eyeliner *n.* 眼线笔，眼线膏

F

face cream 面霜；雪花膏
face powder 扑面粉；敷面粉
face-lift 美化；翻新；面部皱纹切除术；面部拉皮术
facial cleanser 洗面奶；洁面乳
facial mask 面膜
facultative anaerobe 兼性厌氧菌；兼性厌氧；兼性厌氧微生物
false eyelash 假睫毛
fenugreek *n.* 葫芦巴（种子用于南亚食物调味）
fermentation *n.* 发酵
ferric ammonium ferrocyanide 亚铁氰化铁铵
ferric ferrocyanide 氰亚铁化亚铁
ferrocene *n.* 二茂铁
fibrillin-rich microfibril 富含原纤维蛋白的微纤维
fibroblast *n.* 成纤维细胞
fibronectin *n.* 纤连蛋白
filaggrin *n.* 聚丝蛋白
filthy *adj.* 肮脏的；污秽的
fitness *n.* 健身；健康；适合
fixative *n.* 定影剂；定色剂；防（香味）挥发剂；固定剂
flake *n.* 小薄片；（尤指）碎片
flapper *n.* 新潮女郎
flavanol *n.* 黄烷醇
flavonoid *n.* 黄酮类化合物
follicular *adj.* 滤泡的；卵泡的；小囊的
foreseeable *adj.* 可预料的；可预见的；可预知的
formaldehyde-releaser 甲醛释放剂
formate *n.* 甲酸盐
formulation *n.*（药品、化妆品等的）配方；剂型；配方产品
foundation *n.*（化妆打底用的）粉底霜；地基；基础
fragrance *n.* 香味；香气；香水；芳香
fragrant oil 香油；芬芳油；香精
freckle *n.* 雀斑；斑点；黑斑
free radical 自由基
free radical scavenger 自由基清除剂
free radical trapper 自由基捕捉器
free valence 自由价
freeze-dried 冷冻干燥
freeze-dried powder 冻干粉
freeze-thaw 冻融；冷冻-解冻；解冻；冻融法；冻融作用
frequency sweep 频率扫描；扫频
friction *n.* 摩擦；摩擦力；争执；分歧；不和
fringe projection profilometry 条纹投影轮廓术
frizzy hair 卷发
fructose *n.* 果糖
Fucus *n.* 墨角藻属；墨角藻；褐藻；黑角菜属之海藻；岩藻
fullerene *n.* 富勒烯

functional class name　官能团类名
fungal　*adj.* 真菌的；真菌引起的
fungi　*n.* 真菌
fused polycyclic　稠合多环

G

galactomannan　*n.* 半乳甘露聚糖
galactose　*n.* 半乳糖
gauge　*v.*（用仪器）测量；判定；判断（尤指人的感情或态度）；估计
gelatin　*n.* 明胶
general provision　一般规定
genre　*n.* 体裁；类型
geranium　*n.* 天竺葵；老鹳草
glabrene　*n.* 光甘草
glabridin　*n.* 光甘草定
glaze　*n.* 釉；瓷釉
glucaric acid　葡萄糖二酸
β-D-glucopyranoside　β-D-吡喃葡萄糖苷
glutaraldehyde　*n.* 戊二醛
glycerol monostearate　单硬脂酸甘油酯
glycolic acid　乙醇酸
glycoproteins　*n.* 糖蛋白
glycosaminoglycan　*n.* 糖胺聚糖
glyoxyldiureide　*n.* 尿囊素
glyoxylic acid　乙醛酸
Gram-negative　革兰氏阴性
Gram-positive　革兰氏阳性
granular layer　颗粒层
granule　*n.* 颗粒状物；微粒；细粒
guanidine carbonate　碳酸胍；胍碳酸盐
guar gum　瓜尔胶；古尔胶
gum benzoin　安息香树胶

H

hair care　护发
hair conditioner　护发素
hair dye　染发剂
hair follicle　毛囊
hair frizzing　头发卷曲
hair lacquer　毛发定型剂
hair spray　发胶
hair straightening　拉直头发
hair wax　发蜡
haircare　*n.* 头发护理
hand sanitizer　手部消毒剂；洗手液
hapticity　*n.* 扣数
heather　*n.* 帚石楠
hedonic　*adj.* 享乐的；享乐主义的
hematopoietic　*adj.* 造血的
hemostatic　*adj.* 止血的
henna　*n.* 散沫花染剂（棕红色，尤用于染发）
herpes virus　疱疹病毒
heterocycle　*n.* 杂环
heterocyclic dye　杂环染料
heterologously　*adv.* 异源地
heteropolyatomic　*adj.* 杂多原子的
hexamethyldisiloxane　*n.* 六甲基二硅醚；六甲基二硅氧烷
hierarchical cluster analysis　系统聚类分析；层次聚类分析
high-end product　高端产品
highlighter　*n.* 荧光笔；高光色
histidine　*n.* 组氨酸
histological　*adj.* 组织（学）的
histology　*n.* 组织学
homeostasis　*n.* 内稳态；体内平衡；内环境稳定；稳定性；稳态
homoatomic　*adj.* 同原子的
homogeneity　*n.* 同质性；同质；同种
homogenisation　*n.* 均化；同质化；均一化；均质化
homogenizer　*n.* 均化（匀浆）器；均质机
homopolymer　*n.* 均聚物
huckleberry　*n.* 美洲越橘；美洲越橘树
human trial　人体试验
humectant　*n.* 保湿剂
hyaline cartilage　透明软骨
hyaluronan synthase 2　透明质酸合成酶2
hyaluronic acid synthase 1　透明质酸合酶1
hyamine　*n.* 季铵盐
hybrid material　混杂材料；杂化材料
hydrate　*n.* 水合物
hydride　*n.* 氢化物
hydrogenation　*n.* 氢化
hydrolysis　*n.* 水解；水解作用
hydrolytic　*adj.* 水解的
hydrophilia　*n.* 亲水性；吸水性
hydrophilic　*adj.* 亲水（性）的

hydrophilicity *n.* 亲水性
hydrophilic-lipophilic balance 亲水亲油平衡
hydrophobic *adj.* 疏水的
hydroquinone *n.* 氢醌；对苯二酚
hydroquinone glycoside 对苯二酚糖苷
hydroxide *n.* 氢氧化物
hydroxy *n.* 羟基；氢氧根的
4-hydroxyanisole 4-羟基茴香醚
hydroxyester *n.* 羟基酯
hydroxyethyl-behenamidopropyl-dimonium chloride 羟乙基苯胺丙基二氯化铵
hygiene *n.* 卫生
Hymenobacter *n.* 薄层菌属
hyperpigmentation *n.* 色素沉着过度
hypnotic *n.* 催眠药；安眠药
hypoallergenic *adj.* 低敏感性的；低过敏的

I

ichthammol *n.* 鱼石脂
imminent *adj.* 迫在眉睫的；即将发生的；临近的
immunohistological *adj.* 免疫组织的
immunomediate *adj.* 免疫介导的
immunosuppression *n.* 免疫抑制
implant *n.*（植入人体中的）移植物；植入物
implicit *adj.* 含蓄的；不直接言明的；成为一部分的；内含的；完全的；无疑问的
in silico 生物信息学；预测基因
in vitro 在生物体外
indigoid dye 靛蓝染料
inflammation *n.* 发炎；炎症
ingredient *n.* 成分；原料；因素；要素
inhalation *n.* 吸入；吸入剂；吸入药
inhale *v.* 吸入；吸气
innate *adj.* 与生俱来的；天生的；先天的
innate immunity 先天免疫
inorganic oxoacid 无机含氧酸
insanitary *adj.* 不卫生的；不洁的
intercellular cement 胞间黏合质；细胞间胶质
inter-corneocyte cohesion 角质细胞间内聚力
interleukin-6 白细胞介素 -6
interpretative *adj.* 解释的；作为说明的
intervene *v.* 介入；出面；插嘴；打断；阻碍；干扰
intervertebral disc 椎间盘
intimate *adj.* 亲密的；密切的；个人隐私的
ionic chromatography 离子色谱
iontophoretic permeability coefficient 离子导入渗透系数
irish moss 爱尔兰苔藓；角叉菜；爱尔兰藓
irregular polymer 非规整聚合物
Isochrysis *n.* 等鞭金藻属
isododecane *n.* 异十二烷
isoform *n.* 亚型；同工型；同型；异构体
isoliquiritigenin *n.* 异甘草素
isoliquiritin *n.* 异甘草苷
isoparaffinic *adj.* 异构化烷烃的
isopropyl lanolate 羊毛酸异丙酯
isopropyl myristate 肉豆蔻酸异丙酯；十四酸异丙酯
isopropyl palmitate 十六酸异丙酯；棕榈酸异丙酯

J

jargon *n.* 行话
jasmine *n.* 茉莉
jawline *n.* 下颌的轮廓；下巴的外形
jojoba oil 荷荷巴油；霍霍巴油
jurisdiction *n.* 司法权；审判权；管辖权

K

keratin *n.* 角蛋白
keratin filament 角蛋白丝；角蛋白纤维
keratin-based peptide 角蛋白基肽
keratinocyte *n.* 角蛋白形成细胞
keratinocyte proliferation 角质形成细胞增殖
keratolytic *adj.* 角质层分离的；角蛋白溶解的
ketoconazole *n.* 酮康唑
ketone *n.* 酮
kohl *n.* 黑色眼影粉（尤指东方人用的）
kojic acid 曲酸
konjac sponge 魔芋海绵

L

lace *n.* 网眼织物；花边；蕾丝
lactate *n.* 乳酸盐
ladder polymer 梯形聚合物
lake *n.* 色淀
lamellar *adj.* 薄片状的；层式的；成薄层的；多层（片）的
Lamellar *adj.* 薄片状的；层式的；成薄层的
lamiaceae *n.* 唇形科

lanolin *n.* 羊毛脂
lard *n.* 猪油
large pore 大孔；粗大毛孔
latex *n.* 乳胶
latex glove 乳胶手套
lauric acid 月桂酸
laurtrimonium chloride 月桂基三甲基氯化铵
lavender oil 薰衣草油
Lawsonia alba 散沫花；指甲花
lecithin *n.* 卵磷脂；磷脂酰胆碱
legal liability 法律责任
legible *adj.* 易读的；可读的
lemon peel oil 柠檬皮油
Lemongrass *n.* 柠檬草；香茅
lentigines *n.* 雀斑痣；（尤指）老年斑
lesion *n.*（因伤病导致皮肤或器官的）损伤；损害
leuphasyl *n.* 五肽
lexicon *n.* 全部词汇；词汇表；字典
licochalcone A 甘草查尔酮A
licorice *n.* 甘草精
licuraside *n.* 异甘草素葡萄糖洋芫荽糖苷
ligand *n.* 配体
lightener *n.* 美白剂
linoleic acid 亚油酸
lip balm 润唇膏；护唇膏
lip gloss 唇彩；亮唇膏
lip liner 唇线笔；唇线
lipid peroxidation 膜脂质过氧化；脂质过氧化作用
lipid rancidity 油脂酸败
liposome *n.* 脂质体
liposuction *n.* 吸脂术；脂肪抽吸（术）
lipoxygenase *n.* 脂加氧酶；脂氧化酶
lipstick *n.* 口红；唇膏
liquid foundation 粉底液；粉底；液体粉底；粉底霜；湿粉
liquid paraffin 液体石蜡
liquiritin *n.* 甘草苷
local tolerance 局部耐受性
locant *n.* 位次
lotion *n.* 洗剂；化妆水；水粉剂
low-end market 低端市场
lumican *n.* 光蛋白聚糖
Lycopersicon esculentum 番茄

M

maceration *n.* 浸软；浸渍
macromolecule *n.* 大分子
maculae *n.* 缺陷；斑点；暗斑；气门斑（蜱螨）
magnesium-1-ascorbyl-2-phosphate 1-抗坏血酸-2-磷酸镁
makeup *n.* 化妆品；天性；性格；组成；构成
malachite *n.* 孔雀石
Malassezia furfur 糠秕马拉色菌
maltese cross 黑十字花样
mammalian *n.* 哺乳类动物
mammalian tyrosinase 哺乳动物酪氨酸酶
mandate *n.* 授权；命令
mandatory *adj.* 强制性的；强制的
mandelic acid 扁桃酸
manganese violet 锰紫
manifold *adj.* 多的；多种多样的；许多种类的
mannitol *n.* 甘露醇；甘露糖醇
MAPK phosphorylation MAPK磷酸化
mascara *n.* 睫毛膏；染睫毛油
masculine *adj.* 男子的；男性的
mask *n.* 面具；面罩；假面具；护肤膜；面膜
matrix metalloproteinase 基质金属蛋白酶
mature *adj.* 成熟的
Mediterranean basin 地中海盆地
melanin *n.* 黑色素
melanin granule 黑色素颗粒
melanocyte *n.* 黑素细胞
melanogenesis *n.* 黑色素生成
melanoma *n.* 黑（色）素瘤
melasma *n.* 黄褐斑
menthol *n.* 薄荷醇
mequinol *n.* 甲氧苯酚；对甲氧酚
mercurialentis *n.* 汞中毒性晶状体变色
meridional *adj.* 子午线的；经向的
mesocarp *n.* 中果皮
metabolic *adj.* 代谢的；新陈代谢的
metabolism *n.* 新陈代谢；代谢
metabolite *n.* 代谢物；代谢产物
metabolize *v.* 新陈代谢
metallocene *n.* 茂金属
metal-metal bond 金属-金属键
methyl cellulose 甲基纤维素

methylene glycol 亚甲二醇
methylparaben *n.* 羟苯甲酯；对羟基苯甲酸甲酯
mica *n.* 云母
microdermabrasion *n.* 微晶磨皮术；微晶磨皮；微晶换肤术；微晶换肤仪
microfiber cloth 超细纤维布
microfiltration *n.* 微滤
micropropagation *n.* 微体繁殖
mid-end market 中端市场
milk thistle 水飞蓟；乳蓟
mink oil 水貂油
minoxidil *n.* 米诺地尔
mitt *n.* 接球手套；棒球手套；手
moisturising agent 保湿剂
moisturizer *n.* 润肤霜；润肤膏
monocyclic *adj.* 单环的
monograph *n.* 专论；专题文章；专著
monolaurin *n.* 甘油一月桂酸酯；单月桂酸甘油酯
monomer *n.* 单体
mononuclear complex 单核配合物
moraceae *n.* 桑科
Morus alba 桑树；桑科植物
mould *n.* 霉；霉菌
mucilage *n.* 黏液；黏质；胶浆剂
mucilaginous *adj.* 黏的；黏液质的
muco adhesiveness 黏液黏附性
mucopolysaccharide *n.* 黏多糖；糖胺聚糖
mucosa *n.* 黏膜
mucous *adj.* 黏液的；黏的；分泌黏液的
mulberry *n.* 桑树；桑葚；深紫红色
multielement *n.* 多元素；多元件
musk *n.* 麝香
musk xylene 二甲苯麝香；二麝香
musky *adj.* 有麝香味的；麝香的
mutagen *n.* 诱变剂
mutagenic *adj.* 诱变的
mutagenicity *n.* 诱变性
mutation *n.* 突变；（生物物种的）变异
mycosis *n.* 真菌病
myristamine oxide 肉豆蔻胺氧化物

N

nacreous *adj.* 珍珠母的；珍珠（质）的；珍珠似的；有光泽的
nail polish 指甲油
nanofiltration *n.* 纳米过滤
naphthenic *adj.* 环烷（烃）的
naphthoflavone *n.* 萘黄酮
National Medical Products Administration 国家药品监督管理局
natural moisturizing factor 天然保湿因子
nausea *n.* 恶心；恶心想吐
nephrotoxicity *n.* 肾毒性；中毒性肾损害
neurotransmitter *n.* 神经递质
neurotransmitter inhibitor peptide 神经递质抑制肽
neutrophil *n.* 中性粒细胞；嗜中性粒细胞
niacinamide *n.* 尼克酰胺；烟酰胺
Nicotiana sylvestris 美花烟草
nicotinamide adenine dinucleotide phosphate 烟酰胺腺嘌呤二核苷酸磷酸
night cream 晚霜
niosome *n.* 泡囊；非离子表面活性剂囊泡
nitrate *n.* 硝酸盐
nitric acid 硝酸
nitrite *n.* 亚硝酸盐
nitroaromatic amine 硝基芳香胺
nitrocellulose *n.* 硝化纤维素；硝酸纤维素
nitrosamine *n.* 亚硝胺
nitrosating agent 亚硝化剂
nitrous acid 亚硝酸
nomenclature *n.* 命名法
non-cytotoxic 非细胞毒性
non-exhaustive 非详尽无遗的
non-ionic 非离子（式）的
non-linear polymer 非线型聚合物
nonoxynol *n.* 壬苯醇醚
nonpolar *adj.* 非（无）极性的
nonprescription *n.* 非处方药
non-setting facial mask 不定型面膜
nonsteroidal *adj.* 非甾体化合物的；非类固醇的
norartocarpetin *n.* 降桂木生黄亭
nordihydroguaiaretic acid 去甲二氢愈创木酸
notification person 通知人
nucleic acid 核酸
nutmeg *n.* 肉豆蔻

O

oakmoss *n.* 橡树苔；橡木苔
oat bran 燕麦麸
objective evaluation 客观评价

obligation　*n.* 职责；责任；（已承诺的或法律等规定的）义务
occlusive　*adj.* 闭塞的；咬合的
ocher　*n.* 赭石
octahedral　*n.* 八面体
octinoxate　*n.* 桂皮酸盐
octocrylene　*n.* 氰双苯丙烯酸辛酯；奥克立林
odoriferous　*adj.* 有气味的；散发气味的
ointment　*n.* 药膏；软膏；油膏
oleic acid　油酸
oleosome　*n.* 油质体
olfaction　*n.* 嗅觉
olfactory sense　嗅觉
opaqueness　*n.* 不透明性
opioid　*n.* 阿片样物质；阿片类[物质]
optesthesia　*n.* 视觉
Opuntia humifusa　仙人掌
organelle　*n.* 细胞器
organoleptic　*adj.* 影响（或涉及）器官（尤指味觉、嗅觉或视觉器官）的；感官的
organometallic compound　金属有机化合物
organosulfate　*n.* 有机硫酸盐
oscillation　*n.* 摆动；摇摆；振动；浮动；振幅
oxidase　*n.* 氧化酶
oxidation number　氧化数
8-oxo-7,8-dihydro-2′-deoxyguanosine　8-氧代-7,8-二氢-2′-脱氧鸟苷
8-oxoguanine glycosylase　8-氧桥鸟嘌呤糖基化酶
oxy subunit　氧亚基
oxybenzone　*n.* 羟苯甲酯

P

p53 protein　p53蛋白
palm　*n.* 棕榈；棕榈树
palmitic acid　棕榈酸
palmitoyl pentapeptide-4　棕榈酰五肽-4
p-aminophenol　*n.* 对氨基苯酚
panelist　*n.* 小组成员
p-anisidine　对氨基苯甲醚；对甲氧基苯胺
panthenol　*n.* 泛醇
papain　*n.* 木瓜蛋白酶
para-aminobenzoic acid　对氨基苯甲酸
paraben　*n.* 尼泊金（苯甲酸酯类）
paraffin wax　石蜡
parent compound　母体化合物
parent hydride　母体氢化物
parfum　*n.* 香精；（法）香水
partner　*n.* 搭档；合伙人；同伴；舞伴；伙伴；配偶
passion fruit juice　西番莲汁；百香果汁
patchouli　*n.* 广藿香精油；广藿香
pathogen　*n.* 病原体；病原物
pathogenic　*adj.* 致病性的；致病的；病原的
pathogen　*n.* 病原体
pearlised agent　珠光剂
pendant and sessile drop methodology　悬挂式和固定式跌落法
penetration　*n.* 穿透；渗透
pentaerythritol　*n.* 季戊四醇
pentaerythrityl tetraacrylate　季戊四醇四丙烯酸酯
pentaerythrityl triacrylate　季戊四醇三丙烯酸酯
peppermint leaf oil　薄荷叶油
percutaneous　*adj.* 经皮的；通过皮肤的
perfume　*n.* 香水；芳香；香味；馨香
periorbital wrinkle　鱼尾纹
permanent color　永久性染色；固定的颜色
permanent dyeing　永久性染色
permanent wave　卷发；烫发
permanent waving　烫发
permeation　*n.* 渗透；通透
permed hair　烫发
peroxide　*n.* 过氧化物；过氧化氢
persistent　*adj.* 持久的；持续的；坚持不懈的
persistent pigment darkening　持续性色素沉着
personal care　个人护理；个人护理用具；个人护理用品
personal deodorant　个人除臭剂
persulfate salt　过硫酸盐
pertaining to　关于；有关；适合；属于
petal　*n.* 花瓣
petrolatum　*n.* 矿脂
petroleum jelly　蜡膏
pharmacokinetic　*n.* 药代动力学
phenobarbital　*n.* 苯巴比妥
phenolic　*adj.* 酚的
phenyl dimethicone　苯基聚二甲基硅氧烷
phenylmercury　*n.* 苯汞；苯基汞
phenyl trimethicone　苯基聚三甲基硅氧烷

phosphate n. 磷酸盐
phosphatidyl-inositol 3-kinase activation 磷脂酰肌醇3-激酶激活
phospholipid n. 卵磷脂；磷脂类；磷脂
phosphoric acid 磷酸
photoaging n. 光老化
photocarcinogenesis n. 光致癌
photolyase n. 光裂合酶；光修复酶
photomutagenicity n. 光致突变性
photophobia n. 畏光；羞明；恐光症
phthalocyanine dye 酞菁染料
Phyla nodiflora 过江藤
pig pancreas 猪胰腺
pigment n. 颜料；色素
pigment symptom 色素症状
pinhole n. 针刺的孔；针孔
pink grapefruit peel oil 粉红葡萄柚皮油
Pinus n. 松属；松树；松科松属；松柏科松属
piroctone olamine 吡啶酮乙醇胺盐
pityriasis versicolor 汗斑；花斑癣；杂色糠疹；花斑糠疹；变色糠疹
placebo n. 安慰剂
placebo-controlled 安慰剂对照；安慰剂对照组；安慰剂控制
placenta n. 胎盘
plaque n. 匾牌；牙菌斑
plastic surgery 整形手术；整形外科
plasticizer n. 增塑剂；粉末增塑剂
platelet n. 血小板
plethora n. 过多；过量；过剩
poliovirus n. 脊髓灰质炎病毒；小儿麻痹病毒
poly(lactic acid) 聚乳酸
polyethylene glycol 聚乙二醇
polyhedral symbol 多面体符号
polyhydroxy compound 多羟基化合物
polyhydroxyalkanoate 聚羟基乙醇酸酯
polymer n. 聚合物
polymethyl methacrylate 有机玻璃；聚甲基异丁烯酸；聚甲基丙烯酸甲酯
polypeptide n. 多肽
polysaccharide n. 多糖；聚糖
polysorbate-80 吐温类乳化剂-80
polysorbate n. 聚山梨醇酯；聚山梨酸酯；吐温类乳化剂
polyvinylpyrrolidone n. 聚乙烯吡咯烷酮
pomace n.（水果榨汁后的）果渣；油渣
pomade n. 发油
pomegranate n. 石榴
Pongamia pinnata (L.)-Indian beech tree 印度山毛榉
Populus nigra 黑杨；欧洲黑杨；钻天杨
posterior adj. 在后面的；在后部的
post-natal 产后的；分娩后的
potassium cetyl phosphate 十六烷基磷酸钾
potency n. 影响力；支配力；效力
pour femme 女士香水
pour homme 男士香水
pourability n. 倾倒性；可浇注性；流动性
P. pastoris 酵母菌
p-phenylene diamine 对苯二胺
pragmatic adj. 实用的；讲求实效的；务实的
pragmatically adv. 实用主义地；讲究实效地
prerequisite n. 先决条件；前提；必备条件
preservative n. 防腐剂；保鲜剂
primer n. 底漆；底层涂料；打底妆
profilometry n. 轮廓仪；轮廓分析；轮廓测定法；轮廓形貌测量；轮廓术
prohibited substance 违禁物质；禁用物质
prokaryotic adj. 原核生物的
proliferation n. 增生；增殖
proline n. 脯氨酸
prone to 易于
propagation n. 传播；扩展；培养
propanediol n. 丙二醇
propanoic acid 丙酸
propyl gallate 没食子酸丙酯
propylene glycol 丙二醇
propylparaben n. 尼泊金丙酯；对羟基苯甲酸丙酯
propyltrimonium n. 丙基三甲基铵
protease n. 蛋白酶；蛋白[水解]酶
protein carbonylation 蛋白质羰基化
protein phosphorylation 蛋白质磷酸化
proteoglycan n. 蛋白聚糖；蛋白多糖
provitamin n. 维生素原
prune n. 干梅子；西梅干；李子干
pruritus n. 瘙痒[症]
prussian blue 普鲁士蓝
psoriasis n. 牛皮癣；银屑病
puffiness n. 浮肿；自夸；虚胖；膨胀；肿胀

punch biopsies 穿孔活检
purview *n.* 权限；（组织、活动等的）范围
pustule *n.* 脓疱；脓包
putrid *adj.* 腐烂的；腐臭的
putty *n.* 腻子
pyrimidine dimer 嘧啶二聚体
pyrrolidone carboxylic acid 吡咯烷酮羧酸
pyruvate *n.* 丙酮酸；丙酮酸盐
pyruvic acid 丙酮酸

Q

qualifier *n.* 修饰词
quaternisation *n.* 季铵盐化
quaternium ammonium compound 季铵化合物

R

radiance *n.* 容光焕发；红光满面；（散发出来的）光辉
rancidity *n.* 酸败；油脂酸败
randomize *v.* 使随机化；（使）作任意排列
raspberry *n.* 树莓
raspberry extract 树莓精华；覆盆子提取物
reactive oxygen species 活性氧物种；活性氧类
recasting 重铸；重塑；改正
recipe *n.* 配方；食谱；秘诀；方法；烹饪法；诀窍
red beet juice 红甜菜汁
red rose water 红玫瑰水
red petrolatum 红色凡士林
reflectance confocal microscopy 反射共焦显微术
regimen *n.* 养生；养生之道；生活规则
registration person 注册人
rejuvenation *n.* 更新；复苏
relaxer *n.* 直发膏；顺发剂
reprotoxic *adj.* 对生殖系统有毒性的
resin *n.* 树脂；合成树脂
resorcinol *n.* 间苯二酚
restricted substance 限用物质
retinoic acid 视黄酸；维甲酸；维生素A酸
retinoid *n.* 类视黄醇
retinol *n.* 视黄醇
retinyl palmitate 棕榈酸视黄酯；棕榈酰视黄酯
revitalize *v.* 使恢复；使更强壮
rheological *adj.* 流变的
rheometer *n.* 流变仪
rhinestone *n.* 水钻；莱茵石（用于仿钻石首饰）
rhinitis *n.* 鼻炎（感染或过敏引起）
rhizome *n.* 根茎；根状茎
rice bran oil 米糠油
rice bran protein 米糠蛋白
rice peptide 大米肽
rigid *adj.* 固执的；僵化的
roman chamomile oil 罗马洋甘菊油
rosacea *n.* 酒渣鼻；红鼻头
rosaceae *n.* 蔷薇科
rouge *n.* 胭脂

S

S. aureus 金黄色葡萄球菌
S. muticum 海黍子；马尾藻
Saccharomyces *n.* 酵母属；酵母菌属；酵母菌
safety evaluation 安全性评价；安全评价
saffron *n.* 藏红花粉（用作食用色素）；橘黄色
sage *n.* 圣人；鼠尾草（可用作调料）
sagging eyelid 眼睑或眼皮下垂
sahel *n.* 萨赫勒；萨赫勒地区；萨赫尔
salicylate *n.* 水杨酸盐
salicylic acid 水杨酸；邻羟基苯甲酸
salicylism *n.* 水杨酸反应
sallow complexion 面色萎黄
salve *n.* 药膏；软膏；油膏
sandalwood *n.* 檀香木
Sargassum muticum extract 马尾藻提取物
Saxifragaceae *n.* 虎耳草科；虎耳草属
scatter *v.* 散射；分散；散开；撒播；四散
S. cerevisiae 酿酒酵母
sebaceous gland dysfunction 皮脂腺功能障碍
seborrheic dermatitis 脂溢性皮炎
sebum *n.* 皮脂
sedative *n.* 镇静剂
seizure *n.* 没收；起获；充公；没收的财产
selenium sulfide 硫化硒；二硫化硒
senile lentigo 老年性雀斑
serum *n.* 血清；乳清
shampoo *n.* 洗发水；香波；洗发剂
shea butter 乳木果油
shellac *n.* 虫胶；紫胶；紫胶片
shiitake *n.* 香菇
shimmer *n.* 微光；闪烁的光

shrub *n.* 灌木
sienna *n.* 黄土；褐土
signal peptide 信号肽
silica microcapsule 二氧化硅微胶囊
silicate *n.* 硅酸盐
silicic acid 硅酸
silicone brush 硅胶刷
silicone oil 硅油
silk fibroin peptide 丝素肽
silkworm *Bombyx mori* 家蚕
Simmondsia chinensis 希蒙德木；油蜡树
singlet oxygen quencher 单线态氧猝灭剂
sketch *n.* 素描；速写；草图；简报；概述
skin bleaching 漂白皮肤
skin firmness 皮肤紧致度
skin radiance 皮肤光泽
skin roughness 皮肤粗糙度
skin thickness 蒙皮厚度
skin whitener 皮肤美白剂
skincare *n.* 护肤品
skin-restoring 皮肤修复
slurry *n.*（土、煤末或水泥混合而成的）泥浆；稀泥
snake venom 蛇毒；蛇毒蛋白；蛇毒素
sodium alginate 褐藻酸钠；海藻多糖；海藻酸钠
sodium ascorbyl phosphate 抗坏血酸磷酸酯钠；维生素C磷酸酯钠盐
sodium cetyl sulfate 十六烷基硫酸钠
sodium coco sulfate 椰油醇硫酸酯钠
sodium dodecyl sulfate 十二烷基硫酸钠
sodium lactate 乳酸钠
sodium laureth sulfate 十二烷基醚硫酸钠
sodium lauryl sulfate 十二烷基硫酸钠
sodium myreth sulfate 肉豆蔻醇聚醚硫酸钠
sodium polyacrylate 聚丙烯酸钠
solar simulator 太阳模拟器
solubiliser *n.* 增溶剂
somatic *adj.* 体细胞的；躯体的；体壁的
soothing *adj.* 安慰性的；减轻（痛苦）的；起镇定作用的
sorbitan monostearate 山梨醇酐单硬脂酸酯
sorbitan oleate 山梨醇油酸酯
sorbitol *n.* 山梨醇
sorbitol-6-phosphate 2-dehydrogenase 山梨醇-6-磷酸2-脱氢酶
source-based nomenclature 基于来源的命名法
soy oligopeptide 大豆低聚肽
Spathodea campanulata (L.)-African tulip tree 火焰树，非洲郁金香树
spearmint leaf oil 留兰香叶油
special cosmetic 特殊用途化妆品
spermicide *n.* 杀精子剂
sphingomyelin *n.* 鞘磷脂；神经鞘磷脂
sphingosine *n.* 鞘氨醇
spice *n.*（调味）香料
spinning brush 旋转刷
spiro polycyclic 螺多环
spiro polymer 螺环聚合物
split-face 左右面
sports cosmetic 运动化妆品
spot lightening cream 祛斑霜
spreadability *n.* 铺展性；蔬菜汁的延展性；覆盖性；涂抹性
squalane *n.* 鲨烷；角鲨烷
squalene *n.*（角）鲨烯；三十碳六烯
square-planar 平面正方形
square-pyramidal 四角锥形
stearamidopropyl dimethylamine 硬脂酰胺丙基二甲胺
stearic acid 硬脂酸；十八烷酸
steartrimonium chloride 硬脂基三甲基氯化铵
stearyl alcohol 十八烷醇
stem cell 干细胞
steppogenin *n.* 草大戟素；2’, 4’, 5, 7-四羟基黄烷酮
stereo descriptor 立体描述符
stereoisomer *n.* 立体异构体
sterilize *v.* 灭菌；消毒
stickiness *n.* 黏稠度；黏性
stiff *adj.* 不易弯曲（或活动）的；硬的；挺的；僵硬的
stipulation *n.* 规定；约定；合同
stoichiometric or compositional name 化学计量或组成名称
stratum corneum 角质层；角化层
streptococcus *n.* 链球菌
stretchability *n.* 可拉伸性；可延性；拉伸性
structure-based nomenclature 基于结构的命名法
sturdy *adj.* 结实的；坚固的
styrene *n.* 苯乙烯
sub-chronic 亚慢性的

subjective evaluation 主观评价
subordinated regulations 附属法规
substantiation *n.* 实证；证实；证明；具体化；实任制
substituent *n.* 取代基
subtilisin *n.* 枯草杆菌蛋白酶
subunit *n.* 亚单元；亚基；亚单位
sulfate *n.* 硫酸盐
sulfonamide *n.* 磺胺
sulfone *n.* 砜
sulfoxide *n.* 亚砜
sulfur dye 硫化染料；硫化染料染色
sulfuric acid 硫酸
sun protection factor (SPF) 防晒系数
sun-block 防晒霜
sunblock/suntan lotion 防晒霜；防晒油
suncream *n.* 防晒霜
sunscreen *n.* 防晒霜；防晒油
superficial peel 表面剥离
superoxide dismutase 超氧化物歧化酶
surfactant *n.* 表面活性剂
suspending agent 助悬剂；悬浮剂
swelling *n.* 膨胀；肿胀；肿胀处；浮肿处
syndecan-1 黏结蛋白聚糖-1
synovial joint fluid 滑膜关节液
Syringa vulgaris 紫丁香；欧丁香
systematic name 系统命名法；系统名称

T

tactile sense 触觉；触感
talc steatite 滑石
tallow *n.* 牛脂；动物油脂
tattoo *n.* 文身；（在皮肤上刺的）花纹
tautomerization *n.* 互变异构化
tea tree oil 茶树油
tear-free 无泪
telogen effluvium 休止期脱发；休止期落发；静止期脱发
teratogenic *adj.* 致畸形的
teratogenicity *n.* 致畸胎性；致畸性；致畸
terminal tyrosine 末端酪氨酸
terminally blocked ethoxylate 末端封闭乙氧基化物
terpene *n.* 萜烯；萜
tertiary butylhydroquinone 叔丁基对苯二酚
tetrahedral *adj.* 四面体的；有四面的
tetrahexyldecyl ascorbate 四己基癸醇抗坏血酸酯
Theobroma oil 可可油
theobromine *n.* 可可碱；咖啡碱
therapeutic *adj.* 治疗的；治病的；有助于放松精神的
thermolysin *n.* 嗜热菌蛋白酶
thermoplastic resin 热塑性树脂
thiocyanate *n.* 硫氰酸盐；硫氰酸酯
thiomersal *n.* 硫柳汞钠；硫柳汞
thixotropy *n.* 触变性；摇溶性；触变剂；搅溶性
thrombospondin-1 血小板应答蛋白-1
tingling *n.* 刺痛感
titanium dioxide 二氧化钛
tocopherol *n.* 生育酚；维生素E
tocopherol acetate 生育酚乙酸酯
toilet water 花露水
toluene *n.* 甲苯
toner *n.* 爽肤水；呈色剂；色粉
tonic *n.* 补品；补药；护发液；护肤液
tonnage *n.* 吨位；吨数
touchup *n.* 润色；修补画面
toxicokinetic *adj.* 毒代动力学的；毒物动力学的
traditional name 传统名称
tragacanth gum 黄芪胶
tragacanthin *n.* 西黄蓍胶素；黄耆质；黄耆糖
transdermal *n.* 经皮给药的；透过皮肤吸收的
transepidermal *adj.* 经皮的
transesterification *n.* 酯交换；转酯基作用
transparency *n.* 透明度；透明；透明性；显而易见
trauma *n.* 创伤；损伤；痛苦经历；挫折
tretinoin *n.* 维甲酸；维生素A酸
triarylmethane dye 三芳基甲烷染料
trichorrhexis nodosa 结节性脆发病
triclosan *n.* 三氯羟基二苯醚；三氯苯氧氯酚；三氯生
triethylhexanoin *n.* 三异辛酸甘油酯
trigger *v.* 触发；引起；发动；开动；起动
trigonal bipyramidal 三角双锥的
tripeptide-10 三肽-10
triplicate *n.* 一式三份
trivial name 俗名
tumor necrosis factor-α 肿瘤坏死因子-α
turmeric *n.* 姜黄；姜黄根粉
tyrosinase *n.* 酪氨酸酶

U

ubiquinone *n.* 泛醌；辅酶Q
ubiquitous *adj.* 无处不在的；十分普遍的
ultramarine blue 群青蓝
ultra-pure water 超纯水
ultra-violet (UV) filter 紫外线（UV）过滤器
umber *n.*（油漆中用的）棕土；赭土
umbilical cord 脐带
urate oxidase 尿酸氧化酶
urea *n.* 尿素；脲
5-ureidohydantoin 尿囊素
uric acid 尿酸
uricase *n.* 尿酸氧化酶
urinary tract disorder 尿路疾病
ursolic acid 乌苏酸；熊果酸

V

valencia orange peel oil 瓦伦西亚橘皮油
vanilla *n.* 香草
variance *n.* 方差；变化幅度；差额
vegetable glycerin 蔬菜甘油；植物甘油
venom *n.* 毒液
verbascoside *n.* 毛蕊花糖苷
verbatim *adj./adv.* 逐字的（地）；一字不差的（地）
violation *n.* 违反；违法；违章；越轨；侵犯；破坏；违例；犯规
viper tropidolaemus wagleri 瓦氏毒蛇
viscoelastic region 黏弹性区
viscoelasticity *n.* 黏弹性
viscometer *n.* 黏度计
viscosity *n.* 黏度；（液体的）黏性
vitreous fluid 玻璃体液

W

W/O emulsifier W/O型乳化剂
water-in-oil 油包水
water-resistant performance 防水性能
wet wipe 湿巾；湿纸巾
white mulberry 白桑树
whitening agent 美白成分；增白剂
wintergreen leaf oil 冬青叶油
witch hazel 金缕梅酊剂（用于治疗皮肤创伤）；金缕梅

X

xanthan gum 黄原胶
xenon arc 氙弧
xylene *n.* 二甲苯

Y

yeast *n.* 酵母；酵母菌
Yucca herbal extract 丝兰草本提取物

Z

zeta potential Zeta电位
zig-zag *adj.* 锯齿状的
zinc pyrithione 吡啶硫酮锌
Ziziphus nummularia 铜钱枣
zwitterionic *adj.* 两性离子的